Electrical Principles IV

Electrical Principles IV

A textbook covering the Level IV syllabus of the Business and Technician Education Council

D C Green
MTech, CEng, MIERE
Senior Lecturer in Telecommunication Engineering
Willesden College of Technology

Longman Scientific & Technical
an imprint of
Longman Group Limited
Longman House, Burnt Mill, Harlow
Essex CM20 2JE, England
Associated companies throughout the world

First published in Great Britain by Pitman Publishing Limited 1983
Reprinted by Longman Scientific & Technical 1986

ISBN 0-582-98853-5

Produced by Longman Singapore Publishers (Pte) Ltd
Printed in Singapore.

Preface

A wide range of electrical principles is usually considered to be desirable knowledge for any student of either heavy-current or light-current electrical engineering. For this reason "Principles" are studied at each level of the majority of certificate/diploma courses run under the auspices of the UK Business and Technician Education Council (BTEC). For the study of electrical/electronic principles at higher certificate/diploma level, BTEC has produced a large unit from which colleges are invited to select material to make up one or more 60 hour units. This book has been written to cover *all* of the material contained in this unit plus some extra topics which, although not included by BTEC, were thought to be of considerable importance. These extra topics include some network theory, non-linear circuits and the use of Laplace transforms to solve transient problems. It is hoped that most, if not all, of the requirements of the majority of college's principles units will have been satisfied.

The book has been written on the assumption that the reader will have studied Electrical Principles to the standard attained by the level III units of BTEC. This is particularly true of Chapters 6 and 7 where prior knowledge of electric and magnetic fields and of energy has been taken for granted in order to keep those chapters within reasonable proportions.

A large number of worked examples are provided throughout the book and each chapter concludes with a number of exercises. Answers to the numerical exercises will be found at the rear of the book.

Acknowledgement is made to BTEC for its permission to use the content of the BTEC unit in an appendix to this book. (The Council reserves the right to amend the content of its unit at any time.)

D C G

The following abbreviations for other titles in this series are used in this book:

[RSIII] Radio Systems III
[EIV] Electronics IV
[TSII] Transmission Systems II
[DT & S] Digital Techniques and Systems

Contents

1 Alternating Current Theory

The calculation of the currents and voltages at points in an electrical circuit can be carried out using phasor diagrams. This is the approach normally employed in introductory treatments and it is assumed that the reader will be familar with phasor diagrams. It is generally more convenient, however, particularly with the more complex circuits, to employ the **j operator** or **complex algebra**.

Revision of Complex Algebra

The symbol j represents the square root of minus 1, i.e. $j=\sqrt{-1}$. From this it is evident that $j^2=-1$, $j^3=-j$, and, of course, $j^4=+1$.

j also represents an angle of 90° so that $j=1\underline{/90^\circ}$, $j^2=1\underline{/180^\circ}$, $j^3=1\underline{/270^\circ}$, and $j^4=1\underline{/0^\circ}$.

A quantity $r=a+jb$ is known as a **complex number**; a is the **real part** and b is the **imaginary part** of that number. The term "imaginary" has its origins in the history of mathematics when the meaning of j was not clearly understood. In this book the imaginary term will often be referred to as the j part. The number $a+jb$ represents the phasor sum of a distance a in the positive real direction, which is normally taken as being the reference direction, and a distance b in the direction 90° anticlockwise to this reference. This is shown by Fig. 1.1.

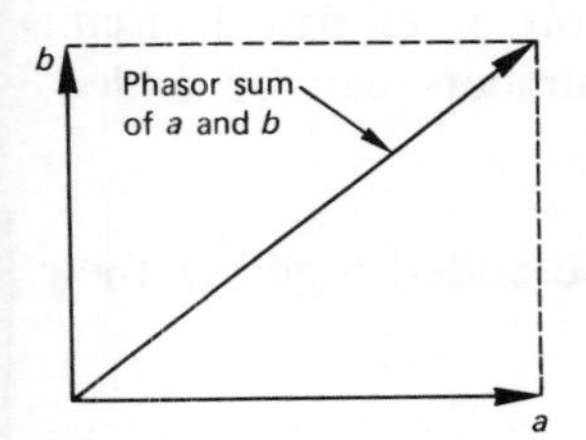

Fig. 1.1 Complex number $a+jb$

Any complex number can be represented on an **Argand Diagram** (see Fig. 1.2). Thus $(2+j3)$ is the point marked as A and $(-1-j4)$ is the point labelled as B.

Consider the point A:

$$A=2+j3=\sqrt{(2^2+3^2)}\underline{/\tan^{-1}3/2}=3.61\underline{/56.3^\circ} \qquad (1.1)$$

If the complex number A had originally been quoted as $3.61\underline{/56.3^\circ}$ it would be possible to obtain the values of its real and imaginary parts. The real part is obtained by multiplying the magnitude of the number by the cosine of its angle, i.e.

$$a=3.61\cos 56.3^\circ=2$$

The imaginary part is obtained by writing

$$b=3.61\sin 56.3^\circ=3$$

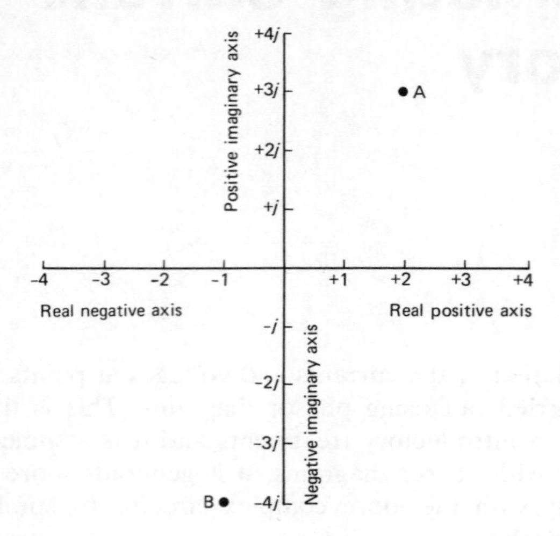

Fig. 1.2 Argand diagram

This process is known as *resolving* the complex number into its real and imaginary parts. A complex number written as $a+jb$ is said to be in its **rectangular form,** and when written as $r\underline{/\theta^\circ}$ it is in its **polar form**. The **conjugate** of a complex number has the same real and imaginary parts but the imaginary part has the opposite sign. Thus the conjugate of $a+jb$ is $a-jb$.

Manipulation of Complex Numbers

Before complex algebra can be used in the solution of electrical circuit problems it is necessary to know how complex numbers can be added, subtracted, multiplied and divided.

a) **Addition** When two complex numbers are to be added together their real and j parts must be added separately. Thus

$$(a+jb)+(c+jd)=a+c+j(b+d)$$

For example,

$$(4+j3)+(2+j5)=6+j8$$

$$(6-j2)+(-2+j4)=4+j2$$

Two numbers given in polar form cannot be directly added together; before addition can take place both of the numbers must be resolved into their real and imaginary parts. Consider the sum of the two numbers $25\underline{/60^\circ}$ and $40\underline{/-35^\circ}$. Resolving each number into its real and imaginary components gives

$$25\cos 60^\circ+j25\sin 60^\circ=12.5+j21.65$$

and

$$40\cos(-35^\circ)+j40\sin(-35^\circ)=32.77-j22.94$$

The sum is

$$12.5+32.77+j(21.65-22.94) \quad \text{or} \quad 45.27-j1.29$$

Expressed in polar form this is

$$\sqrt{(45.27^2+1.29^2)}\underline{/\tan^{-1}-1.29/45.27}=45.29\underline{/-1.63^\circ}$$

b) **Subtraction** The subtraction of one complex number from another follows the same rules as addition. Thus

$$(a+jb)-(c+jd)=a-c+j(b-d)$$

For example,

$$(6+j8)-(4-j10)=2+j18$$

As before, two complex numbers expressed in their polar forms must be changed into their rectangular form before subtraction can be carried out.

c) **Multiplication** The multiplication of two complex numbers can be carried out with the numbers expressed in *either* their polar *or* their rectangular form. In the rectangular form,

$$\begin{aligned}(a+jb)(c+jd)&=ac+jad+jbc+j^2bd\\&=ac-bd+j(ad+bc)\end{aligned}$$

In the polar form

$$R\underline{/\theta^\circ}\times S\underline{/\varphi^\circ}=RS\underline{/\theta^\circ+\varphi^\circ}$$

Thus the rule is: multiply the magnitudes and add the angles to obtain the product of the two complex numbers.

Example 1.1

Multiply together the complex numbers $10-j10$ and $5\underline{/30^\circ}$.

Solution

a) Working in rectangular form

$$5\underline{/30^\circ}=4.33+j2.5$$

and

$$\begin{aligned}(10-j10)(4.33+j2.5)&=43.3+j25-j43.3-j^225\\&=68.3-j18.3 \quad (Ans)\end{aligned}$$

In polar form,

$$\sqrt{(68.3^2+18.3^2)}\underline{/\tan^{-1}-18.3/68.3}=70.7\underline{/-15^\circ} \quad (Ans)$$

b) Working in polar form

$$10-j10=\sqrt{(10^2+10^2)}\underline{/\tan^{-1}-10/10}=14.14\underline{/-45^\circ}$$

and

$$14.14\underline{/-45^\circ}\times 5\underline{/30^\circ}=70.7\underline{/-15^\circ} \quad \text{(as before)} \quad (Ans)$$

d) **Division** The division of one complex number by another can be carried out using either the polar or the rectangular forms of the two numbers.

When the rectangular form is used, a procedure known as **rationalization** must be employed. A complex number is rationalized when both its numerator and its denominator are multiplied by the conjugate of the denominator. Thus,

$$\frac{a+jb}{c+jd}=\frac{(a+jb)(c-jd)}{(c+jd)(c-jd)}$$

$$=\frac{ac+bd+j(bc-ad)}{c^2+d^2+j(cd-cd)}$$

$$=\frac{ac+bd+j(bc-ad)}{c^2+d^2}$$

The reason the rationalization process is used is that it produces an entirely real denominator. It should be noted that the denominator of a rationalized expression *always* consists of the square of the real part *plus* the square of the imaginary part of the original denominator.

When a complex number in polar form is to be divided by another complex number, also in polar form, the division is easily carried out by taking the quotient of their magnitudes and the difference between their angles, i.e.

$$\frac{R\angle\theta^\circ}{S\angle\varphi^\circ}=\frac{R}{S}\angle\theta^\circ-\varphi^\circ$$

Example 1.2

Divide $10-j10$ by $5\angle 30^\circ$ by working in rectangular form.

Solution

$$5\angle 30^\circ=4.33+j2.5$$

Therefore,

$$\frac{10-j10}{4.33+j2.5}=\frac{(10-j10)(4.33-j2.5)}{4.33^2+2.5^2}$$

$$=\frac{43.3-25-j(43.3+25)}{4.33^2+2.5^2}=\frac{18.3-j68.3}{25}$$

$$=0.73-j2.73 \quad (Ans)$$

Expressed in polar form,

$$\sqrt{(0.73^2+2.73^2)}\angle\tan^{-1}\frac{-2.73}{0.73}=2.83\angle-75^\circ \quad (Ans)$$

Example 1.3

Divide $10-j10$ by $5\angle 30^\circ$ by working in polar form

Solution

$$10-j10=14.14\angle-45^\circ$$

Therefore,

$$\frac{14.14\angle -45^\circ}{5\angle 30^\circ} = 2.83\angle -75^\circ \quad \text{(as before)} \quad (Ans)$$

e) **Exponential form of complex numbers** Two important relationships are given in equations (1.2) and (1.3).

$$e^{j\theta} = \cos\theta + j\sin\theta \tag{1.2}$$

$$e^{-j\theta} = \cos\theta - j\sin\theta \tag{1.3}$$

Multiplying both sides of equation (1.2) or (1.3) by r gives

$$r(\cos\theta + j\sin\theta) = r\angle\theta = re^{j\theta} = e^{p}e^{j\theta} = e^{p+j\theta} \tag{1.4}$$

Also, adding equations (1.2) and (1.3)

$$e^{j\theta} + e^{-j\theta} = 2\cos\theta$$

or

$$\cos\theta = \frac{e^{j\theta} + e^{-j\theta}}{2} \tag{1.5}$$

Subtracting equation (1.3) from equation (1.2)

$$\sin\theta = \frac{e^{j\theta} - e^{-j\theta}}{2j} \tag{1.6}$$

The relationships expressed by equations (1.2) to (1.6) are particularly important in the study of transmission lines and networks.

Example 1.4

Express the complex number $e^{-(4+j2)}$ in the form $r\angle\theta$.

Solution

$$e^{-(4+j2)} = e^{-4-j2} = 0.018\angle -2 = 0.018\angle -114.6^\circ \quad (Ans)$$

Example 1.5

Express the complex number $5+j10$ in exponential form.

Solution

$$5+j10 = 11.18\angle 1.107\text{(rad)} = e^{2.414+j1.107} \quad (Ans)$$

Application of Complex Quantities to Electric Circuits

1 *Resistance* When a sinusoidal voltage $V\sin\omega t$ is applied across a linear resistance R the current that flows is given by Ohm's Law:

$$i = \frac{V}{R}\sin\omega t = I\sin\omega t$$

The current is in phase with the voltage and so it does not possess an imaginary component. The resistance R of the circuit is

$$R = \frac{V}{I} = R + jO = R \tag{1.7}$$

2 *Inductance* The self-induced voltage in an inductance is equal to the inductance times the rate of change of the current:

$$V = -L\,di/dt$$

For a sinusoidal current $i = I \sin \omega t$ and

$$\frac{di}{dt} = \omega I \cos \omega t = \omega I \sin(\omega t + 90°)$$

The voltage across the inductor, $v = \omega LI \sin(\omega t + 90°)$ leads the current by 90°. The reactance X_L of the inductance is the ratio of the peak (or r.m.s.) values of the voltage and the current. Hence

$$X_L = \frac{V}{I} = j\omega L \tag{1.8}$$

3 *Capacitance* When a sinusoidal voltage $v = V \sin \omega t$ is applied across a capacitance C a charge $q = CV \sin \omega t$ will be supplied to the capacitance. Current is the rate of change of charge and so

$$i = \frac{dq}{dt} = \omega CV \cos \omega t = \omega CV \sin(\omega t + 90°).$$

The reactance X_C of the capacitance is the ratio of the peak (or r.m.s.) values of the voltage and the current, and hence,

$$X_C = \frac{V}{I} = \frac{1}{j\omega C} = -\frac{j}{\omega C} \tag{1.9}$$

4 *Resistance and Inductance in Series* Referring to Fig. 1.3. $V_R = IR$ and $V_L = Ij\omega L$. The applied voltage V is the phasor sum of V_R and V_L and this is written as

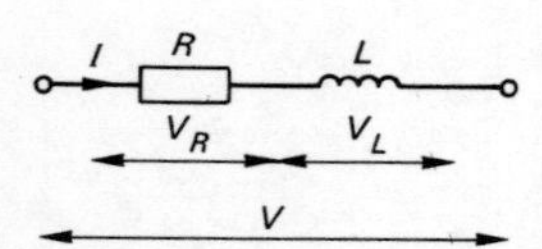

Fig. 1.3 Resistance and inductance in series

$$V = I(R + j\omega L)$$

The impedance Z of the circuit is the ratio of the applied voltage to the circuit current, i.e.

$$Z = V/I = R + j\omega L \tag{1.10a}$$

$$= \sqrt{(R^2 + \omega^2 L^2)} \angle \tan^{-1} \frac{\omega L}{R} \tag{1.10b}$$

Example 1.6

A current of 500 mA flows in an inductor of 60 mH inductance and 10 Ω resistance. Calculate the voltage across the inductor at a frequency of 1000 Hz.

Solution The reactance of the inductor is

$$X_L = 2\pi \times 1000 \times 60 \times 10^{-3} = 377\ \Omega$$

The impedance of the circuit is

$$Z = 10 + j377\ \Omega$$

Therefore,

$$\text{Voltage across inductor} = IZ = 0.5(10 + j377)$$
$$= 5 + j188.5$$
$$= 188.56\underline{/88.5°}\text{ V} \quad (Ans)$$

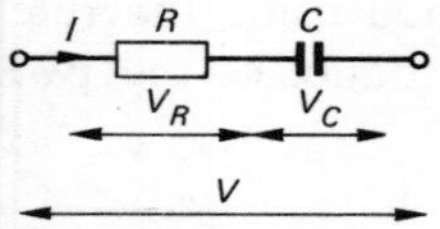

Fig. 1.4 Resistance and capacitance in series

5 *Resistance and Capacitance in Series* From Fig. 1.4,

$$V = V_R + V_C = I(R + 1/j\omega C)$$
$$Z = V/I = R + 1/j\omega C \tag{1.11a}$$
$$= \sqrt{(R^2 + 1/\omega^2C^2)}\,\underline{/\tan^{-1} 1/\omega CR} \tag{1.11b}$$

It should be noted from both equations (1.10) and (1.11) that the reactance is replaced by either $j\omega L$ or $1/j\omega C$ and then the components are written down as a series addition. The principle can be extended to cater for any number of components connected in series. Thus, for a circuit containing resistance, inductance and capacitance connected in series,

$$Z = R + j(\omega L - 1/\omega C) \tag{1.12a}$$
$$= \sqrt{[R^2 + (\omega L - 1/\omega C)^2]}\,\underline{/\tan^{-1}\frac{\omega L - 1/\omega C}{R}} \tag{1.12b}$$

Example 1.7

Calculate the voltage applied across the circuit given in Fig. 1.5 if, at a frequency of 2000 Hz, 2 V are dropped across the 1000 Ω resistance.

Solution Since the current is common to all components it is taken to be the reference. The current is equal to 2/1000 = 2 mA.

$$X_L = 2\pi \times 2 \times 10^3 \times 200 \times 10^{-3} = j2513\ \Omega$$
$$X_C = 1/(2\pi \times 2 \times 10^3 \times 10^{-8}) = -j7958\ \Omega$$

Therefore,

$$V = IZ = 2 \times 10^{-3}(1000 + j2513 - j7958)$$
$$= 2 \times 10^{-3}(1000 - j5445)$$
$$= 2 \times 10^{-3}(5536\underline{/-79.6°})$$
$$= 11.1\underline{/-79.6°}\text{ V} \quad (Ans)$$

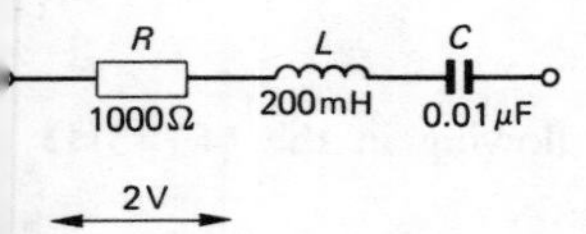

Fig. 1.5

6 *Resistance and Capacitance in Parallel* The total current I supplied to the parallel combination of a resistance R and a capacitance C in parallel (Fig. 1.6) is the phasor sum of the currents I_C and I_R flowing in each component. Since the components are connected in parallel they will have the same voltage developed across them and so the voltage is taken as the reference. Therefore

$$I = I_R + I_C = \frac{V}{R} + \left(V\Big/\frac{1}{j\omega C}\right) = \frac{V}{R} + jV\omega C$$

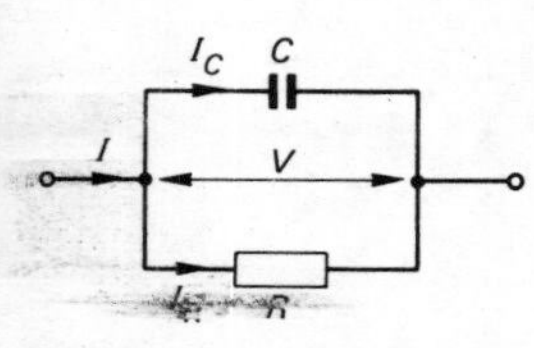

Fig. 1.6 Resistance and capacitance in parallel

The impedance Z of the circuit is the ratio voltage/current and hence,

$$Z = \frac{V}{I} = \frac{1}{\frac{1}{R} + j\omega C} = \frac{R}{1 + j\omega CR} \tag{1.13a}$$

Rationalizing,

$$Z=\frac{R(1-j\omega CR)}{1+\omega^2C^2R^2} \tag{1.13b}$$

The impedance of the parallel circuit can be determined using the rule developed for two resistors in parallel, i.e. divide the product of the two impedances by their sum. Thus

$$Z=\frac{R\times 1/j\omega C}{R+(1/j\omega C)}=\frac{R}{1+j\omega CR} \tag{1.13a}$$

Series–Parallel Circuits

A generalized approach to the solution of circuits containing components connected both in series and in parallel is not possible. Once each inductance and capacitance in the circuit has been replaced by its complex impedance the methodology is similar to that employed for the solution of purely d.c. resistive circuits. The approach required will be illustrated by a number of examples.

Example 1.8

Calculate the impedance of the circuit given in Fig. 1.7 at a frequency of 79.6 Hz.

Solution

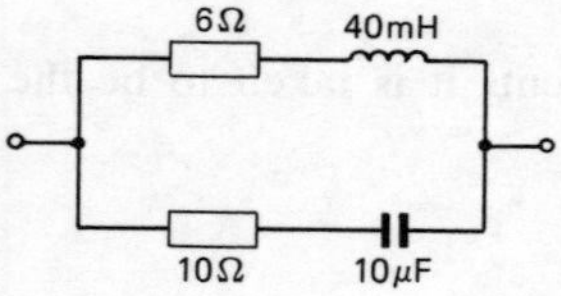

$$X_L=2\pi\times 79.6\times 40\times 10^{-3}=j20\ \Omega$$

$$X_C=1/(2\pi\times 79.6\times 10^{-5})=-j200\ \Omega$$

Fig. 1.7

Therefore,

$$Z=\frac{(6+j20)(10-j200)}{(6+j20)+(10-j200)}=23.14\underline{/71.1^\circ}\ \Omega \quad (Ans)$$

Example 1.9

For the circuit shown in Fig. 1.8 determine the current flowing in the $(4+j5)\ \Omega$ impedance.

Solution The current I flowing in the $(20+4)\ \Omega$ impedance is $25/(20+j4)$.

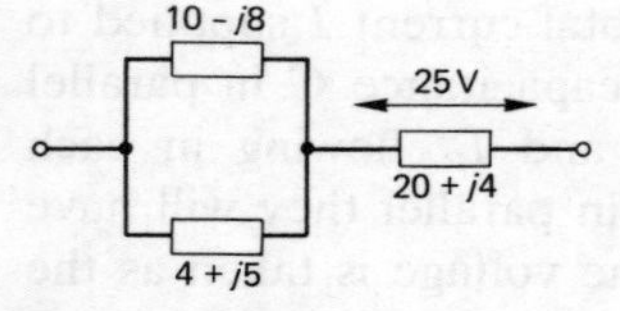

Fig. 1.8

Therefore, the current in the $(4+j5)\ \Omega$ impedance is†

$$I=\frac{25}{20+j4}\times\frac{10-j8}{(10-j8)+(4+j5)}=\frac{250-j200}{(20+j4)(14-j3)}$$

$$=1.096\underline{/-37.9^\circ}\ \text{A} \quad (Ans)$$

† $Z_T=Z_1Z_2/(Z_1+Z_2)$ $\quad V=IZ_1Z_2/(Z_1+Z_2)$
$I_1=V/Z_1=IZ_2/(Z_1+Z_2)$ $\quad I_2=V/Z_2=IZ_1/(Z_1+Z_2)$

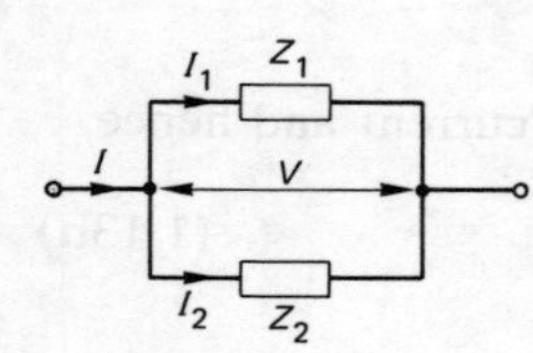

Fig. 1.9 Calculations of the current in one branch of a parallel circuit

Example 1.10

For the circuit shown in Fig. 1.10 calculate *a*) the impedance of the circuit, *b*) the current that flows when a voltage of 100 sin ωt is applied to the circuit, and *c*) the effective capacitance of the circuit, if the frequency is 796 Hz.

Solution

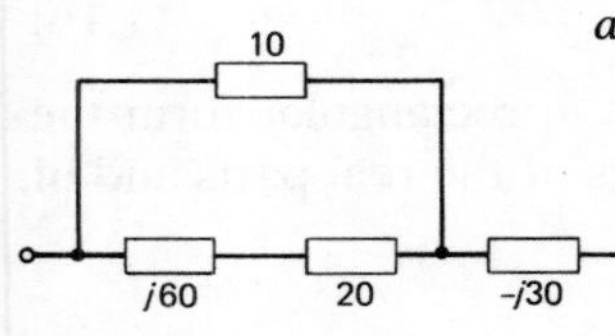

Fig. 1.10

$$a)\ Z = -j30 + \frac{10(20+j60)}{10+20+j60} = -j30 + \frac{200+j600}{30+j60}$$

$$= -j30 + \frac{60-j120+j180+360}{45} = -j30 + \frac{420+j60}{45}$$

$$= -j30 + 9.33 + j1.33 = 9.33 - j28.67$$

$$= 30.15\underline{/-72^\circ}\ \Omega \quad (Ans)$$

$$b)\ \text{The current } I = V/Z = \frac{100}{30.15\underline{/-72^\circ}} = 3.32\underline{/72^\circ}\ \text{A} \quad (Ans)$$

$$c)\ -j28.67 = \frac{-j}{2\pi \times 796 \times C}$$

$$\text{or} \quad C = 1/(2\pi \times 796 \times 28.67) = 6.97\ \mu F \quad (Ans)$$

Power in a Circuit

The instantaneous power dissipated in a circuit is the product of the instantaneous values of the current flowing in the circuit and the voltage applied across it. In general, the current and the voltage will not be in phase with one another. If their phase difference is θ then the instantaneous power is given by

$$p = V_m \sin \omega t \times I_m \sin(\omega t + \theta)$$

A trigonometric identity is

$$2 \sin A \sin B = \cos(A - B) - \cos(A + B)$$

Using this, with $A = \omega t + \theta$ and $B = \omega t$, gives

$$p = \frac{V_m I_m}{2}[\cos \theta - \cos(2\omega t + \theta)]$$

The average power dissipated over a complete cycle is

$$P = \frac{1}{2\pi}\int_0^{2\pi} vi d\omega t = \frac{V_m I_m}{4\pi}\int_0^{2\pi}[\cos \theta - \cos(2\omega t + \theta)]\, d\omega t$$

$$= \frac{V_m I_m}{4\pi}\left[\cos \theta \cdot \omega t - \frac{\sin(2\omega t + \theta)}{2}\right]_0^{2\pi}$$

$$= \frac{V_m I_m}{4\pi}[2\pi \cos \theta] = \frac{V_m I_m}{2}\cos \theta$$

or

$$P = VI \cos \theta \tag{1.14}$$

$$= |I|^2 \times \text{real part of impedance } V/I \tag{1.15}$$

Thus, the *mean* power dissipated is the product of the r.m.s. values of the current and the voltage and a term, $\cos\theta$, known as the **power factor** of the circuit. Note carefully that the power factor is *only* equal to $\cos\theta$ when both the current and the voltage are of sinusoidal waveform. The general definition of power factor, which is true for any waveform, is given by

$$\text{Power factor} = \text{power/volt–amps} \tag{1.16}$$

When the current and the voltage are expressed in rectangular form the power dissipated is given by the sum of the products of the real parts and of the imaginary parts.

Example 1.11

Calculate the power dissipated when a voltage $V = (5 + j2)$ volts is developed across an impedance carrying a current of $(20 + j15)$ mA.

Solution

$$V = 5 + j2 = 5.385\,\underline{/21.8°}\ \text{V} \qquad I = 20 + j15 = 25\,\underline{/36.9°}\ \text{mA}$$

From equation (1.14),

$$P = 5.385 \times 25 \times 10^{-3} \cos 15.1° = 130\ \text{mW} \quad (\textit{Ans})$$

Alternatively,

$$P = (5 \times 20) + (2 \times 15) = 100 + 30 = 130\ \text{mW} \quad (\textit{Ans})$$

Admittance, Conductance and Susceptance

The work involved in the solution of a parallel circuit is very often eased if the **admittance** of the circuit is determined and used. The admittance, symbol Y, of a circuit is the reciprocal of its impedance Z, i.e. $Y = 1/Z$. In general, the admittance of a circuit will be a complex quantity; the real part of an admittance is known as the **conductance** G, and the imaginary part is known as the **susceptance** B. Thus

$$Y = G + jB \tag{1.17}$$

The unit for admittance, conductance and susceptance is the siemen, S.

The use of admittances in the solution of a parallel circuit is convenient because *admittances in parallel are summed to obtain their total admittance.* Consider the circuit shown in Fig. 1.11 The total current I is the sum of the individual branch currents I_1, I_2 and I_3. Thus

$$I = (V/Z_1) + (V/Z_2) + (V/Z_3) = VY_1 + VY_2 + VY_3$$

The total admittance Y_T of the circuit is the ratio

(total current)/(total voltage)

and so

$$Y_T = I/V = Y_1 + Y_2 + Y_3$$

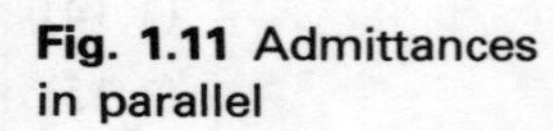

Fig. 1.11 Admittances in parallel

The power dissipated in the circuit is equal to V^2G_T, where G_T is the real part of Y_T.

Example 1.12

An inductor of inductance 50 mH and resistance 100 Ω is connected in parallel with a 200 Ω resistor. Calculate the admittance of the circuit at a frequency of $10^4/\pi$ Hz. Calculate the power dissipated when a voltage of 12 V is applied across the circuit at the same frequency.

Solution The reactance of the inductor is

$$2\times10^4\times50\times10^{-3}=1000\ \Omega$$

and so the impedance of the inductor is $100+j1000\ \Omega$. Therefore the admittance of the inductor is

$$\frac{1}{Z}=\frac{1}{100+j1000}=\frac{100-j1000}{10^4+10^6}$$
$$=9.9\times10^{-5}-j9.9\times10^{-4}\ \mathrm{S}$$

The admittance of the parallel resistor is $1/200=5\times10^{-3}$ S. The total admittance is

$$Y=5.099\times10^{-3}-j9.9\times10^{-4}\ \mathrm{S}\quad (Ans)$$

Power dissipated is

$$V^2G=144\times5.099\times10^{-3}=734\ \mathrm{mW}\quad (Ans)$$

Example 1.13

Calculate the impedance of the circuit shown in Fig. 1.12.

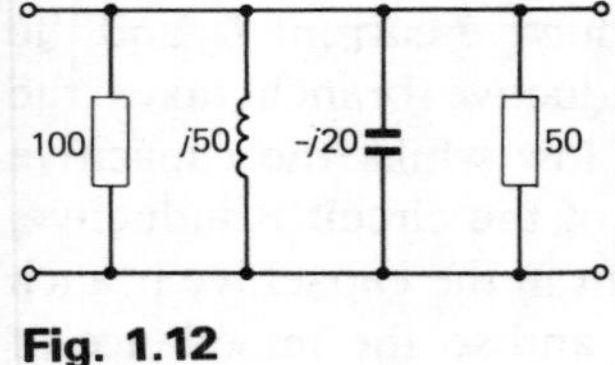

Fig. 1.12

Solution

$$Y=(1/100)+(1/50)+j(1/20-1/50)=0.03+j0.03\ \mathrm{S}$$

Therefore,

$$Z=\frac{1}{Y}=\frac{1}{\sqrt{(0.03^2+0.03^2)}\,/\tan^{-1}1}=23.57\,\underline{/-45^\circ}\quad (Ans)$$

Resonant Circuits

A **resonant circuit** consists of an inductor and a capacitor connected either in series or in parallel with one another. Some resistance is always present because of the unavoidable resistance of the inductor windings, but often the resistance of the capacitor is negligibly small. Resonant circuits are widely used, particularly in radio engineering, because of their ability to be tuned to select a required narrow band of frequencies from a range of applied frequencies.

The Series-Resonant Circuit

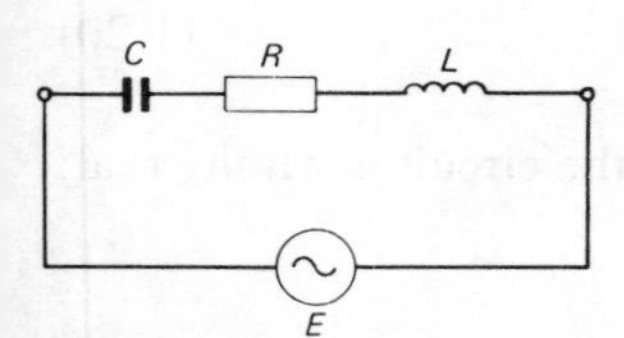

Fig. 1.13 Series-resonant circuit

A **series-resonant circuit** consists of an inductor and a capacitor connected in series as shown by Fig. 1.13. The impedance of the circuit at a particular frequency $\omega/2\pi$ Hz is

$$Z=R+j[\omega L-(1/\omega C)] \tag{1.18a}$$

$$=\sqrt{\left[R^2+\left(\omega L-\frac{1}{\omega C}\right)^2\right]}\,\underline{/\tan^{-1}\frac{\omega L-(1/\omega C)}{R}} \tag{1.18b}$$

The impedance of the circuit varies with change in frequency because of the $[\omega L-(1/\omega C)]$ term. At one particular frequency, known as the **resonant frequency** f_0, the inductive and capacitive reactances cancel out, i.e. $\omega_0 L = 1/\omega_0 C$.

Then, $\omega_0^2 = 1/LC$, $\omega_0 = 1/\sqrt{(LC)}$ or

$$f_0 = 1/2\pi\sqrt{(LC)} \tag{1.19}$$

Note that the resonant frequency is independent of the resistance of the circuit.

At the resonant frequency the impedance of the circuit is merely the resistance R and so the current flowing in the circuit has its maximum value of $I_0 = E/R$. At all other frequencies the impedance of the circuit is greater than the resonant value since $[\omega L-(1/\omega C)] \neq 0$. The impedance at any frequency *can* be obtained with the aid of equation (1.18) but a more convenient method involves the use of the Q-factor (p. 14). However, it should be evident that at frequencies above resonance the impedance of the circuit will be inductive, while the impedance will be capacitive at frequencies below resonance.

The Parallel-resonant Circuit

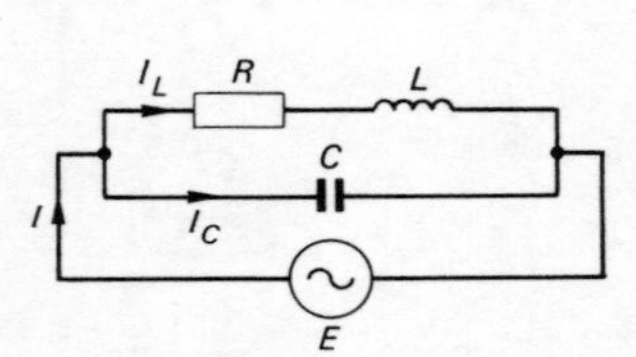

Fig. 1.14 Parallel-resonant circuit

A **parallel-resonant circuit** consists of an inductor and a capacitor connected in parallel as shown in Fig. 1.14. R is the resistance of the inductor and the capacitor is assumed to have a negligible resistance. The total current supplied to the circuit is the phasor sum of the inductive current I_L and the capacitive current I_C. At low frequencies the inductive branch takes the larger current because the inductive reactance is low whilst the capacitive reactance is high. This means that the impedance of the circuit is inductive. At high frequencies the converse is true: the current in the capacitive branch is larger than the current in the inductive branch and so the impedance of the circuit is capacitive.

At a particular frequency, known as the resonant frequency, the impedance of a parallel-tuned circuit is purely resistive and it is then generally known as the **dynamic resistance** R_d.

The admittances of the inductive and capacitive branches are, respectively, given by

$$Y_L = 1/(R+j\omega L) = \frac{R-j\omega L}{R^2+\omega^2 L^2} \quad \text{and} \quad Y_C = j\omega C$$

The total admittance of the circuit is $Y_T = Y_L + Y_C$, or

$$Y_T = \frac{R}{(R^2+\omega^2 L^2)} + j\left(\omega C - \frac{\omega L}{R^2+\omega^2 L^2}\right) \tag{1.20}$$

At the resonant frequency $\omega_0/2\pi$ the admittance of the circuit is wholly real; therefore,

$$\omega_0 C - \frac{\omega_0 L}{R^2+\omega_0^2 L^2} = 0$$

$$R^2 C + \omega_0^2 L^2 C = L$$

$$\omega_0^2 = \frac{L}{L^2C} - \frac{R^2C}{L^2C} = \frac{1}{LC} - \frac{R^2}{L^2}$$

and

$$f_0 = \frac{1}{2\pi}\sqrt{\left(\frac{1}{LC} - \frac{R^2}{L^2}\right)} \tag{1.21}$$

Very often the second term, R^2/L^2, is negligibly small compared with the first term and then equation (1.21) reduces to the same form as that derived for the series resonant frequency, i.e.

$$f_0 = 1/2\pi\sqrt{(LC)}$$

If the resistance R_C of the capacitive arm is *not* negligibly small then the resonant frequency will be given by

$$f_0 = \frac{1}{2\pi}\sqrt{\frac{1}{LC}\left(\frac{L - CR_L^2}{L - CR_C^2}\right)} \tag{1.22}$$

Substituting for ω_0 in the real part of equation (1.20) gives

$$Y_T = G_T = \frac{R}{R^2 + [(1/LC) - (R^2/L^2)]L^2} = \frac{R}{R^2 + (L/C) - R^2} = CR/L$$

Therefore, the dynamic resistance R_d of a parallel-tuned circuit is

$$R_d = 1/G_T = L/CR\ \Omega \tag{1.23}$$

When a series- or a parallel-tuned circuit is at resonance its impedance is purely resistive and so the circuit then has unity power factor.

Example 1.14

An inductor has a series resistance of 100 Ω and an inductance of 40 mH. *a)* Calculate its conductance and susceptance at a frequency of 5000/π Hz. *b)* Calculate the value of capacitance to be connected (i) in series and (ii) in parallel with the inductor for the circuit to have unity power factor. For each case calculate the circuit's impedance.

Solution

a) $Y_L = \dfrac{1}{100 + j400} = \dfrac{100 - j400}{170 \times 10^3}$

$= (0.588 - j2.353) \times 10^{-3}$ S (*Ans*)

b) (i) The series-connected capacitor must have a reactance of $-j400\ \Omega$. Therefore,

$400 = 1/10^4 C_s$ or $C_s = 1/4 \times 10^6 = 0.25\ \mu F$ (*Ans*)

b) (ii) The parallel-connected capacitor must have a susceptance of $+j2.353 \times 10^{-3}$ S. Therefore,

$2.353 \times 10^{-3} = 10^4 C_p$ or $C_p = 0.235\ \mu F$ (*Ans*)

The resistance of the capacitive arm of a parallel-tuned circuit may not always be negligibly small. An interesting case occurs when the resistances

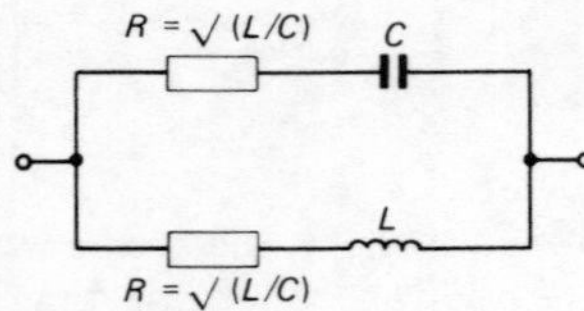

Fig. 1.15 Constant resistance network

of the inductive and the capacitive arms are equal to one another and also to $\sqrt{(L/C)}$ (Fig. 1.15). Then

$$Y_T = \frac{1}{R+(1/j\omega C)} + \frac{1}{R+j\omega L}$$

$$= \frac{(R+j\omega L)+[R+(1/j\omega C)]}{[R+(1/j\omega C)](R+j\omega L)} = \frac{2R+j[\omega L-(1/\omega C)]}{R^2+(L/C)+j[\omega LR-(R/\omega C)]}$$

Since $R=\sqrt{(L/C)}$

$$Y_T = \frac{2R+j[\omega L-(1/\omega C)]}{R\{2R+j[\omega L-(1/\omega C)]\}} = 1/R \tag{1.24}$$

This means that the impedance $Z=1/Y_T$ of the circuit is constant at R ohms at *all* frequencies.

Q-factor

Ideally, components such as inductors and capacitors would have zero resistance and hence dissipate zero power. In practice, of course, all components possess some self-resistance although the resistance of a capacitor is usually small enough to be ignored. Losses in inductors and capacitors will be discussed in Chapter 3.

The losses in a component, or in a circuit, can be expressed in terms of its ***Q*-factor**. The definition of Q-factor is given by equation (1.25).

$$Q = \frac{2\pi \times \text{maximum energy stored}}{\text{energy dissipated per cycle}} \tag{1.25}$$

(at resonance if a tuned circuit)

1 *Inductor* A practical inductor has an inductance L and a resistance R. The maximum energy stored in the magnetic field due to the inductance is

$$\tfrac{1}{2}LI^2_{max} \quad \text{or} \quad \tfrac{1}{2}L(I\sqrt{2})^2 = LI^2 \text{ joules}$$

The energy dissipated per cycle in the resistance is I^2R/f joules. Substituting into equation (1.25),

$$Q = \frac{2\pi LI^2}{I^2R/f} = \frac{2\pi fL}{R} = \frac{\omega L}{R} \tag{1.26}$$

i.e. the Q-factor is the ratio reactance/resistance.

2 *Capacitor* The losses in a capacitor C can be represented by a resistance R connected in series with the capacitor. The maximum energy stored in the electric field due to the capacitance is

$$\tfrac{1}{2}CV^2_{C(max)} = \tfrac{1}{2}C(V_C\sqrt{2})^2 = CV_C^2 \text{ joules}$$

The voltage V_C developed across the capacitance C is equal to the product of the current I and the reactance $1/\omega C$ of the capacitor. Thus the maximum energy stored may be written as

$$CI^2/(\omega C)^2 \quad \text{or} \quad I^2/\omega^2 C \text{ joules}$$

The energy dissipated per cycle in the resistance is I^2R/f and so

$$Q = \frac{2\pi I^2/\omega^2 C}{I^2 R/f} = 2\pi f/\omega^2 CR = 1/\omega CR \qquad (1.27)$$

This equation may also be written as reactance/resistance.

3 *Series-resonant Circuit* In a series-resonant circuit energy is continually being transferred between the electric field of the capacitance and the magnetic field of the inductor. Only at resonance are the two fields equal and so it is customary to consider the Q-factor of a resonant circuit at its resonant frequency. The maximum energy stored in a series-resonant circuit is then LI^2 joules and the energy dissipated per cycle is I^2R/f_0 joules. Therefore,

$$Q = \frac{2\pi LI^2}{I^2 R/f_0} = \frac{\omega_0 L}{R} = 1/\omega_0 CR \quad (\text{since } \omega_0 L = 1/\omega_0 C) \qquad (1.28)$$

The Q-factor of a series-tuned circuit is the same as that of the inductor alone if the capacitor is lossless and there is no added resistance.

From equation (1.19), $\omega_0 = 1/\surd(LC)$ and substituting into equation (1.28),

$$Q = \frac{L}{R\surd(LC)} = \frac{1}{R}\sqrt{\frac{L}{C}} \qquad (1.29)$$

Equation (1.29) demonstrates clearly that for the maximum Q-factor the resistance must be as small as possible *and* the L/C ratio should be as high as possible.

The current I_0 flowing in a series-tuned circuit at resonance is $I_0 = E/R$, where E is the applied voltage. This current flows in both the inductance and the capacitance and develops voltages, I_0X_L and I_0X_C, respectively, across them. Therefore,

$$V_C = I_0/\omega_0 C = E/R\omega_0 C = QE \qquad (1.30)$$

Example 1.15

A series-resonant circuit is connected across a 10 V, 2 MHz supply having an internal impedance of 5 Ω. Calculate the values of inductance and capacitance required to give a capacitor voltage of 250 V at the resonant frequency. Assume the resistance of the inductor to be 5 Ω.

Solution The required Q-factor is $250/10 = 25$ and hence

$$25 = \frac{1}{10}\sqrt{\frac{L}{C}} \quad \text{or} \quad \surd L = 10 \times 25 \times \surd C$$

Thus $2\pi \times 2 \times 10^6 = 1/250C$ and so

$$C = 1/(1000\pi \times 10^6) = 318 \text{ pF} \quad (\textit{Ans})$$

Also $L = 1/(4\pi^2 \times 4 \times 10^{12} \times 318 \times 10^{-12}) = 19.9\ \mu\text{H}$ (*Ans*)

4 *Parallel-resonant Circuit* The maximum energy stored in a parallel-tuned circuit is LI_L^2 joules, where I_L is the current flowing in the inductive branch. The energy dissipated per cycle at resonance is I_L^2R/f_0 joules. Therefore,

$$Q = \frac{2\pi LI_L^2}{I_L^2R/f_0} = \omega_0 L/R \tag{1.31}$$

This is the same expression as obtained for the series-tuned circuit and it supposes that the capacitor possesses negligible resistance.

The dynamic impedance of a parallel-tuned circuit is L/CR Ω. Multiplying both the numerator and the denominator by ω_0 gives

$$R_d = \omega_0 L/\omega_0 CR = Q\omega_0 L = Q/\omega_0 C \tag{1.32}$$

When a parallel-tuned circuit has a current I supplied to it a voltage IR_d will be developed across its terminals. The current I_C flowing in the capacitive branch is equal to this voltage divided by the reactance of the capacitor, i.e.

$$I_C = V/X_C = \frac{IR_d}{1/\omega_0 C} = IL\omega_0 C/CR = I\omega_0 L/R$$

or $$I_C = QI \tag{1.33}$$

Similarly, the current I_L flowing in the inductive branch is

$$I_L = \frac{V}{\sqrt{(R^2+\omega_0^2L^2)}} = \frac{V}{\omega_0 L\sqrt{[1+R^2/(\omega_0 L)^2]}} = \frac{IL/CR}{\omega_0 L\sqrt{[1+(1/Q^2)]}}$$

$$I_L = \frac{QI}{\sqrt{[1+(1/Q^2)]}} \simeq QI \tag{1.34}$$

Equations (1.33) and (1.34) show that the current flowing in either branch of a parallel-tuned circuit is Q times greater than the total current supplied to the circuit.

Example 1.16

A parallel-tuned circuit consists of an inductance L of 50 μH and 10 Ω connected in parallel with a loss-free capacitor at a frequency of $10^7/2\pi$ Hz. The circuit is connected in series with a 1 mA constant current source. Calculate *a*) the current in the inductor, *b*) the dynamic resistance and *c*) the voltage across the circuit.

Solution

$$Q = \omega_0 L/R = 10^7 \times 50 \times 10^{-6}/10 = 50$$

a) $I_L = QI = 50$ mA (*Ans*)
b) $R_d = Q\omega_0 L = 50 \times 500 = 25$ kΩ (*Ans*)
c) $V = IR_d = 25$ V (*Ans*)

Effective *Q*-factor

In most cases the voltage source applied across a series-resonant circuit will not have an internal resistance of very nearly zero ohms. The source

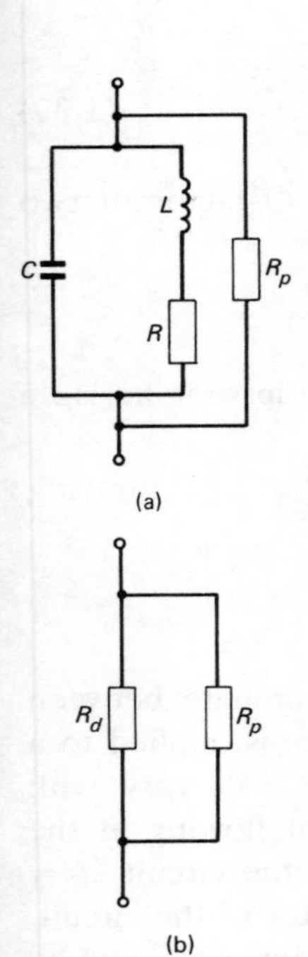

Fig. 1.16 Effective Q-factor

impedance will then be in series with a series-tuned circuit or in parallel with a parallel-tuned circuit. The Q-factor of the series circuit will be reduced

from $Q=\omega_0 L/R$ to $Q=\omega_0 L/(R+R_s)$ or

$$Q_{eff}=\frac{\omega_0 L}{R}\left(\frac{1}{1+(R_s/R)}\right)=\frac{Q}{[1+(R_s/R)]} \tag{1.35}$$

Similarly, for a parallel circuit consider Fig. 1.16a which shows such a circuit shunted by a resistor R_p. At the resonant frequency the tuned circuit can be represented by its dynamic resistance R_d (see Fig. 1.16b). The two resistances are in parallel and their total resistance, the **effective dynamic resistance, $R_{d(eff)}$**, is

$$R_{d(eff)}=R_d R_p/(R_d+R_p)$$

Therefore,

$$Q_{eff}=R_{d(eff)}/\omega_0 L=Q\omega_0 L R_p/(Q\omega_0 L+R_p)\omega_0 L=QR_p/(R_d+R_p)$$

$$Q_{eff}=\frac{Q}{1+(R_d/R_p)}$$

Example 1.17

A parallel-tuned circuit has a 100 pF capacitor connected in parallel with an inductor of Q-factor 100. The circuit is resonant at 1 MHz. Calculate its effective Q-factor when a resistance of 20 kΩ is connected in parallel.

Solution From equation (1.32)

$$R_d=\frac{100}{2\pi\times 1\times 10^6\times 100\times 10^{-12}}=10^6/2\pi\ \Omega$$

Hence

$$Q_{eff}=\frac{100}{1+(10^6/2\pi\times 20\times 10^3)}=11.2 \quad (Ans)$$

Q-Factors in Series and in Parallel

When the losses of a capacitor are not negligibly small, the overall Q-factor of the circuit will depend upon the Q-factors of the individual components. Suppose that the Q-factors of the inductor and the capacitor are, respectively, Q_L and Q_C. The overall factor Q_T is

$$Q_T=\frac{1}{R_T}\sqrt{\frac{L}{C}} \qquad \text{where } R_T=R_L+R_C$$

Hence

$$Q_T=\frac{1}{(\omega_0 L/Q_L)+(1/\omega_0 C Q_C)}\times\sqrt{\frac{L}{C}}$$

$$=\frac{1}{\frac{1}{Q_L}\sqrt{\frac{L}{C}}+\frac{1}{Q_C}\sqrt{\frac{L}{C}}}\times\sqrt{\frac{L}{C}}$$

$$Q_T = \frac{1}{(1/Q_L)+(1/Q_C)} = Q_L Q_C/(Q_L + Q_C) \qquad (1.37)$$

Equation (1.37) can also be applied to find the overall Q-factor of two Q-factors in *parallel.*

Example 1.18

An inductor of Q-factor 80 is connected in parallel with a capacitor having a Q-factor of 500. Calculate the Q-factor of the combination.

Solution From equation (1.37),

$$Q = \frac{80 \times 500}{80 + 500} = 69 \quad (Ans)$$

Selectivity of a Resonant Circuit

The **selectivity** of a resonant circuit is its ability to discriminate between signals at different frequencies. If a constant-voltage source is applied to a series-resonant circuit the current flowing in the circuit will vary with frequency in the manner shown in Fig. 1.17. The current flowing at the resonant frequency depends solely upon the resistance of the circuit ($I_0 = E/R$), and the shape of the curve depends upon the Q-factor of the circuit. Reducing the circuit resistance increases the peak value of the curve, and by increasing the Q-factor, makes the circuit more selective. The selectivity of a resonant circuit is usually expressed in terms of its **3 dB bandwidth**. This is the bandwidth $f_2 - f_1$ over which the current is not less than $1/\sqrt{2}$ times the resonant current. The current/frequency curve is not symmetrical about the resonant frequency if a linear frequency base is employed, since, for example, the current at $0{\cdot}8f_0$ is equal to the current at $1{\cdot}25f_0$. To obtain a symmetrical curve the current should be plotted to a logarithmic frequency base, Fig. 1.17*b*, although if the Q-factor is fairly high the difference between the two curves is small and is often neglected.

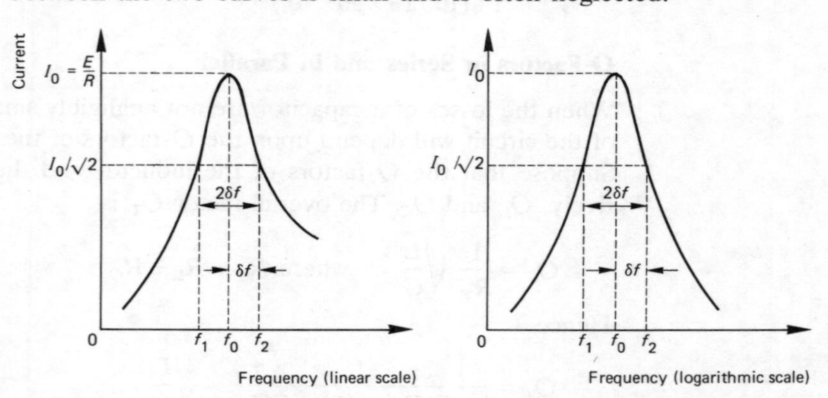

Fig. 1.17 Selectivity of a tuned circuit

If a current is supplied to a parallel-resonant circuit, then the voltage developed across the circuit will vary with change in frequency in the same way as the current in a series-resonant circuit. Thus, Fig. 1.17 will show the voltage/frequency characteristic of a parallel-resonant circuit if the vertical axis is changed to read $V_0 = IR_d$ at resonance, where I is the constant current and R_d the dynamic impedance of the circuit.

The impedance at any frequency $\omega/2\pi$ of a series-tuned circuit is given by equation (1.18*a*) which is repeated here for convenience.

$$Z = R + j[\omega L - (1/\omega C)] \tag{1.38}$$

The current i flowing in the circuit is $i = V/Z$ or

$$i = \frac{V}{R + j[\omega L - (1/\omega C)]} = \frac{V/R}{1 + \dfrac{j\omega_0 L}{R}\left(\dfrac{\omega}{\omega_0} - \dfrac{1}{\omega_0 \omega LC}\right)}$$

$$= \frac{I_0}{1 + jQ\left(\dfrac{\omega}{\omega_0} - \dfrac{\omega_0}{\omega}\right)} = \frac{I_0}{1 + jQ\left(\dfrac{\omega^2 - \omega_0^2}{\omega_0 \omega}\right)}$$

Let $\omega = \omega_0 + \delta\omega_0$. Then $\omega^2 = \omega_0^2 + 2\omega_0\,\delta\omega_0 + (\delta\omega_0)^2$. The last term is negligibly small and hence

$$i \simeq \frac{I_0}{1 + jQ(2\omega_0\,\delta\omega_0/\omega_0^2)} = \frac{I_0}{1 + jQ(2\delta\omega_0/\omega_0)} \tag{1.39a}$$

The term $2\delta\omega_0$ is the bandwidth of the circuit for *any* required number of dB reduction on the maximum current value. Hence equation (1.39*a*) can be re-written as

$$i = \frac{I_0}{1 + (jQB/f_0)} \tag{1.39b}$$

where B is the bandwidth and f_0 is the resonant frequency. Equation (1.39*b*) can be used to obtain a very useful equation for the 3 dB bandwidth of a series-tuned circuit in terms of its resonant frequency and its Q-factor. The magnitude of equation (1.39*b*) is

$$\left|\frac{i}{I_0}\right| = \frac{1}{\sqrt{[1 + (Q^2B^2/f_0^2)]}}$$

3 dB is a current ratio of $1/\sqrt{2}$. Thus for 3 dB reduction,

$$\frac{1}{\sqrt{2}} = \frac{1}{\sqrt{[1 + (Q^2B_{3dB}^2/f_0^2)]}}$$

Hence $1 = Q^2B_{3dB}^2/f_0^2 = QB_{3dB}/f_0$ or

$$B_{3dB} = f_0/Q \tag{1.40}$$

The lower 3 dB frequency f_1 is equal to

$f_0 - (B_{3dB}/2) = f_0 - (f_0/2Q) = f_0[1 - (1/2Q)]$

The upper 3 dB frequency f_2 is equal to $f_0[1 + (1/2Q)]$.

The product $f_1 f_2$ of the two 3 dB frequencies is

$$f_0^2\left(1 - \frac{1}{2Q}\right)\left(1 + \frac{1}{2Q}\right) = f_0^2$$

The impedance of a parallel-tuned circuit can be written as the product of the impedances of the two arms divided by their sum, i.e.

$$Z=\frac{(R+j\omega L)\times 1/j\omega C}{R+j[\omega L-(1/\omega C)]}\simeq\frac{L/C}{R+j[\omega L-(1/\omega C)]}$$

$$=\frac{L/CR}{1+j\dfrac{1}{R}[\omega L-(1/\omega C)]}=\frac{R_d}{1+j\dfrac{1}{R}[\omega L-(1/\omega C)]}$$

This equation is of the same form as after equation (1.38) and following the same steps as before, will lead to a similar result, i.e.

$$Z=\frac{R_d}{1+(jQB/f_0)} \qquad (1.41)$$

Example 1.19

A parallel-tuned circuit is resonant at 120 kHz and has a 3 dB bandwidth of 6 kHz. Calculate the bandwidth over which its impedance is *a*) −6 dB, *b*) −10 dB less than the dynamic impedance.

Solution From equation (1.40)

$$Q=f_0/B_{3dB}=120\times 10^3/6\times 10^3=20$$

a) 6 dB is a voltage ratio of 2:1, hence

$$\left|\frac{Z}{R_d}\right|=\frac{1}{2}=\frac{1}{\sqrt{[1+(Q^2B_{6dB}^2/f_0^2)]}}=\frac{1}{\sqrt{[1+(400\times B_{6dB}^2/144\times 10^8)]}}$$

$$3=400B_{6dB}^2/144\times 10^8$$

$$B_{6dB}=\sqrt{\left(\frac{3\times 144\times 10^8}{400}\right)}=10.39\text{ kHz} \quad (Ans)$$

b) 10 dB is a voltage ratio of 3.162:1, hence

$$\left|\frac{Z}{R_d}\right|=\frac{1}{3.162}=\frac{1}{\sqrt{[1+(400\times B_{10dB}^2/144\times 10^8)]}}$$

$$9=400B_{10dB}^2/144\times 10^8$$

and therefore

$$B_{10dB}=3B_{3dB}=18\text{ kHz} \quad (Ans)$$

Exercises 1

1.1 Explain what is meant by the power factor of a circuit. Determine the voltage across the circuit shown in Fig. 1.18, the current supplied to it, and the phase difference between them at a frequency of $5000/2\pi$ R/S, if the current in the 100 Ω resistor is 10 mA. Also find the total power dissipated in the network and its power factor.

1.2 A coil of inductance 50 mH and *Q*-factor 80 is connected in parallel with a 600 pF loss-free capacitor. This circuit is supplied by a constant current source at the resonant frequency. If the supply current is *I* mA and the voltage developed across the circuit is 25 V calculate *a*) the resonant frequency, *b*) the capacitor current, *c*)

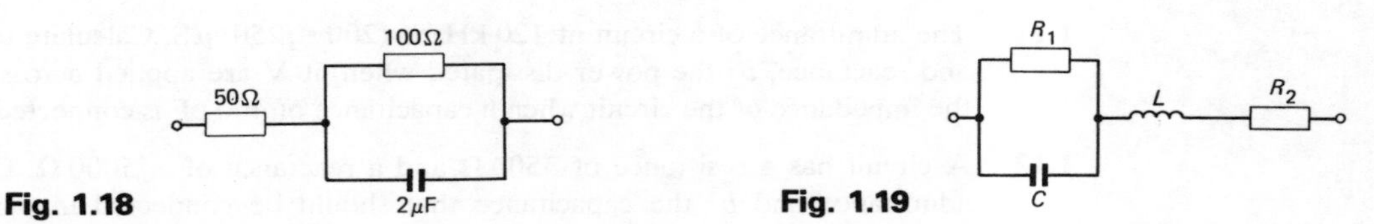

Fig. 1.18 **Fig. 1.19**

the Q-factor of the coil, *d*) the voltage across the circuit when a resistor of 250 kΩ is connected in parallel.

1.3 Derive an expression for the resonant frequency of the circuit shown in Fig. 1.19. If $R_1 = 2000\ \Omega$, $R_2 = 30\ \Omega$, $L = 20$ mH and $C = 0.3\ \mu$F, calculate *a*) the resonant frequency and *b*) the impedance of the circuit at resonance.

1.4 A parallel-tuned circuit is resonant at 100 kHz and has a 3 dB bandwidth of 5 kHz. Calculate the bandwidth over which its impedance is *a*) −6 dB, *b*) −10 dB and *c*) −20 dB relative to the resonant value.

1.5 Three series-tuned circuits are connected in series with one another. The data for the circuits are
Circuit A $R = 10\ \Omega$, LC product $= 2 \times 10^{-10}$, L/C ratio $= 5000$
Circuit B $R = 20\ \Omega$, LC product $= 2 \times 10^{-10}$, L/C ratio $= 5000$
Circuit C $R = 10\ \Omega$, LC product $= 2 \times 10^{-10}$, L/C ratio $= 1250$
For each of these circuits find its capacitance, its inductance, its resonant frequency and its Q-factor. Also find the 3 dB bandwidth of each circuit, and the overall 3 dB bandwidth and resonant frequency.

1.6 Calculate the impedance of the circuit shown in Fig. 1.20. Also find the current that flows when a voltage of 120 V is applied across the circuit.

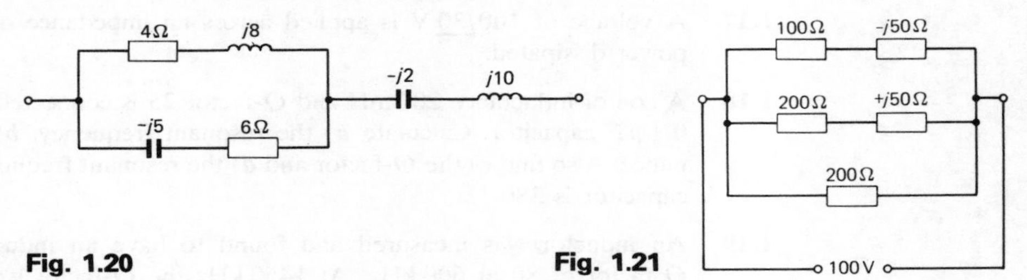

Fig. 1.20 **Fig. 1.21**

1.7 A sinusoidal voltage of 100 V r.m.s. is applied across a circuit whose impedance is $300 + j200\ \Omega$. Calculate *a*) the current flowing in the circuit, *b*) the power dissipated and *c*) the conductance and susceptance of the circuit.

1.8 A sinusoidal signal of 12 V r.m.s. is applied across a circuit whose conductance is 100 μS and whose susceptance is +22 μS. Calculate *a*) the current that flows, *b*) the power dissipated, *c*) the power factor, and *d*) the resistance and reactance of the circuit.

1.9 For the circuit given in Fig. 1.21 calculate *a*) the impedance and the admittance, *b*) the current taken from the supply and *c*) the power dissipated.

1.10 At a frequency of 200 kHz a coil has a resistance of 50 Ω and a reactance of 1200 Ω. Calculate its impedance and its admittance. Also calculate the power dissipated when a 12 V signal is applied across the circuit.

1.11 The admittance of a circuit at 120 kHz is $(200-j250)$ μS. Calculate *a*) its resistance and reactance, *b*) the power dissipated when 50 V are applied across the circuit, *c*) the impedance of the circuit when a capacitance of 400 pF is connected in parallel.

1.12 A circuit has a resistance of 3500 Ω and a reactance of $+j5000$ Ω. Calculate *a*) its admittance and *b*) the capacitance that should be connected in parallel with the circuit in order to make the power factor unity. $\omega = 5000$ R/S.

1.13 An inductor has a resistance of 500 Ω and an inductance of 250 mH at a frequency of 1.59 kHz. Calculate the current that flows in the circuit when 25 V at 1.59 kHz are applied across the circuit.

1.14 For the circuit shown in Fig. 1.22 calculate *a*) the admittance, *b*) the current taken from the supply and *c*) the power dissipated.

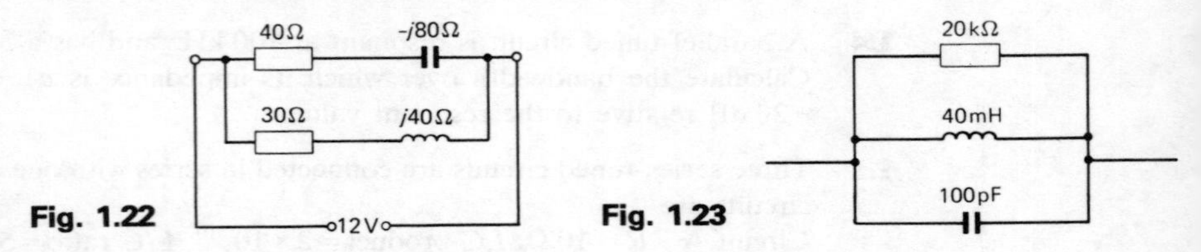

Fig. 1.22 Fig. 1.23

Short Exercises

1.15 An inductor with a Q-factor of 60 is connected in *a*) series, *b*) parallel with a capacitor of $Q = 200$. Calculate the overall Q-factor in each case.

1.16 For the circuit shown in Fig. 1.23 calculate *a*) the dynamic resistance, *b*) the resonant frequency, and *c*) the Q-factor.

1.17 A voltage of $100\angle 30$ V is applied across an impedance of $25\angle 60$ Ω. Calculate the power dissipated.

1.18 A coil of inductance 200 mH and Q-factor 25 is connected in series with a loss free 0·1 μF capacitor. Calculate *a*) the resonant frequency, *b*) the impedance at resonance. Also find *c*) the Q-factor and *d*) the resonant frequency if the Q-factor of the capacitor is 380.

1.19 An inductor was measured and found to have an inductance of 7·5 mH and a Q-factor of 80 at 600 kHz. At 1450 kHz the Q-factor was only 30. Calculate the effective resistance of the inductor at each frequency.

1.20 Calculate the current taken from the supply in the circuit of Fig. 1.24.

1.21 For the circuit given in Fig. 1.25 calculate the impedance.

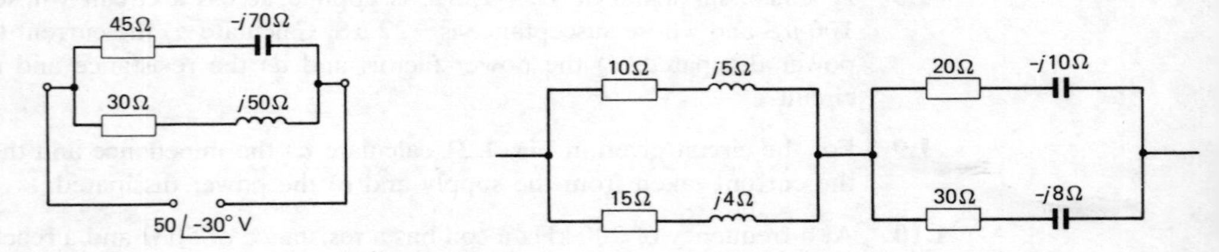

Fig. 1.24 Fig. 1.25

1.22 A coil of inductance 150 mH and Q-factor 25 is connected in series with a capacitor of 0.03 μF. Calculate *a*) the resonant frequency, *b*) the impedance at resonance, *c*) the Q-factor, *d*) the impedance at resonance if the capacitor had a Q-factor of 800.

1.23 An inductance of 2 mH and $Q = 50$ is connected in parallel with a 620 pF loss-free capacitor. Calculate *a*) the resonant frequency, *b*) the capacitor current, *c*) the Q-factor of the circuit, when a constant current of 1 mA is supplied to the circuit.

1.24 The impedance of a circuit is $Z = 100(1 + j1200\,\delta f)$ where δf is the deviation from the resonant frequency. Derive the expression for the admittance of the circuit.

1.25 An inductance of 250 μH is connected in series with a capacitor and the circuit formed is resonant at 500 kHz. A voltage at this frequency is then applied across the circuit with the result that 300 mV appears across the capacitor. A 12 Ω resistor connected in series with the circuit causes the capacitor voltage to fall to 200 mV. Calculate *a*) the capacitance, *b*) the resistance and *c*) the Q-factor of the circuit.

1.26 A 10 mH coil has a Q-factor of 100 and is connected in parallel with a 3 nF capacitor. A variable frequency 6 V source is connected across the circuit. Calculate *a*) the resonant frequency, *b*) the minimum current taken from the source, and *c*) the current in the capacitor at the resonant frequency.

1.27 A 20 kΩ resistor, a 25 μH inductance of $Q = 100$ at 2 MHz, and a capacitor are connected in parallel. Calculate *a*) the capacitance to resonate the circuit, *b*) the dynamic impedance of the circuit, *c*) the Q-factor of the circuit.

2 Circuit Theorems

The solution of electric circuits containing several components can always be carried out using Kirchhoff's first and second laws. These are:

Kirchhoff's First Law The algebraic sum of the currents entering any node in a circuit is zero.

Kirchhoff's Second Law The e.m.f. applied to a loop in a circuit is equal to the sum of the potential differences around that loop.

However, some simplification of the work involved can usually be obtained if the method of circulating currents due to Maxwell is used. With this method a current is assumed to circulate around *each* loop in the circuit in a clockwise direction. Then Kirchhoff's second law is applied. If the solution of the problem yields any current with a negative sign, this merely means that the assumed clockwise direction was in error and the current really flows in the opposite direction.

Very often a considerable reduction in the work involved in the solution of a circuit problem is possible if one, or more, of a number of circuit theorems is employed. The most commonly used theorems are those due to *Thevenin* and to *Norton*, the *superposition theorem*, and the *star-to-delta* (or T-to-π) transform theorem. The application of each of these theorems will be discussed in this chapter together with a discussion of the *maximum power transfer theorem*.

Mesh (or Loop) and Nodal Analysis

The current flowing in any part of a network can be determined by drawing the Maxwell circulating currents in the circuit and then applying Kirchhoff's second law to each **mesh** or **loop**. The resulting simultaneous equations must then be solved to obtain the numerical values for the circulating currents. The actual current flowing in any component is then the algebraic sum of the circulating currents in that component.

When the number of **nodes** or junctions in the circuit is less than the number of loops, it will generally prove to be easier to employ **nodal analysis** instead. For a nodal analysis of a circuit the currents entering each node are summed to produce a number of simultaneous equations.

Consider the circuit shown in Fig. 2.1*a* and suppose the current I flowing in the 20 Ω resistor is to be determined.

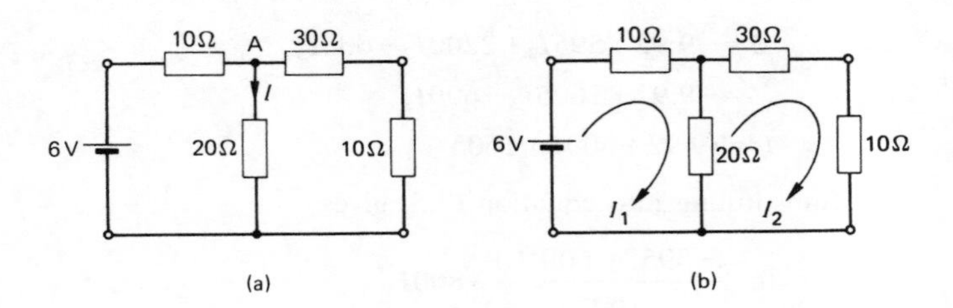

Fig. 2.1

1 Using *mesh analysis*, circulating currents I_1 and I_2 are drawn (see Fig. 2.1*b*). Then, Kirchhoff's second law is applied to each loop:

$$6 = I_1(10+20) - I_2 20 \tag{2.1}$$

$$0 = -I_1 20 + I_2(20+30+10) \tag{2.2}$$

From equation (2.2)

$$I_2 = 20I_1/60 = I_1/3$$

Substituting into equation (2.1) gives

$$6 = I_1 30 - 20I_1/3 \quad \text{or} \quad I_1 = 0.257 \text{ A}$$

Therefore

$$I_2 = 0.257/3 = 0.086 \text{ A} \quad \text{and} \quad I = I_1 - I_2 = 0.171 \text{ A} \quad (Ans)$$

2 Using *nodal analysis*, let the voltage at the junction or node marked as A be V_A. Then

$$\frac{6-V_A}{10} - \frac{V_A}{20} - \frac{V_A}{40} = 0$$

$$0.6 = V_A(1/10 + 1/20 + 1/40) = 0.175\,V_A$$

and $V_A = 3.43$ V. Therefore $I = 3.43/20 = 0.171$ A (as before).

Example 2.1

Calculate using *a*) mesh analysis and *b*) nodal analysis, the current flowing in the 200 Ω resistor in Fig. 2.2.

Solution

a) *Mesh analysis*

$$10 = I_1 605 \quad - I_2 600 \tag{2.3}$$

$$0 = -I_1 600 + I_2 2200 - I_3 600 \tag{2.4}$$

$$0 = \quad - I_2 600 \quad + I_3 800 \tag{2.5}$$

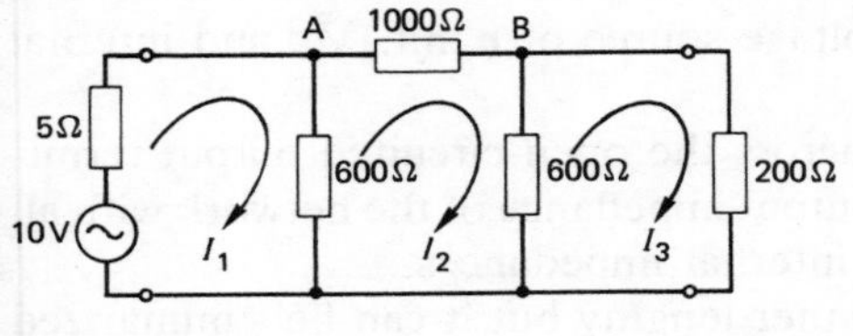

Fig. 2.2

From equation (2.3)

$$I_1 = \frac{10 + 600I_2}{605}$$

and substituting into equation (2.4)

$$0 = -9.92 - 595I_2 + 2200I_2 - 600I_3$$
$$= -9.92 + 1605I_2 - 600I_3$$
$$I_2 = (9.92 + 600I_3)/1605$$

Substituting into equation (2.5) gives

$$0 = \frac{(-5952 - 600^2 I_3)}{1605} + 800I_3$$
$$= -3.71 - 224.3I_3 + 800I_3 = -3.71 + 575.7I_3$$

Therefore $I_3 = 3.71/575.7 = 6.44$ mA (*Ans*)

b) Nodal analysis

$$\frac{10 - V_A}{5} - \frac{V_A}{600} - \frac{V_A - V_B}{1000} = 0 \qquad (2.6)$$

$$\frac{V_A - V_B}{1000} - \frac{V_B}{600} - \frac{V_B}{200} = 0 \qquad (2.7)$$

$$2 - V_A(1/5 + 1/600 + 1/1000) = V_B/1000$$
$$2 - V_A \times 0.203 = 0.001\,V_B \qquad (2.8)$$

Also $0.001\,V_A = V_B(1/1000 + 1/600 + 1/200) = 0.0077\,V_B$

Therefore

$$V_A = V_B \frac{0.0077}{0.001} = 7.7\,V_B \qquad (2.9)$$

Substituting into equation (2.8):

$$2 - (0.203 \times 7.7\,V_B) = 0.001\,V_B$$
$$2 = V_B(0.001 + 0.203 \times 7.7) \quad \text{so} \quad V_B = 1.28 \text{ V}$$

Therefore $I_{200} = V_B/200 = 1.280/200 = 6.4$ mA (as before)

The slight difference between the two answers obtained arises because of the rounding-off of results that has taken place in both of the calculations.

Thevenin's and Norton's Theorems

The two theorems due to Thevenin and Norton are often used to reduce a network to a simpler form and thereby minimize the effort involved in a calculation.

Thevenin's Theorem

Thevenin's theorem states that

The current flowing in an impedance Z_L connected across the output terminals of a linear network will be the same as the current that would flow if Z_L were connected to a voltage source of e.m.f. V_{oc} and internal impedance Z_{oc}.

V_{oc} is the voltage that appears across the open-circuited output terminals of the network and Z_{oc} is the output impedance of the network with all internal sources replaced by their internal impedances.

The statement of the theorem is rather lengthy but it can be summarized by means of the circuits shown in Fig. 2.3.

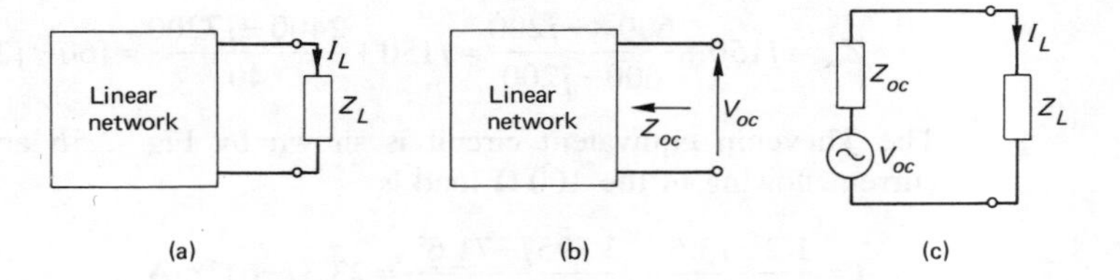

Fig. 2.3 Thevenin's theorem

Example 2.2

Use Thevenin's theorem to determine the current flowing in an impedance of $(250-j500)\ \Omega$ when it is connected across the terminals AB of the network in Fig. 2.4*a*.

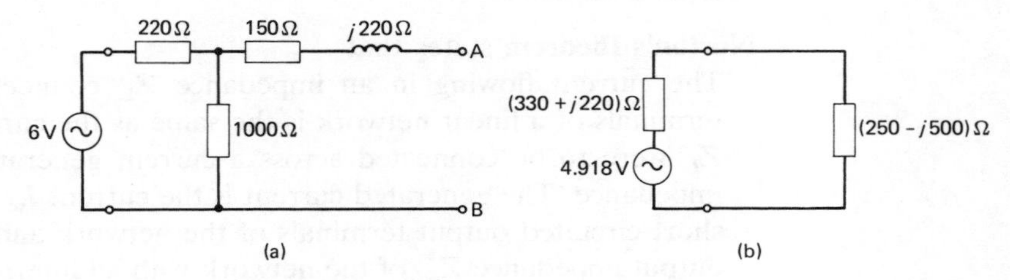

Fig. 2.4

Solution The voltage V_{oc} appearing across the open-circuited terminal AB is $V_{oc}=6\times 1000/(220+1000)=4.918$ V

The output impedance of the network is

$$Z_{oc}=150+j220+\frac{1000\times 220}{1000+220}=(330+j220)\ \Omega$$

The Thevenin equivalent circuit is shown by Fig. 2.4*b* and from this

$$I=\frac{4.918}{580-j280}=7.64\underline{/25.8^\circ}\ \text{mA} \quad (Ans)$$

Example 2.3

For the circuit shown in Fig. 2.5, use Thevenin's theorem to calculate the power dissipated in the 100 Ω load impedance.

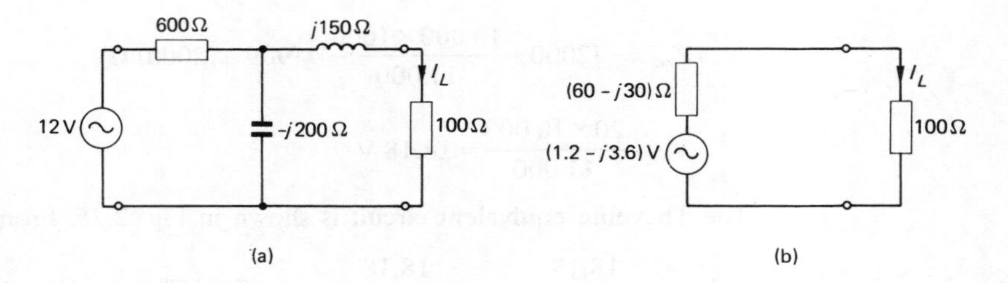

Fig. 2.5

Solution

$$V_{oc}=\frac{12\times -j200}{600-j200}=\frac{48-j144}{40}=(1.2-j3.6)\ \text{V}$$

$$Z_{oc} = j150 + \frac{600 \times -j200}{600 - j200} = j150 + \frac{2400 - j7200}{40} = (60 - j30)\ \Omega$$

The Thevenin equivalent circuit is shown by Fig. 2.5*b* and from this figure the current flowing in the 100 Ω load is

$$I = \frac{1.2 - j3.6}{160 - j30} = \frac{3.795\angle{-71.6^\circ}}{162.79\angle{-10.6^\circ}} = 23.3\angle{-61^\circ}\ \text{mA}$$

The power dissipated in the 100 Ω load is

$$(23.3 \times 10^{-3})^2 \times 100 = 54.3\ \text{mW} \quad (Ans)$$

Norton's Theorem

Norton's theorem states that

The current flowing in an impedance Z_L connected across the output terminals of a linear network is the same as the current that would flow if Z_L were to be connected across a current generator in parallel with an impedance. The generated current is the current I_{sc} that would flow in the short-circuited output terminals of the network and the impedance is the output impedance Z_{oc} of the network with all internal sources replaced by their internal impedances.

Norton's theorem is summarized by Fig. 2.6.

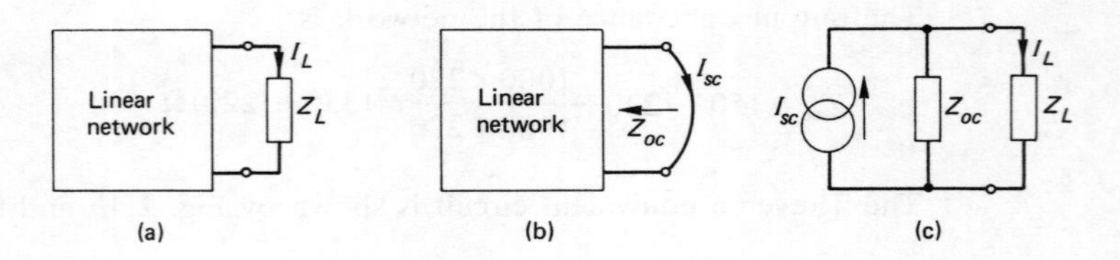

Fig. 2.6 Norton's theorem

Example 2.4

For the network shown in Fig. 2.7*a* calculate the current flowing in a 1000 Ω resistor connected across terminals AB using *a*) Thevenin's theorem and *b*) Norton's theorem.

Solution

a) *Thevenin's Theorem*

$$Z_{oc} = -j2000 + \frac{10\,000 \times 1000}{11\,000} = (909 - j2000)\ \Omega$$

$$V_{oc} = \frac{20 \times 10\,000}{11\,000} = 18.18\ \text{V}$$

The Thevenin equivalent circuit is shown in Fig. 2.7*b*. From this figure,

$$I_L = \frac{18.18}{1909 - j2000} = \frac{18.18}{2765\angle{-46.3^\circ}} = 6.58\angle{46.3^\circ}\ \text{mA} \quad (Ans)$$

b) *Norton's Theorem*

As in *a*) $Z_{oc} = (909 - j2000)\ \Omega$

The impedance "seen" by the voltage source when the load is short-circuited is

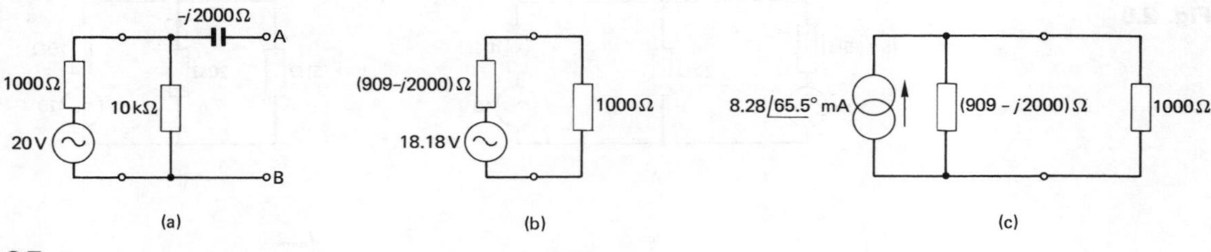

Fig. 2.7

$$1000+\frac{10\,000\times(-j2000)}{10\,000-j2000}=1385-j1923\ \Omega$$

The input current to the network is $20/(1385-j1923)$ and so the short-circuit current I_{sc} is equal to

$$\frac{20}{1385-j1923}\times\frac{10\,000}{10\,000-j2000}=8.28\underline{/65.5^\circ}\text{ mA}$$

The Norton equivalent circuit is shown in Fig. 2.7*c*, from which

$$I_L=8.28\underline{/65.5^\circ}\times\frac{909-j2000}{1909-j2000}=6.58\underline{/46.3^\circ}\text{ mA}\quad (Ans)$$

If Norton's theorem is applied to Fig. 2.7*b*,

$$I_{sc}=18.18/(909-j2000)=18.18\underline{/0^\circ}/2196\underline{/-65.5^\circ}$$
$$=8.28\underline{/65.5^\circ}\text{ mA}$$

as in Fig. 2.7*c*. Alternatively, if Thevenin's theorem is applied to Fig. 2.7*c*,

$$V_{oc}=8.28\underline{/65.5^\circ}\times10^{-3}\times(909-j2000)=18.18\text{ V}$$

as in Fig. 2.7*b*. This means, of course, that either of the two equivalent circuits can be quickly converted into the other.

Superposition Theorem

The **superposition theorem** states that

The current at any point in a *linear* network containing two, or more, voltage or current sources is the sum of the currents that would flow at that point if only one source is considered at a time, all other sources being replaced by impedances equal in value to their internal impedances.

Example 2.5

Calculate the current flowing in the 20 Ω resistor of Fig. 2.8*a* if the two generators operate at the same frequency and their e.m.f.s are *a*) in phase and *b*) in antiphase with one another.

Solution

a) Replace the left-hand generator by its internal impedance to obtain the circuit shown in Fig. 2.8*b*. Either Thevenin's or Norton's theorem can be used to simplify the circuit or the current I_1 can be determined using basic principles.

Using Thevenin's theorem,

$$Z_{oc}=\frac{10(5+j5)}{15+j5}=\frac{50+j50}{15+j5}=4+j2\ \Omega$$

Fig. 2.8

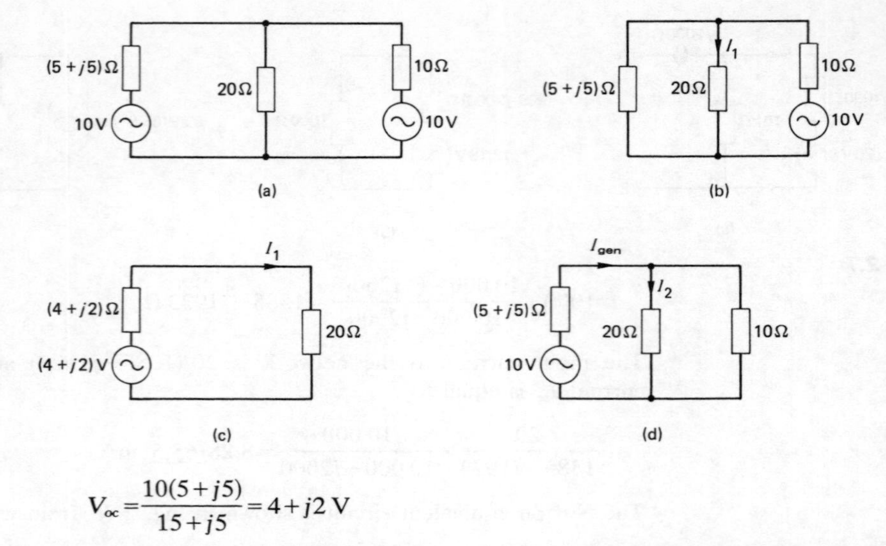

$$V_{oc} = \frac{10(5+j5)}{15+j5} = 4+j2 \text{ V}$$

The Thevenin equivalent circuit is shown in Fig. 2.8*c*. From this figure

$$I_1 = \frac{4+j2}{24+j2} = \frac{(4+j2)(24-j2)}{24^2+2^2} = 0.172+j0.069 \text{ A}$$

b) Now replace the right-hand generator by its internal impedance to obtain the circuit shown in Fig. 2.8*d*. The current I_{gen} supplied by the generator is

$$I_{gen} = \frac{10}{5+j5+\frac{20\times 10}{20+10}} = \frac{10}{11.67+j5}$$

Therefore,

$$I_2 = \frac{10}{11.67+j5} \times \frac{10}{20+10} = \frac{10}{35+j15} = (0.241-j0.103) \text{ A}$$

a) When the two generators are in phase with one another the current I_{20} flowing in the 20 Ω resistor is

$$I_{20} = I_1 + I_2 = 0.413 - j0.034 \text{ A} = 414.4\underline{/-4.7^\circ} \text{ mA} \quad (Ans)$$

b) When the two generators are in antiphase,

$$I_{20} = I_1 - I_2 = -69 + j172 \text{ mA} = 185.3\underline{/112^\circ} \text{ mA} \quad (Ans)$$

The Star–Delta (or T–π) Transformation Theorem

The terms *star* and *delta* refer to two types of network commonly employed in heavy current engineering, particularly in conjunction with three-phase circuits. Star and delta networks are illustrated by Figs. 2.9*a* and *b* respectively. Figs. 2.9*c* and *d* respectively show the T and π networks. It can be seen that the T network is just a star network drawn upside down with Z_2 and Z_3 drawn horizontally. Similarly, the delta and the π networks are really the same. The networks are normally drawn in their T or π forms in light current work.

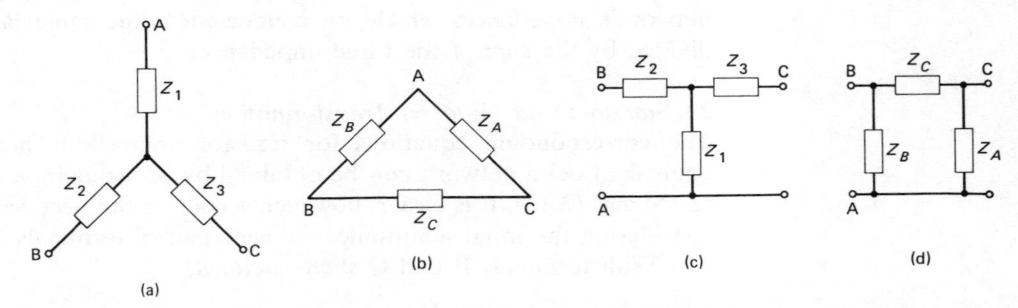

Fig. 2.9 Showing *a*) a star network, *b*) a delta network, *c*) a *T* network, and *d*) a π network

Very often the solution of a circuit can be simplified if a star (T) network is converted into the equivalent delta (π) network or vice versa.

1 *Delta-to-Star* (π*-to-*T) *Transformation*

For the star (T) and delta (π) networks to be equivalent, the impedances "seen" looking into any pair of terminals, with the third terminal open-circuited, must be equal.

i) With terminals C open-circuited:

$$Z_1+Z_2=\frac{Z_B(Z_A+Z_C)}{Z_A+Z_B+Z_C} \tag{2.10}$$

ii) With terminals B open-circuited,

$$Z_1+Z_3=\frac{Z_A(Z_B+Z_C)}{Z_A+Z_B+Z_C} \tag{2.11}$$

iii) With terminals A open-circuited

$$Z_2+Z_3=\frac{Z_C(Z_A+Z_B)}{Z_A+Z_B+Z_C} \tag{2.12}$$

Add equations (2.10) and (2.11),

$$2Z_1+Z_2+Z_3=\frac{2Z_AZ_B+Z_C(Z_A+Z_B)}{Z_A+Z_B+Z_C} \tag{2.13}$$

Subtract equation (2.12) from equation (2.13),

$$Z_1=\frac{Z_AZ_B}{Z_A+Z_B+Z_C} \tag{2.14}$$

Substitute equation (2.14) into equation (2.10),

$$Z_2=\frac{Z_BZ_C}{Z_A+Z_B+Z_C} \tag{2.15}$$

Substitute equation (2.15) into equation (2.12),

$$Z_3=\frac{Z_AZ_C}{Z_A+Z_B+Z_C} \tag{2.16}$$

These three equations can best be remembered by noting that each impedance in the star (T) network is equal to the product of the delta (π)

network's impedances which are connected to the same labelled terminal divided by the sum of the three impedances.

2 *Star-to-Delta* (T-*to*-π) *Transformation*

The corresponding equations for transforming a star network into the equivalent delta network *can* be obtained by re-arranging equations (2.14), (2.15) and (2.16). It is easier, however, to derive the necessary equations by considering the input admittances of each pair of terminals. Thus,

i) With terminals B and C *short-circuited,*

$$Y_A+Y_B=\frac{Y_1(Y_2+Y_3)}{Y_1+Y_2+Y_3} \tag{2.17}$$

ii) With terminals A and C short-circuited,

$$Y_B+Y_C=\frac{Y_2(Y_1+Y_3)}{Y_1+Y_2+Y_3} \tag{2.18}$$

iii) With terminals A and B short-circuited,

$$Y_A+Y_C=\frac{Y_3(Y_1+Y_2)}{Y_1+Y_2+Y_3} \tag{2.19}$$

Clearly these three equations are of the same form as those derived for the delta–star transformation and they will hence lead to similar results.

Adding equations (2.17) and (2.18) and then subtracting equation (2.19) from the sum will give

$$Y_B=\frac{Y_1Y_2}{Y_1+Y_2+Y_3} \tag{2.20}$$

or

$$Z_B=\frac{Y_1+Y_2+Y_3}{Y_1Y_2}=(1/Y_2)+(1/Y_1)+(Y_3/Y_1Y_2)$$

$$=Z_1+Z_2+\frac{Z_1Z_2}{Z_3}$$

Therefore,

$$Z_B=\frac{Z_1Z_3+Z_2Z_3+Z_1Z_2}{Z_3} \tag{2.21}$$

Substitute equation (2.20) into equation (2.18),

$$Y_C=\frac{Y_2Y_3}{Y_1+Y_2+Y_3} \tag{2.22}$$

or $$Z_C=\frac{Z_1Z_3+Z_2Z_3+Z_1Z_2}{Z_1} \tag{2.23}$$

Also $$Y_A=\frac{Y_1Y_3}{Y_1+Y_2+Y_3} \qquad (2.23)$$

and $$Z_A = \frac{Z_1Z_3 + Z_1Z_2 + Z_2Z_3}{Z_2} \quad (2.24)$$

It should be noted, as a memory aid, that an impedance is equal to the sum of the products of pairs of star impedances divided by the star impedance on the opposite side of the network.

Example 2.6

Reduce the circuit given in Fig. 2.10 to a single equivalent resistance in series with a voltage source. Hence calculate the power dissipated in the network.

Fig. 2.10

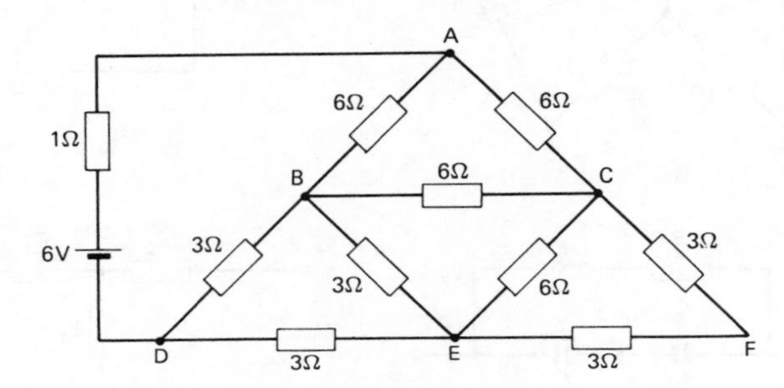

Solution The resistance between terminals C and E is 6 Ω in parallel with 6 Ω or 3 Ω. If the delta circuit A,B,C is transformed into its equivalent star circuit, the component values are

i) From equation (2.14) $Z_1 = \frac{6\times6}{6+6+6} = 2\,\Omega$

ii) From equation (2.15), $Z_2 = 2\,\Omega$, and from equation (2.16), $Z_3 = 2\,\Omega$ also. Similarly, transforming the delta network B,D,E gives

$$Z_1' = \frac{3\times3}{3+3+3} = 1\,\Omega \qquad Z_2' = 1\,\Omega \qquad Z_3' = 1\,\Omega$$

Fig. 2.10 can therefore be re-drawn as shown in Fig. 2.11*a*. This circuit can be still further reduced, since the paths XBY and XCEY are in parallel, to the circuit of Fig. 2.11*b*. Thus, there is a single equivalent resistance of 5 Ω across the terminals AD (*Ans*)

Therefore, the current flowing is 6/6 = 1 A and the power dissipated is
$1^2 \times 5 = 5$ W (*Ans*)

Example 2.7

Determine an expression for the frequency at which the bridged-T network shown in Fig. 2.12*a* has zero output voltage.

Solution Let $Z_C = R_1 + j\omega L$ and $Z_A = Z_B = 1/j\omega C$ and convert the network into its equivalent T network.

i) From equation (2.16)

$$Z_3 = \frac{(R_1 + j\omega L)/j\omega C}{R_1 + j[\omega L - (2/\omega C)]} = \frac{(L/C) - (jR_1/\omega C)}{R_1 + j[\omega L - (2/\omega C)]}$$

This is also Z_2, since $Z_A = Z_B$.

Fig. 2.11

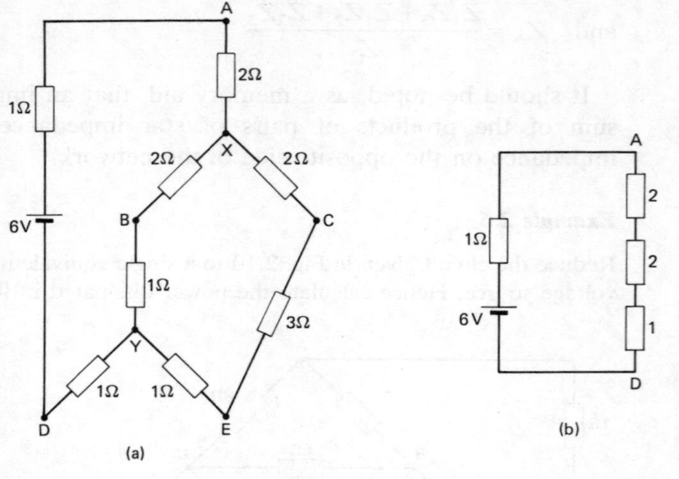

Fig. 2.12 Bridged T network

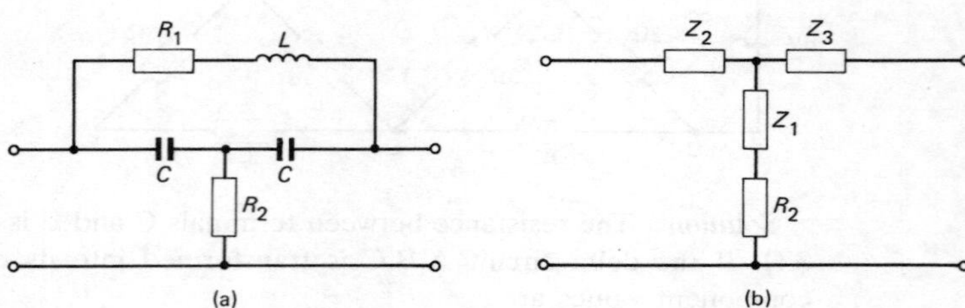

ii) From equation (2.14)

$$Z_1 = \frac{(1/j\omega C)\times(1/j\omega C)}{R_1 + j[\omega L - (2/\omega C)]} = \frac{-1/\omega^2C^2}{R_1 + j[\omega L - (2/\omega C)]}$$

The network will have zero output voltage at the frequency at which the shunt path has zero impedance. (The values of Z_2 and Z_3 are unimportant.) Hence, from Fig. 2.12*b*,

$$R_2 - \frac{1/\omega_0^2C^2}{R_1 + j[\omega_0 L - (2/\omega_0 C)]} = 0$$

$$R_1R_2 + jR_2[\omega_0 L - (2/\omega_0 C)] = 1/\omega_0^2C^2$$

Equating the real terms,

$$R_1R_2 = 1/\omega_0^2C^2 \qquad \omega_0^2 = 1/R_1R_2C^2$$

$$\text{or} \quad f_0 = \frac{1}{2\pi C\sqrt{(R_1R_2)}}\text{Hz} \tag{2.25}$$

Equating the imaginary parts,

$$R_2[\omega_0 L - (2/\omega_0 C)] = 0 \qquad \omega_0^2 = 2/LC$$

$$\text{or} \quad f_0 = \frac{1}{2\pi}\sqrt{\frac{2}{LC}}\text{Hz} \tag{2.26}$$

Both equation (2.25) and equation (2.26) must be satisfied for the network to have a null in its output voltage/frequency characteristic at frequency f_0 (in practice, a fairly sharp minimum is obtained).

Maximum Power Transfer Theorem

The **maximum power transfer** theorem states that

The maximum power is transferred from a source to a load when the load impedance is the conjugate of the source impedance.

Fig. 2.13 shows a voltage source of e.m.f. E_S and internal impedance $R_S + jX_S$ connected to a load of impedance $R_L + jX_L$. The current I_L flowing in the load is

$$I_L = \frac{E_S}{R_S + R_L + j(X_S + X_L)} \tag{2.27}$$

and the load power P_L is $|I_L|^2 R_L$ or

$$P_L = \frac{E_S^2 R_L}{(R_S + R_L)^2 + (X_S + X_L)^2} \tag{2.28}$$

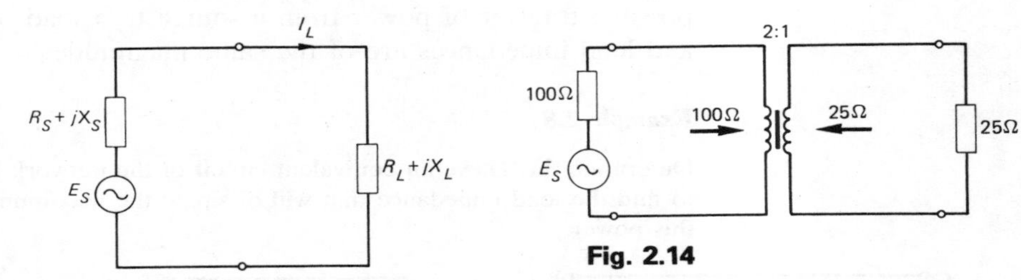

Fig. 2.13

Fig. 2.14

For the load power to be a maximum it is clear that the total reactance of the circuit must be zero, so that $X_L = -X_S$. For this condition

$$P_L = E_S^2 R_L / (R_S + R_L)^2$$

Differentiating with respect to R_L and equating the result to zero gives

$$\frac{dP_L}{dR_L} = \frac{(R_S + R_L)^2 E_S^2 - 2E_S^2 R_L (R_S + R_L)}{(R_S + R_L)^4} = 0 \quad \text{or} \quad R_S^2 = R_L^2$$

The condition for maximum power transfer is therefore

$$R_S = R_L \quad \text{and} \quad X_S = X_L \tag{2.29}$$

If a *purely resistive source* and a *purely resistive load* are considered, equation (2.28) will apply with $X_S = X_L = 0$, and following the same steps leads to the conclusion that for maximum power transfer

$$R_S = R_L \tag{2.30}$$

Very often it is only possible to control the effective magnitude of the load impedance, its angle remaining unchanged. This is the case, for example, when a transformer is used for impedance transformation (see Fig. 2.14). Re-writing equation (2.28),

$$P_L = \frac{E_S^2 |Z_L| \cos \varphi_L}{(R_S + |Z_L| \cos \varphi_L)^2 + (X_S + |Z_L| \sin \varphi_L)^2}$$

For P_L to be a maximum

$$\frac{dP_L}{d\,|Z_L|}=E_S^2\cos\varphi_L[(R_S+|Z_L|\cos\varphi_L)^2+(X_S+|Z_L|\sin\varphi_L)^2]$$
$$-E_S^2|Z_L|\cos\varphi_L[2(R_S+|Z_L|\cos\varphi_L)\cos\phi_L$$
$$+2(X_S+|Z_L|\sin\varphi_L)\sin\varphi_L]=0$$

(numerator only). Hence

$$R_S^2+2R_S\,|Z_L|\cos\varphi_L+|Z_L|^2\cos^2\varphi_L+X_S^2+2X_S\,|Z_L|\sin\varphi_L+|Z_L|\sin^2\varphi_L$$
$$=2R_S\,|Z_L|\cos\varphi_L+2\,|Z_L|^2\cos^2\varphi_L+2X_S\,|Z_L|^2\sin\varphi_L+2\,|Z_L|\sin^2\varphi_L$$

or $R_S^2+X_S^2=|Z_L|^2(\cos^2\varphi_L+\sin^2\varphi_L)$ and so

$$|Z_S|=|Z_L| \tag{2.31}$$

Thus, if the angle of the load impedance cannot be altered, the maximum possible transfer of power from a source to a load occurs when the source and load impedances are of the same magnitude.

Example 2.8

Determine the Thevenin equivalent circuit of the network given in Fig. 2.15. Use it to find the load impedance that will dissipate the maximum power and the value of this power.

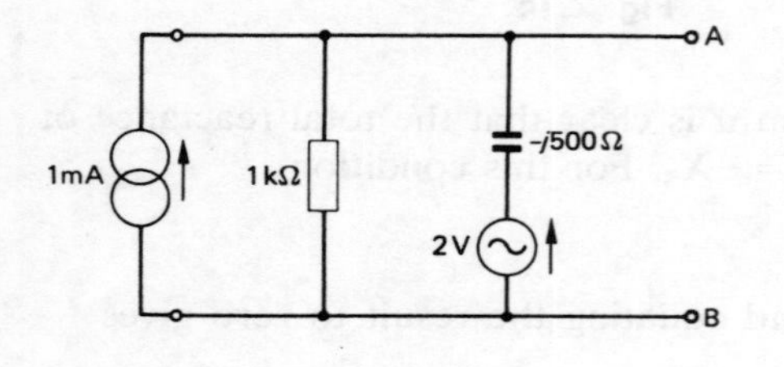

Fig. 2.15

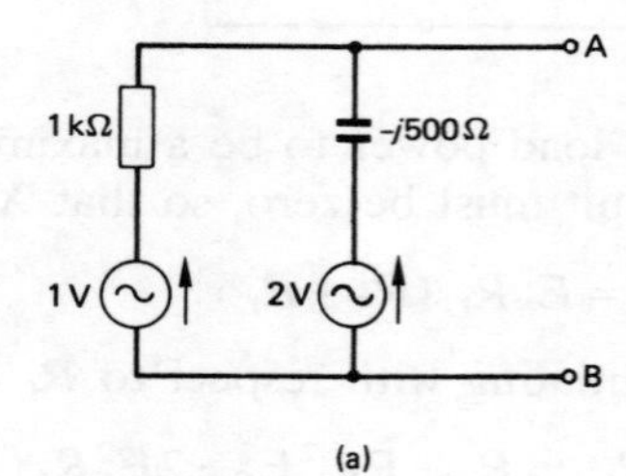

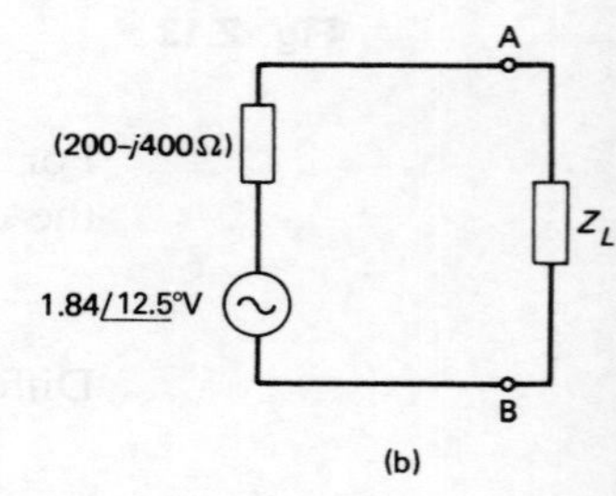

Fig. 2.16

Solution The constant-current source can be converted into a voltage source by the use of Thevenin's theorem. The voltage source has an e.m.f. of $1\times10^{-3}\times1000=1$ V and an impedance of 1000 Ω. The Thevenin equivalent circuit is shown in Fig. 2.16*a*.

Using the superposition theorem

$$V_1=\frac{1\times -j500}{1000-j500}=\frac{-j}{2-j}\text{ V}$$

$$V_2=\frac{2\times1000}{1000-j500}=\frac{4}{2-j}\text{ V}$$

The voltage V_{AB} across the terminals AB is

$$V_{AB}=\frac{4-j}{2-j}=1.84\angle 12.5^\circ\text{ V}$$

$$Z_{oc}=\frac{1000\times(-j500)}{1000-j500}=\frac{-j1000}{(2-j)}=200-j400\ \Omega$$

The Thevenin equivalent circuit is given in Fig. 2.16*b*. The load impedance for maximum power is $(200+j400)\ \Omega$ (*Ans*)

With this value of load impedance connected across terminals A and B

$$I_L = 1.84\underline{/12.5°}/400 = 4.6\underline{/12.5°}\ \text{mA}$$

The load power is

$$(4.6\times10^{-3})^2\times200 = 4.23\ \text{mW} \quad (Ans)$$

Example 2.9

A voltage source has an e.m.f. of 10 V and an internal impedance of $(600+j100)$ ohms. The source is to supply power to a 20 Ω load resistor. Determine the turns ratio of the transformer needed to connect load to source in order to obtain the maximum load power. Calculate the value of this maximum load power, assuming the transformer losses are negligibly small.

Solution The transformer will only match the magnitudes of the source and load impedances. Since

$$\text{Impedance ratio} = (\text{turns ratio})^2$$

$$n = \sqrt{\left[\frac{\sqrt{(600^2+100^2)}}{20}\right]} = 5.51:1 \quad (Ans)$$

The current supplied by the source is

$$\frac{10}{600+j100+(5.51^2\times20)} = \frac{10}{1207+j100}$$

Therefore, $|I| = 8.23$ mA and so the maximum load power P_L is

$$P_L = (8.26\times10^{-3})^2\times607 = 41.1\ \text{mW} \quad (Ans)$$

Exercises 2

2.1 State Thevenin's theorem. Determine the Thevenin equivalent circuit for Fig. 2.17. Thence find the load impedance that will dissipate the maximum power and the value of that power.

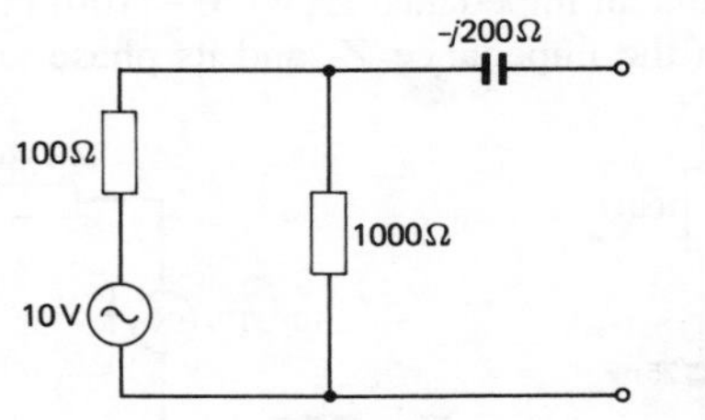

Fig. 2.17

2.2 A voltage source has an e.m.f. of 15 V and an internal impedance of $(500+j30)\ \Omega$. The source is connected to an 8 Ω load by a transformer whose losses may be neglected. Calculate the transformer turns ratio for the maximum load power to be dissipated and the value of this power.

2.3 A voltage source has an e.m.f. of $20 \sin 10^3 t$ volts and an internal impedance of 600 Ω. A 1 μF capacitance is effectively connected across the terminals of the source. The voltage source is connected across an inductor of 60 mH inductance and 50 Ω resistance. Calculate the magnitudes of the current in, and the voltage across, the inductor.

2.4 For the circuit in Fig. 2.18 calculate *a*) the two components which when connected in series, *b*) the two components which, when connected in parallel, across the terminals AB, will dissipate the maximum power.

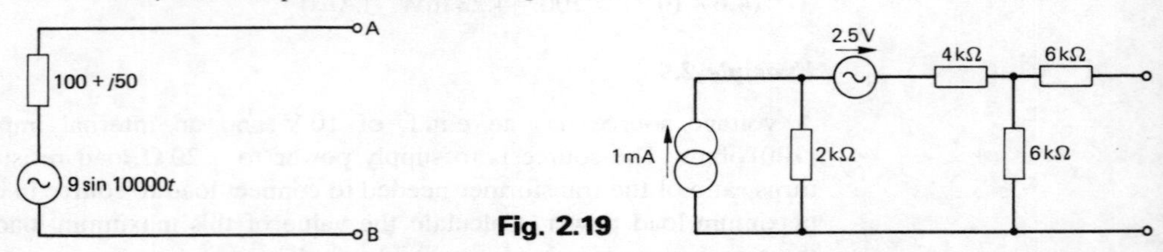

Fig. 2.18

Fig. 2.19

2.5 Obtain the Thevenin and Norton equivalent circuits of the network shown in Fig. 2.19.

2.6 Use both the star-delta transformation and Thevenin's theorem to reduce the circuit shown in Fig. 2.20 to simpler form. Then determine the value of R_L for it to dissipate the maximum possible power and the value of this power.

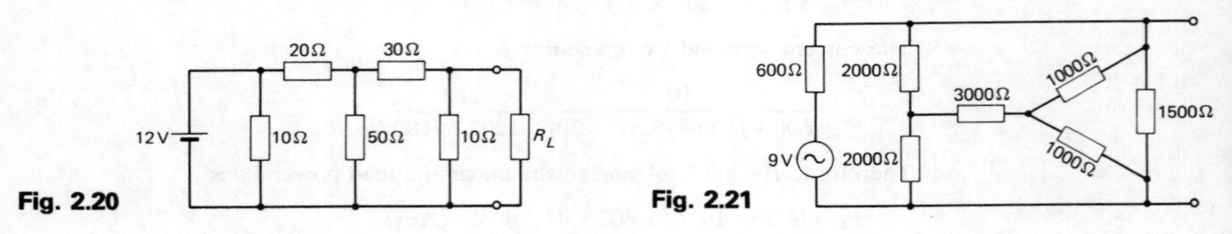

Fig. 2.20

Fig. 2.21

2.7 Repeat exercise 2.6 using the circuit of Fig. 2.21.

2.8 Calculate the power dissipated in the 100 Ω load resistor in Fig. 2.22.

2.9 Two voltages $V_1 = (100 - j100)$ volts and $V_2 = (100 + j100)$ volts having internal impedances of $Z_1 = (30 + j30)$ ohms and $Z_2 = (50 + j0)$ ohms respectively are connected in parallel across an impedance $Z_3 = (20 + j100)$ ohms. Calculate the value of the current flowing in the impedance Z_3 and its phase angle relative to V_2.

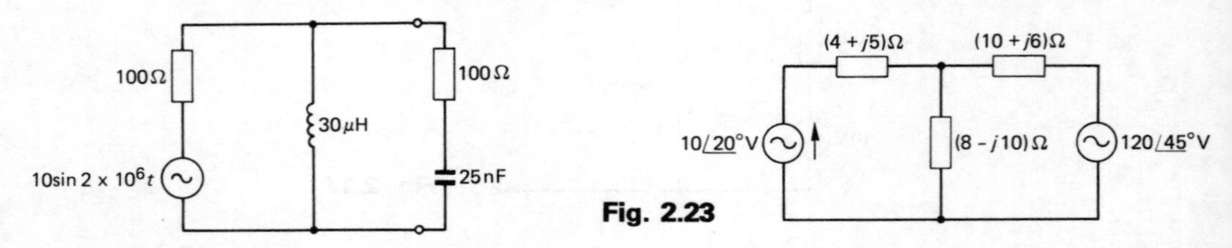

Fig. 2.22

Fig. 2.23

2.10 Calculate the current flowing in each of the impedances shown in Fig. 2.23.

2.11 In the circuit of Fig. 2.24 a current of 3 A in phase with V_1 flows in the unknown impedance Z_3. Use Kirchhoff's laws to find the value of Z_3 and also the currents flowing in the other impedances.

2.12 Use Maxwell's circulating currents to calculate the current I_1 in the circuit given in Fig. 2.25.

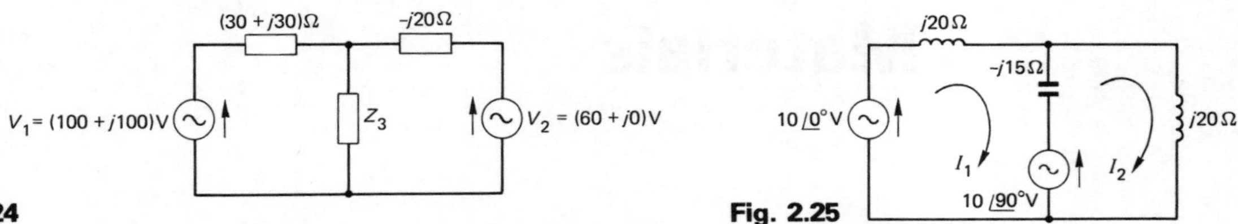

Fig. 2.24

Fig. 2.25

2.13 Repeat exercise 2.12 using nodal analysis.

Short Exercises

2.14 A voltage source is connected to a variable resistor. When the resistance is varied the voltage across it varies in the manner shown by the table.

R (Ω)	100	200	400
V (V)	7	10	12.73

Calculate the maximum power the source is able to deliver to a resistive load.

2.15 Three 10 Ω resistors are connected in delta. Calculate the components of the equivalent star network.

2.16 Three 10 Ω resistors are star connected. Calculate the components of the equivalent delta network.

2.17 A source of impedance $(6000-j125)$ Ω at a frequency of $1000/2\pi$ Hz is to become $(50-j20)$ Ω. Calculate the turns ratio of the necessary matching transformer.

2.18 A source of impedance $(6000-j125)$ Ω at a frequency of $1000/2\pi$ Hz is to be connected to a resistive load of 6000 Ω. Calculate the value of the component that should be connected in series with the load in order that the load should dissipate the maximum possible power.

2.19 A source of 6 V e.m.f. and internal impedance 600 Ω is shunted by a capacitance of 20 pF. Obtain the Thevenin equivalent circuit if the frequency is 20 kHz.

2.20 Calculate the current in the 100 Ω resistor in Fig. 2.26. Use the superposition theorem.

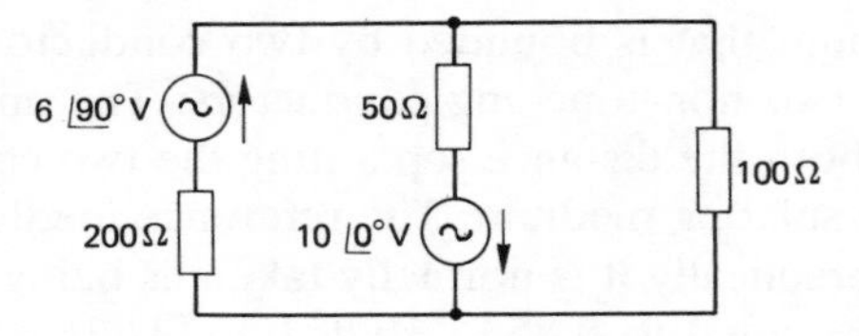

Fig. 2.26

2.21 Repeat 2.20 using mesh analysis.

2.22 Repeat 2.20 using nodal analysis.

2.23 A star network has $Z_1 = 20$ Ω, $Z_2 = (10+j10)$ Ω and $Z_3 = (20-j5)$ Ω. Calculate the values of the components of the equivalent delta circuit.

2.24 The Thevenin equivalent circuit of a network is calculated to consist of a voltage generator of $15.4\underline{/60°}$ volts in series with a resistor of 1 kΩ, a capacitor of reactance $-j100$ Ω and a load impedance Z_L. What should be the value of Z_L for it to dissipate the maximum possible power? What is the value of this power?

3 Magnetic and Dielectric Materials

The passive components which are widely used in electrical and electronic circuits are the resistor, the inductor and the capacitor. Although the first-named component, the resistor, is used to provide a pre-determined amount of resistance into a circuit with consequent power dissipation, the other two components should, ideally, dissipate zero energy.

Unfortunately, all the dielectric and magnetic materials available for use possess inherent losses and so their practical performance must inevitably fall short of the ideal. Dielectric materials are also used to insulate conductors in cables and these should also have the minimum possible loss.

Dielectric Materials and Capacitors

The insulating material between the plates of a capacitor or around the conductors of a telephone cable is known as the **dielectric**. A wide variety of different materials have been used as dielectrics but all of them exhibit various imperfections. The various parameters of a dielectric material will be discussed and then some typical figures for various materials will be given.

Permittivity

Capacitance is a measure of how much electric energy can be stored by an insulating medium that is bounded by two conductors. Capacitance *always* exists between two non-touching conductors. The amount of stored energy depends upon both the distance separating the two conductors and upon the nature of the insulating medium. The reference medium is, strictly speaking, a vacuum but practically it is normally taken as being air. Air is said to have a **permittivity** ε_0 equal to 8.854×10^{-12} F/m [$1/(36\pi \times 10^9)$ F/m]. Any other medium will increase the capacitance between the two conductors because its permittivity will be greater than ε_0. Thus the capacitance of two parallel plates of area A m^2 separated by a distance d metres will possess a capacitance of $\varepsilon_0 A/d$ farads when air is the dielectric material. When some other dielectric material is employed the capacitance will be increased to $\varepsilon_0 \varepsilon_r A/d$ farads where ε_r is a dimensionless quantity known as the **relative permittivity** of the medium.

Some typical figures for the relative permittivity of commonly employed dielectrics are given in Table 3.1.

Table 3.1 Relative permittivities of common dielectrics

Air	1.0006	Paper	2–4
Aluminium oxide	7	Polyethylene	2.3
Bakelite	4–8	Polypropylene	2.5
Ceramic	20–100	Polystyrene	2.55
Glass	4–10	Rubber	2
Mica	2.5–8	Tantalum oxide	10–20
Mylar	3	Teflon	2
Oil	2.5	Water	80

Table 3.2 Dielectric strength

Dielectric material	*E* (V/m)	*Dielectric material*	*E* (V/m)
Air	2×10^6	Oil	6×10^6
Glass	$8–20 \times 10^6$	Rubber	30×10^6
Mica	65×10^6		

Electric Field Strength

The **electric field strength**, often known as the **dielectric strength**, of a dielectric is the electric field in volts/metre at which the dielectric breaks down. Some typical figures are given in Table 3.2.

Leakage Current

The **leakage current** of a dielectric is determined by its *insulation resistance.* When a direct voltage is applied across a dielectric, a current will flow which will reach a steady value after a time determined by a time constant *CR*. Obviously, the insulation resistance should be as high as possible in order to minimize the leakage current.

Dielectric Absorption

When a capacitor is charged up from a constant voltage source and is then discharged, and then the capacitor is left on open-circuit for some time, it is found that a new, but smaller, charge accumulates. This effect arises because some of the original charge was absorbed by the dielectric. The practical result of this is that there must always be a difference between the rates at which a capacitor can be charged and discharged. This means that the effective value of the capacitance falls with increase in the frequency of an applied voltage. **Dielectric absorption** is normally quoted as a percentage and Table 3.3 gives a few typical values.

Loss Angle and Power (Dissipation) Factor

When a capacitance is first charged up and then is discharged it is always found that the energy that can be taken out of the capacitance is less than

Table 3.3 Dielectric absorption values

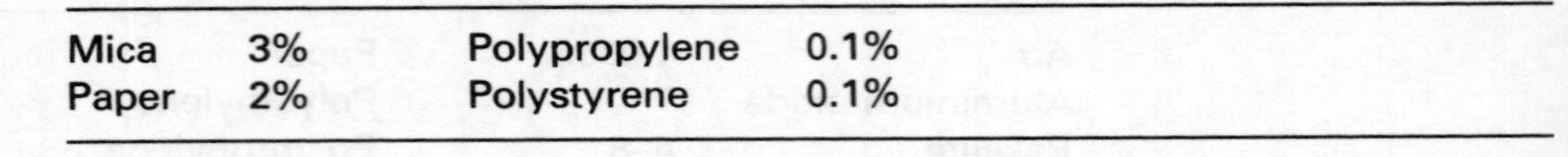

Mica	3%	Polypropylene	0.1%
Paper	2%	Polystyrene	0.1%

R_s C

(a)

C

R_p

(b)

Fig. 3.1 Representation of capacitor losses *a*) by series resistance, *b*) by parallel resistance

the energy that was originally put into it. The missing energy is lost because of a combination of the following reasons:

a) Surface leakage.

b) Leakage paths within the dielectric itself because of various imperfections.

c) The finite insulation resistance of the dielectric.

d) The work done in storing and then releasing charge.

It is neither possible nor particularly desirable to distinguish between these various effects and so their combined effect is usually taken into account by assuming that all the energy loss is caused by I^2R or V^2/R power dissipation in a resistance. This *equivalent loss resistance* may be considered to be connected *either* in series *or* in parallel with the (assumed) perfect capacitance. This is shown by Fig. 3.1*a* and *b*. The series loss resistance has only a low resistance value, typically less than 0.1 Ω, while the parallel loss resistor is of very much higher value, usually 0.5 MΩ or more.

1 The phasor diagram representing the current and voltages in the case of the SERIES LOSS RESISTOR is shown by Fig. 3.2. The phase difference between the sinusoidal applied voltage V and the current I that flows is the angle θ. Since the applied voltage is sinusoidal, the **power factor** of the dielectric is equal to $\cos\theta$.

The **loss angle** ψ of the dielectric is the angle by which θ falls short of 90° (the zero loss case), i.e.

$$\psi = (90-\theta)° \tag{3.1}$$

The loss angle is always a very small angle, certainly less than 1°, and this means that

$$\text{Power factor} = \cos\theta = \sin\psi = \psi \tag{3.2}$$

Thus, the power factor of a dielectric is very nearly equal to its loss angle.

Referring to Fig. 3.2; since ψ is small $\sin\psi \simeq \tan\psi \simeq \psi$ and hence

$$\sin\psi = \frac{IR_s}{I/\omega C} = \omega CR_s \quad \text{or}$$

$$R_s = \psi/\omega C \tag{3.3}$$

Further, from equation (1.27) the Q-factor of R_s in series with C is $Q = 1/\omega CR_s$, hence

$$Q = 1/\psi = 1/(\text{power factor}) \tag{3.4}$$

The modern practice is to quote the quality of a dielectric in terms of another quantity known as the **dissipation factor** D. This parameter is defined as being the reciprocal of the Q-factor. This means, of course, that

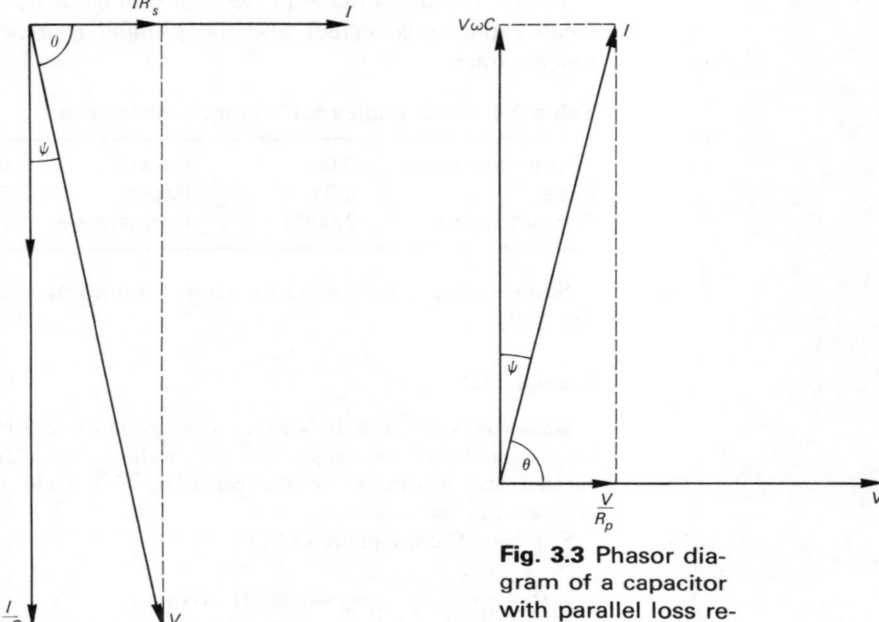

Fig. 3.2 Phasor diagram of a capacitor with series loss resistance

Fig. 3.3 Phasor diagram of a capacitor with parallel loss resistance

provided the loss angle is small the dissipation and power factors are very nearly equal to one another.

Example 3.1

At a frequency of 5000 rad/sec the losses of a 0.1 μF capacitor can be represented by a 0.1 Ω resistor. Calculate *a*) the power factor, *b*) the Q-factor, *c*) the dissipation factor, and *d*) the power dissipated in the capacitor when a 2 V r.m.s. sinusoidal voltage at a frequency of 5000 rad/sec is applied across the capacitor.

Solution

a) Power factor $= \omega CR_s = 5000 \times 0.1 \times 10^{-6} \times 0.1 = 50 \times 10^{-6}$ (*Ans*)

b) $Q = 1/(\text{power factor}) = 20\,000$ (*Ans*) *c*) dissipation factor = 20 000 (*Ans*)

d) The current flowing in the capacitor is

$$I = \frac{2}{\sqrt{[0.1^2 + 1/(5000 \times 0.1 \times 10^{-6})^2]}} \simeq \frac{2}{2000} = 1 \text{ mA}$$

and the power dissipated is

$$I^2R_s = (1 \times 10^{-3})^2 \times 0.1 = 0.1\ \mu\text{W} \quad (Ans)$$

2 The phasor diagram of the currents and voltage in the PARALLEL LOSS RESISTANCE circuit of Fig. 3.1*b* is given in Fig. 3.3. Once again the loss angle is the angle by which θ falls short of 90°. From the phasor diagram,

$$\sin\psi \simeq \tan\psi = \frac{V/R_p}{V\omega C} = 1/\omega CR_p \text{ or since } \sin\psi = \psi$$

$$R_p = 1/\psi\omega C \tag{3.5}$$

In general, the series representation of capacitor losses should be used for small capacitance values and the parallel representation for large capacitance values.

Table 3.4 Loss angles for common dielectrics

Aluminium oxide	0.06	Ceramic	0.005
Mica	0.01	Paper	0.01
Polypropylene	0.0003	Polystyrene	0.003

Some typical loss angles for some commonly used dielectrics are listed in Table 3.4.

Example 3.2

A sinusoidal voltage of 10 V r.m.s. at a frequency of $3000/2\pi$ Hz is applied across a 1 μF capacitor of loss angle 2.5×10^{-4} radians. Calculate the equivalent series and parallel loss resistances of the capacitor. For each resistor calculate the power dissipated in the capacitor.

Solution From equation (3.3)

$$R_s = \frac{2.5 \times 10^{-4}}{3000 \times 1 \times 10^{-6}} = 0.083\ \Omega \quad (Ans)$$

From equation (3.5)

$$R_p = 1/(2.5 \times 10^{-4} \times 3000 \times 1 \times 10^{-6}) = 1.33\ \text{M}\Omega \quad (Ans)$$

For the series loss resistance

$$I = V\omega C = 10 \times 3000 \times 1 \times 10^{-6} = 30\ \text{mA}$$

and the power dissipated is

$$(30 \times 10^{-3})^2 \times 0.083 = 75\ \mu\text{W} \quad (Ans)$$

For the parallel loss resistance the power dissipated is

$$P = 10^2/1.33 \times 10^6 = 75\ \mu\text{W} \quad (Ans)$$

All capacitors possess some inductance and a leakage current and hence their equivalent circuit should really be somewhat more complex (see Fig. 3.4). In this circuit L is the inductance of the capacitor, R_s' is a resistor that

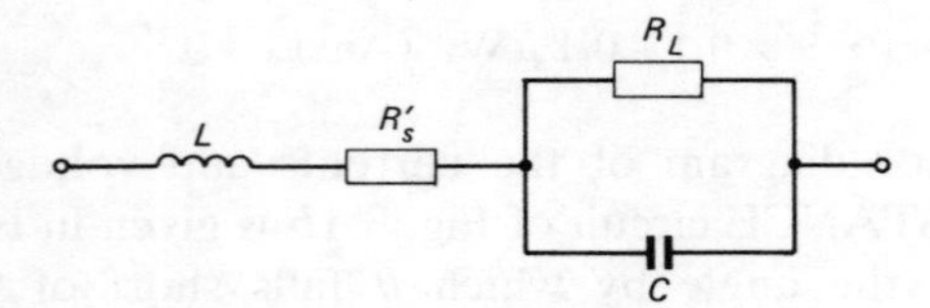

Fig. 3.4 Complete equivalent circuit of a capacitor

represents all the real losses of the capacitor (i.e. the lead, plate and contact resistances), R_L represents the dielectric losses, and C is the actual capacitance. Much of the self-inductance arises from the connecting leads and the manner in which they are connected to the component proper.

The impedance of Fig. 3.4 is

$$Z = R'_s + j\omega L + \frac{R_L/j\omega C}{R_L + (1/j\omega C)}$$

$$= R'_s + \frac{R_L}{1+\omega^2 C^2 R_L^2} + \frac{j[\omega L - \omega R_L^2 C + \omega^3 R_L^2 L C^2]}{1+\omega^2 C^2 R_L^2}$$

The real part of this expression is the *equivalent series loss* resistance R_s of the capacitor. Thus

$$R_s \simeq R'_s + (1/\omega^2 C^2 R_L)$$

The Mechanical and Thermal Characteristics of Dielectric Materials

The choice of a dielectric material for a particular application will be decided by the mechanical and thermal characteristics of the various alternatives. Depending upon the intended use the main requirement may be for mechanical strength, for resistance to attack by acids, or for high thermal resistance. The insulator used for the conductors in a telephone cable must be able to withstand repeated bending but its resistance to a direct pressure need not be high.

Once the main dielectric used in both capacitors and cables was a combination of air and paper. It is obvious that the mechanical strength of such a dielectric is very small and, of course, the resistance to water is very limited. Nowadays the most commonly used dielectric for cables is some kind of plastic and this is used for many types of capacitor also.

The plastic materials are of low density, have a high resistance to attack by chemicals, are of high thermal resistivity, and are relatively easy to fabricate into any required shape. The main disadvantage of plastics is their low mechanical strength and elasticity and their relatively high thermal expansion coefficient. Polyethylene has a high resistance to most chemicals and solvents and it remains both tough and flexible over quite a wide range of temperatures. There are two types: one of which, known as low-density, will soften if immersed in boiling water, while the other, known as high density, does not soften in boiling water. Polypropylene becomes softer at a higher temperature than polyethylene and it is particularly good for use in any environment where it will be subjected to repeated bending, since it is highly resistant to cracking. Polystyrene is of high resistance to most acids but is very vulnerable to some others. It is easily moulded into a required shape and is very light in weight.

The particular characteristic of rubber, real or synthetic, is its ability to show considerable deformation when subject to stress and yet return to its original shape and dimensions immediately the stress is removed. Rubber is not rigid and its mechanical strength is not high.

Bakelite is a very hard and rigid material that has a high thermal resistivity and good resistance to acids, oils, etc. Ceramics are also of high strength and rigidity and also exhibit a high resistance to abrasion and to wear. These attributes also apply to aluminium oxide.

Magnetic Materials

The materials which exhibit the phenomenon of magnetism are chiefly iron and steel. A variety of alloys of iron and steel, such as nickel and cobalt, are also widely used as magnetic materials. When a piece of iron and steel is placed in a magnetic field it will become magnetized by a process known as *induction.* The iron provides a path of smaller reluctance for the lines of magnetic flux and so they tend to concentrate within the iron. This is illustrated by Fig. 3.5.

The degree to which the magnetic field is concentrated in the iron is expressed by the **relative permeability** μ_r of the iron. Relative permeability is defined as being the ratio of the magnetic flux density produced in the iron (or other material) to the flux density produced in air (strictly speaking a vacuum) by the same magnetic field. The permeability of air is known as the **absolute permeability** of free space μ_0 and it is equal to $4\pi \times 10^{-7}$ H/m.

The permeability μ of a magnetic material is then equal to the product $\mu_r\mu_0$. The relative permeability μ_r of a magnetic material may have a value of several thousand but it is not a constant quantity. μ_r is equal to the ratio of the magnetizing force (p. 92) to the flux density and the relationship between these two parameters is not a linear one.

Fig. 3.6 shows a typical B–H curve characteristic from the slope of which values of μ_r can be obtained. Permeability curves for various magnetic materials are available from their manufacturers.

Hysteresis

If a magnetic material is taken through a complete cycle of magnetization and the corresponding values of magnetizing force and flux density are plotted, the **hysteresis loop** of the material will be obtained.

Consider Fig. 3.7*a*. If the material is initially demagnetized and a positive magnetizing force is applied, in a number of discrete steps, the flux density follows the line marked as OA. If, then, the magnetizing force is removed step-by-step it is found that the fall in the flux density does *not* follow the path OA but instead it follows the path AB. It can be seen that, when the magnetizing force has been reduced to zero, some **residual magnetism** remains, represented by the distance OB. This residual magnetism is known as the *remanent flux density.*

To reduce the flux density to zero a negative magnetizing force OC must be applied to the material. This force is known as the *coercive force.*

Further increase in the negative magnetizing force will increase the flux density in the opposite direction to before, see path CD. If, now, the magnetizing force is removed the material will again exhibit some residual magnetism since the flux density will only reduce to the value OE when the magnetizing force has reduced to zero.

If the material is taken into **magnetic saturation** by either, or both, the positive and negative magnetizing forces, then *a*) the remanent flux density is known as the **remanence** and *b*) the coercive force is called the **coercivity**. These terms are illustrated by Fig. 3.7*b*.

Typical hysteresis loops for a permanent magnet material and for an electromagnet material are shown in Fig. 3.8*a* and *b*. It should be noted that a material which is suitable for use as a permanent magnet has a wide

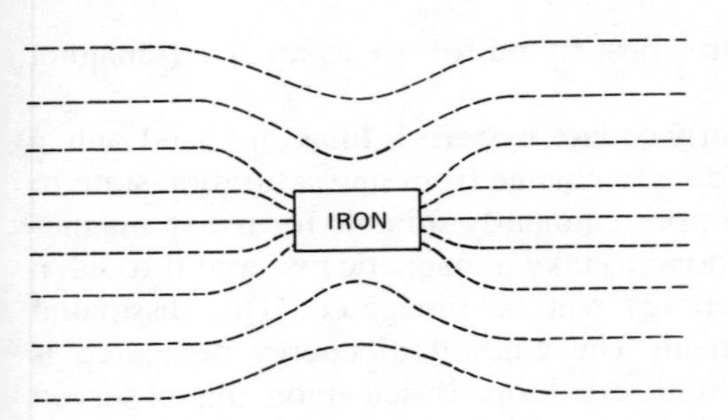

Fig. 3.5 Magnetic field distorted by the presence of an iron bar

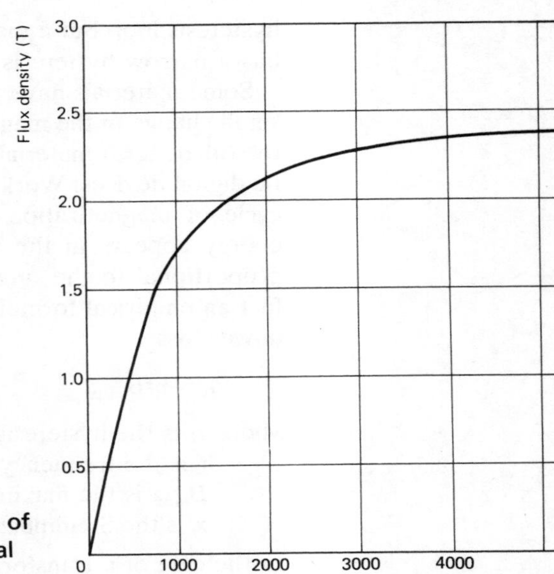

Fig. 3.6 B–H curve of a magnetic material

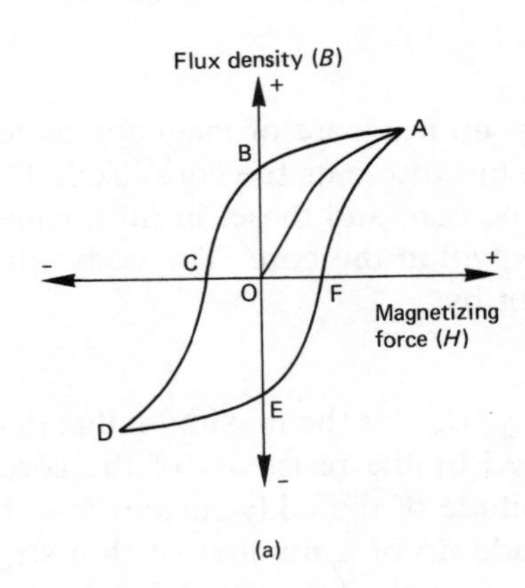

Fig. 3.7 Hysteresis loops

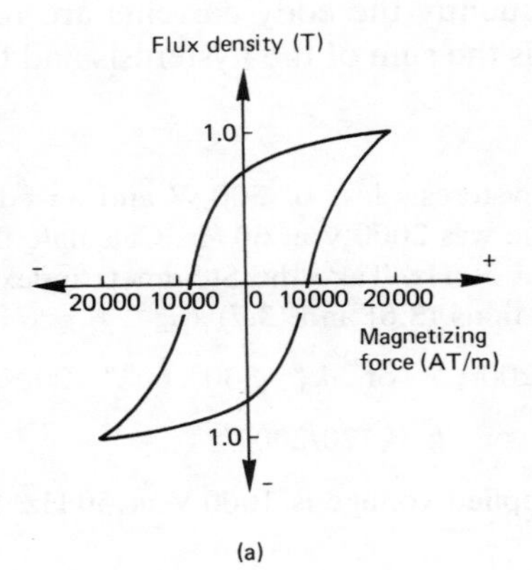

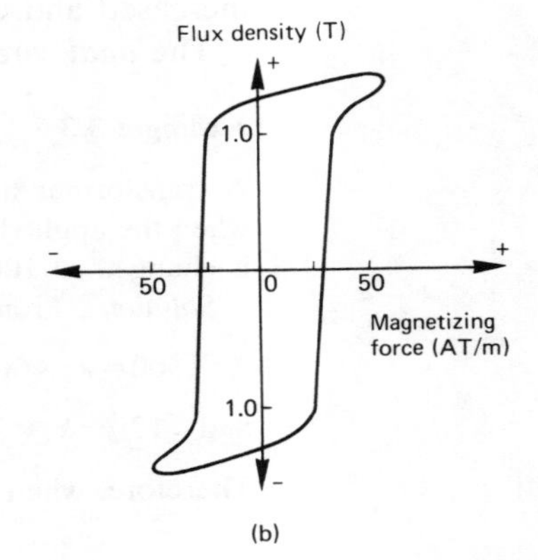

Fig. 3.8 Hysteresis loops of *a*) a permanent magnet material and *b*) an electromagnet material

hysteresis loop but a material that is best suited for use as an electromagnet has a narrow hysteresis loop.

Some materials have a very nearly **square hysteresis loop** and need only a small change in the magnetizing force to change from one saturation state to the other. Such materials are therefore eminently suitable for use in magnetic digital devices. Work must be done to take a magnetic material through a cycle of magnetization and so energy will be dissipated. This dissipated energy appears in the form of heat. The amount of energy dissipated is proportional to the *area* of the hysteresis loop. Based upon this observed fact an empirical formula has been obtained by Steinmetz for the **hysteresis power loss**:

$$P_H = \eta f B_{max}^x \qquad (3.6a)$$

where η is the hysteresis coefficient of the material,
f is the frequency,
B_{max} is the maximum flux density,
x is the Steinmetz index and this is often taken as being equal to 1.6.

In the case of a transformer core, B_{max} is proportional to voltage/frequency (provided the change in frequency is fairly small) and then

$$P_H = kf^{(1-x)}V^x \text{ watts} \qquad \text{where } k \text{ is a constant} \qquad (3.6b)$$

Eddy Current Loss

The alternating flux set up in a core of magnetic material not only cuts any wires wound around it but *also* cuts the core itself. This means that the flux induces voltages into the core and these, in turn, cause currents—known as **eddy currents**—to flow within the core. The eddy currents produce an I^2R power loss that is given by

$$P = kf^2B_{max}^2 \qquad (3.7)$$

where f is the frequency, B_{max} is the maximum flux density in the core and k is a constant determined by the resistivity of the core material.

To reduce the magnitude of the eddy current loss the core of an inductor or a transformer is made up of a number of thin strips, known as laminations, or a *ferrite* core is employed. In either case the resistance of the core is increased and consequently the eddy currents are reduced in magnitude.

The *total core loss* is the sum of the hysteresis and the eddy current losses.

Example 3.3

A transformer had a hysteresis loss of 300 W and an eddy current loss of 120 W when the applied voltage was 2000 V at 60 Hz. Calculate the total loss if this voltage is changed to 1000 V at 50 Hz. Take the Steinmetz index as 1.6.

Solution From equations (3.6) and (3.7)

$$300 = k_1 \times 60^{-0.6} \times 2000^{1.6} \quad \text{or} \quad k_1 = 300 \times 60^{0.6} \times 2000^{-1.6}$$

and $\quad 120 = k_2 \times 2000^2 \quad \text{or} \quad k_2 = 120/2000^2$

Therefore, when the applied voltage is 1000 V at 50 Hz, hysteresis loss is

$$P_H = (300 \times 60^{0.6} \times 2000^{-1.6})50^{-0.6} \times 1000^{1.6} = 110.4\ \text{W}$$

Eddy current loss is

$$P_c = \frac{120}{2000^2} \times 1000^2 = 30\ \text{W}$$

The total iron loss is

$$P = 110.4 + 30 = 140.4\ \text{W} \quad (Ans)$$

Properties of Magnetic Materials

Materials that possess the ability to be magnetized are known as **ferromagnetic** materials and these, in turn, are either suited to use as permanent magnets or as electromagnets.

1 *Permanent magnet* materials are those which have high values of *both* remanence *and* coercivity and (unfortunately) a high hysteresis loss. Because of these features, once such a material has become magnetized, it will retain its magnetism for a long while after the magnetizing force has been removed.

Most permanent magnet materials consist of some kind of iron or steel alloy. In the past, alloys containing cobalt, tungsten and chromium have been used but most modern materials consist of some particular alloy of iron, and one or more of aluminium, nickel, cobalt and copper.

2 Materials suited for use in an *electromagnet* should possess the following properties: *a*) high electrical resistivity, *b*) high permeability, *c*) low remanence, *d*) low coercivity, and *e*) minimum hysteresis loss. Also of importance is the flux density at which the material saturates and the constancy of the magnetic properties with the passage of time.

Suitable materials are often nickel-iron alloys, sometimes with the addition of either some copper or some cobalt. Other electromagnetic materials are various alloys of cobalt and iron. Table 3.5 compares some of the more popular alloys.

Table 3.5 Electromagnetic materials

Magnetic material	*Typical permeability*	*Resistivity*
Iron-silicon	3 500	high
Nickel-iron	100 000	low
Nickel-iron-copper	80 000	medium
Nickel-iron-copper-chromium	75 000	high
Nickel-iron cobalt	10 000	high

Ferrites

Ferrites are a class of non-metallic material which have a very high resistivity and a high permeability. Because of their high resistivity the problems associated with eddy currents are largely overcome. The ferrite material is a compound of iron oxide, zinc, and nickel particles held together by a binding agent and formed by a high temperature process. Typically, the permeability may be somewhere in the range 200–10 000.

Exercises 3

3.1 Fig. 3.9 shows the equivalent circuit of an iron-cored inductor. Derive an expression for the equivalent series resistance and reactance of the inductor. At a particular frequency the reactance of the inductor is 800 Ω, $R_s = 50$ Ω and $R_p = 30$ kΩ. Calculate the equivalent resistance and reactance of the inductor.

3.2 Two capacitors of 0.4 μF each and power factor 3×10^{-4} and 4.5×10^{-4} are connected first in series and then in parallel with one another. For each connection calculate the total capacitance and power factor.

3.3 A 0.1 μF capacitor has a loss angle of 2.5×10^{-4} rad. If the frequency of the sinusoidal supply voltage is $8000/2\pi$ Hz calculate *a*) the reactance of the capacitor, *b*) its series loss resistance, and *c*) its parallel loss resistance.

3.4 A capacitance of 0.159 μF is connected to a 20 V supply at 2 kHz. The power dissipated in the capacitor is then 200 μW. Calculate *a*) the parallel loss resistance, *b*) the dissipation factor, and *c*) the *Q*-factor.

3.5 Explain with the aid of a phasor diagram the meaning of the term loss angle when applied to a dielectric. A 12 V signal at 4 kHz is applied to a capacitor. If the reactance of the capacitor is 1200 Ω calculate *a*) the capacitance, *b*) the capacitor current, and *c*) if the dissipation factor is 4×10^{-4} calculate the power dissipated and the equivalent loss resistance.

3.6 A capacitor may be represented at 1 MHz by a capacitance of 110 pF in series with a resistance of 0.05 Ω. Calculate *a*) the power factor, *b*) the *Q*-factor, *c*) the dissipation factor, and *d*) the parallel loss resistance of the capacitor.

3.7 A 0.5 H inductor has a resistance of 10 Ω. Its losses at a frequency of $5000/2\pi$ Hz may be represented by a parallel resistor of 20 kΩ. Determine the effective series resistance and series inductance of the inductor. A current of 100 mA flows in this component at the same frequency. Calculate the power dissipated in the inductor.

3.8 Fig. 3.10 shows a parallel plate capacitor with two dielectrics arranged as shown. Prove that the effective capacitance is given by

$$C = \frac{\varepsilon_0 \varepsilon_1 \varepsilon_2 d}{\varepsilon_1 (l_2 - l_1) + \varepsilon_2 l_1}$$

Derive an expression for the potential across one of the dielectrics when a voltage V is applied between the plates. If $V = 3500$ V, $\varepsilon_1 = 4$, $\varepsilon_2 = 6$, $l_2 = 10$ mm and $l_1 = 4$ mm calculate the voltage gradient in each dielectric.

3.9 A 100 kVA 6600/330 V single-phase transformer has a hysteresis loss of 400 W and an eddy current loss of 180 W when a voltage of 6600 V at 60 Hz is applied to the primary winding. Calculate the total iron loss when the primary voltage is 6600 V at 50 Hz. Take the Steinmetz index to be 1.6.

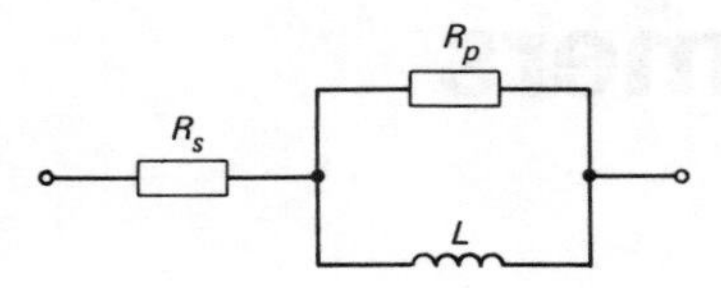

Fig. 3.9

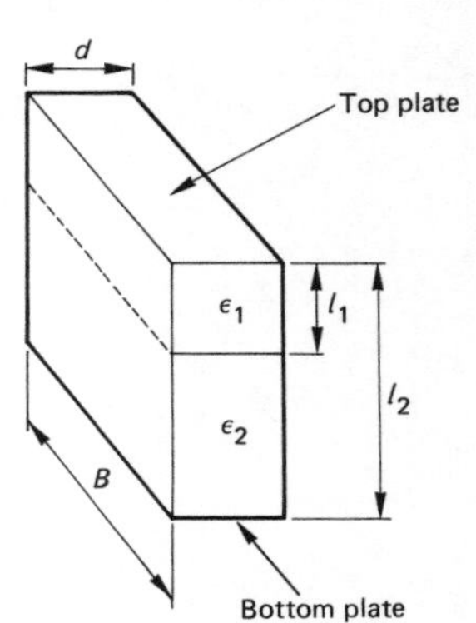

Fig. 3.10

3.10 Explain how the iron losses of a transformer may be separated from one another. A transformer has a hysteresis loss of 220 W and an eddy current loss of 160 W at 50 Hz when a particular voltage is applied to the primary. When the voltage is reduced to 75% of its initial value with the frequency unchanged at 50 Hz the total iron loss is 220 W. Calculate the Steinmetz index for the core material. Also find the total iron loss if the frequency is increased to 60 Hz and the voltage remains at 75% of its initial value.

Short Exercises

3.11 For a certain magnetic material the iron loss is 100 W at 25 Hz and 260 W at 50 Hz. Calculate the iron loss at 100 Hz if the maximum flux density remains constant at all frequencies.

3.12 Explain why magnetic cores are usually laminated.

3.13 Explain, with the aid of a hysteresis loop, what is meant by magnetic hysteresis.

3.14 The iron loss in a magnetic core is 60 W at 30 Hz and 80 W at 50 Hz. Calculate the eddy current and hysteresis losses at *a*) 30 Hz and *b*) 50 Hz.

3.15 A 2 μF capacitor has a loss angle of 1×10^{-4} rad at 1.59 kHz. Calculate its equivalent shunt resistance. The capacitor is connected in parallel with a second capacitor of 3 μF and the loss angle of the combination is then 2×10^{-4} rad. Calculate the equivalent shunt resistance of the second capacitor.

3.16 List the factors that cause a power loss in the core material of a transformer.

4 Coupled Circuits and Transformers

A **transformer** consists essentially of two or more coils of wire coupled together by means of electromagnetic induction. Usually, a step-up, or step-down, of voltage or current occurs. The behaviour of a transformer is profoundly affected by the nature of the former upon which the two coils are wound. If the former is made from some non-magnetic material, which may perhaps be just air, the component is loosely referred to as an **air-cored transformer** or often as "two coupled circuits". When the core is made from some magnetic material, the two windings have a very much increased magnetic coupling and the component is generally known as an **iron-cored transformer**.

Although, essentially, the action of both air-cored and iron-cored transformers is the same it is generally convenient to utilize two different approaches to their analysis.

Mutual Inductance

Consider Fig. 4.1 which shows a coil of wire connected across an alternating voltage source having an e.m.f. of E_p volts. A current I_p flows in the *primary winding* and this sets up a magnetic flux around the winding. As shown by the dotted lines, *some* of the magnetic flux links with the turns of the other (secondary) coil. As the current in the primary changes, the number of flux linkages changes also and consequently an e.m.f. is induced into the secondary winding.

The magnitude of the secondary e.m.f. is given by

$$e = M\frac{dI_p}{dt} \tag{4.1}$$

where M is the *mutual inductance* between the two windings.

The two coils are said to possess a **mutual inductance** of 1 henry (H) when the primary current changing at the rate of 1 A/s induces a voltage of 1 V into the secondary winding.

When the primary current is of sinusoidal waveform, i.e. $i = I_p \sin \omega t$, the rate of change of the primary current is

$$\frac{di_p}{dt} = \omega I_p \cos \omega t = \omega I_p \sin(\omega t + 90°) = j\omega I_p \tag{4.2}$$

Then, from equation (4.1)

$$E_s = j\omega M I_p \tag{4.3}$$

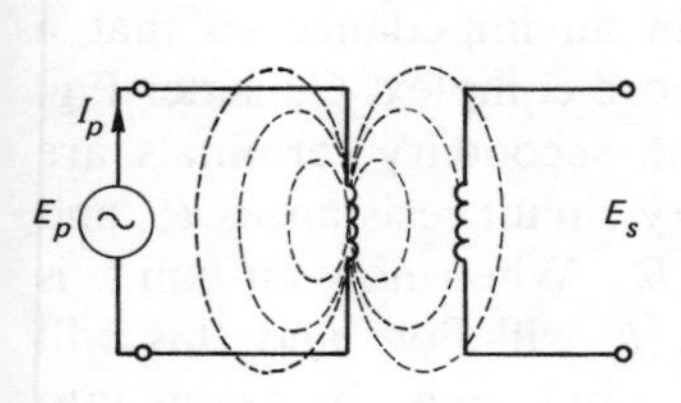

Fig. 4.1 Magnetic flux linking two coils of wire

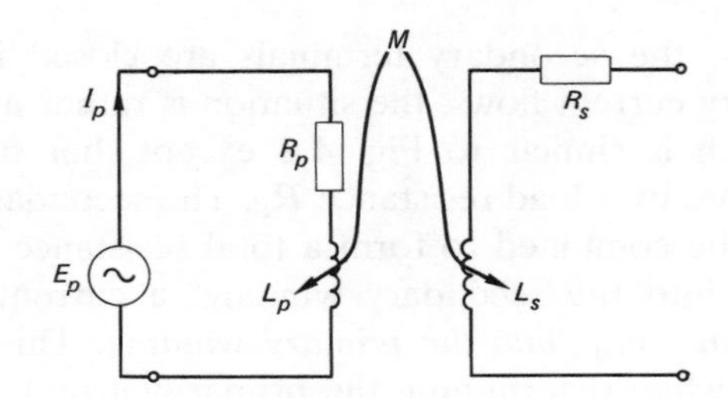

Fig. 4.2 Two coils coupled together by mutual inductance

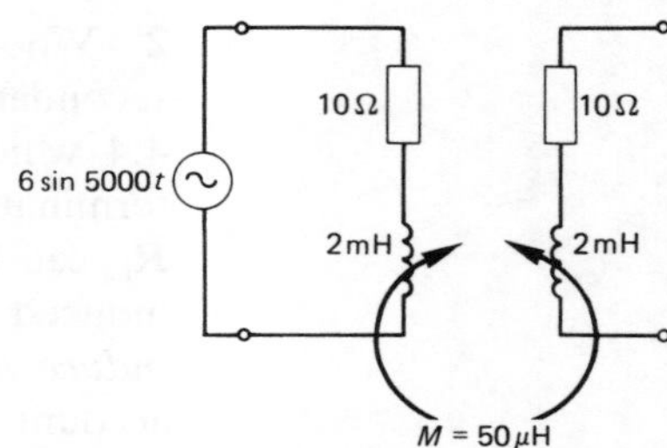

Fig. 4.3

Coupled Circuits

The induced voltage E_s effectively appears in *series* with the secondary winding. Because the flux set up by the primary current also links with the primary winding, there will also be an e.m.f. equal to $j\omega L_pI_p$ induced into the primary winding, where L_p is the inductance of the primary winding.

1 Fig. 4.2 shows two coils, having self-inductances L_p and L_s, which are inductively coupled together by a mutual inductance M. The primary winding has a voltage generator of e.m.f. E_p volts connected across its input terminals. The internal resistance of the source has been included in the primary resistance R_p. The secondary winding is left open-circuited and has a resistance of R_s.

Applying Kirchhoff's voltage law to the primary circuit gives

$$E_p = I_pR_p + L_p\frac{dI_p}{dt}$$

If E_p and I_p are *both* of sinusoidal waveshape this expression can be written as

$$E_p = I_p(R_p + j\omega L_p)$$

Therefore, $I_p = E_p/(R_p + j\omega L_p)$ and, from equation (4.3),

$$E_s = \pm j\omega ME_p/(R_p + j\omega L_p) \qquad (4.4)$$

Example 4.1

Calculate the voltage that appears across the open-circuited secondary winding of the circuit given in Fig. 4.3.

Solution The impedance of the primary circuit is

$$Z_p = 10 + j5000 \times 2 \times 10^{-3} = (10 + j10)\ \Omega$$

Hence, $I_p = E_p/Z_p = 6/(10 + j10)$ and so the secondary voltage is

$$E_s = \pm j\omega MI_p = \frac{\pm j5000 \times 50 \times 10^{-6} \times 6}{10 + j10} = \frac{1.5\angle\pm 90^\circ}{14.14\angle 45^\circ}$$

$$E_s = 106\angle 45^\circ\ \text{mV} \quad \text{or} \quad E_s = 106\angle -135^\circ\ \text{mV} \quad (Ans)$$

2 When the secondary terminals are closed in an impedance so that a secondary current flows, the situation is rather more complex. Consider Fig. 4.4 which is similar to Fig. 4.2 except that the secondary terminals are terminated by a load resistance R_L. The secondary circuit resistances R'_s and R_L can be combined to form a total resistance R_s. When now an e.m.f. is induced into the secondary winding, a current I_s will flow and this will *induce an e.m.f. into the primary winding.* This e.m.f. must be taken into account when determining the primary current.

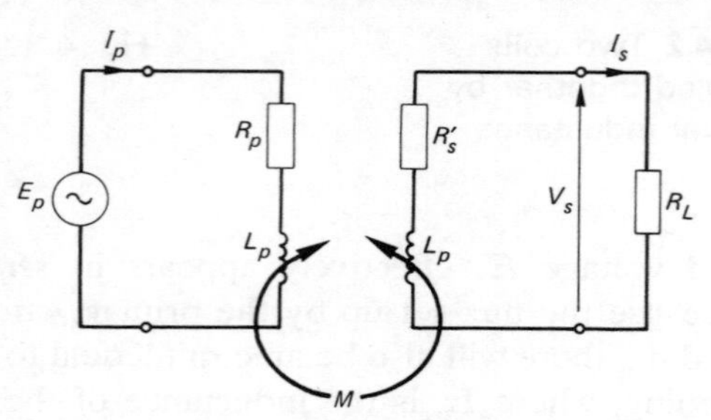

Fig. 4.4 Two coils coupled together by mutual inductance

Applying Kirchhoff's voltage law to both the primary and the secondary circuits of Fig. 4.4. gives

$$E_p = I_p(R_p + j\omega L_p) \pm j\omega MIs \tag{4.5}$$

and $$0 = I_s(R_s + j\omega L_s) \pm j\omega MI_p \tag{4.6}$$

From equation (4.6) $I_s = \mp j\omega MI_p/(R_s + j\omega L_s)$ and substituting for I_s into equation (4.5) gives

$$E_p = I_p\left[R_p + j\omega L_p + \frac{\omega^2 M^2(R_s - j\omega L_s)}{R_s^2 + \omega^2 L_s^2}\right]$$

The *effective* primary impedance $Z_{p(eff)}$ of the circuit is

$$Z_{p(eff)} = E_p/I_p = R_p + \frac{\omega^2 M^2 R_s}{R_s^2 + \omega^2 L_s^2} + j\left(\omega L_p - \frac{\omega^3 M^2 L_s}{R_s^2 + \omega^2 L_s^2}\right) \tag{4.7}$$

The effective primary resistance $R_{p(eff)}$ of the primary winding is the real part of equation (4.7) and it is always greater than the resistance R_p of the primary on its own. The effective reactance $X_{p(eff)}$ of the primary circuit is the imaginary part of equation (4.7) and it is smaller than the reactance of the primary on its own.

The effective Q-factor of the primary circuit is the ratio $X_{p(eff)}/R_{p(eff)}$ and it will always be smaller than the Q-factor of the uncoupled primary circuit.

3 Very often the primary and/or the secondary circuit may be caused to resonate at particular frequencies—not necessarily equal to one another—by means of tuning capacitors. The tuning capacitors can be connected either in series with, or in parallel with, the two windings. Fig. 4.5 shows two windings each tuned by series-connected capacitors C_p and C_s. The analysis of the circuit is very similar to that already given for the circuit of Fig. 4.4 except that now the total reactances of the two windings are

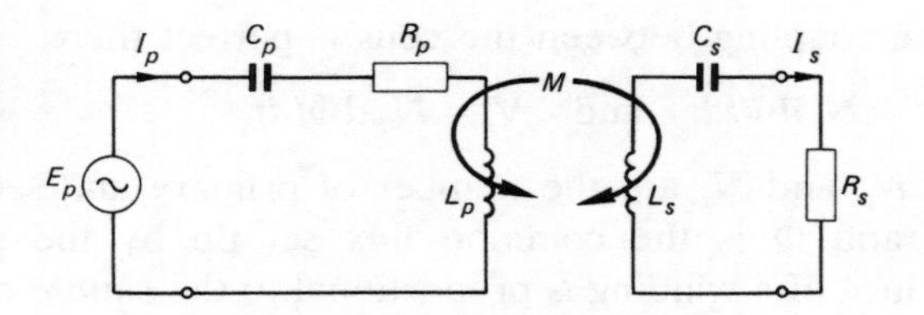

Fig. 4.5 Two series-tuned circuits coupled together by mutual inductance

$$jX_p = [j(\omega L_p - (1/\omega C_p)] \quad \text{and} \quad jX_s = j[\omega L_s - (1/\omega C_s)]$$

respectively. This means that equations (4.5), (4.6) and (4.7) apply with ωL_p replaced by X_p and ωL_s replaced by X_s.

Example 4.2

For the circuit shown in Fig. 4.6 calculate the current in the secondary winding.

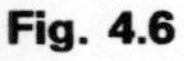
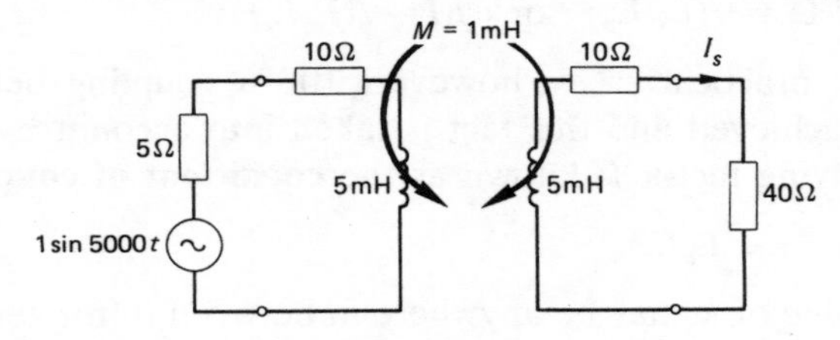

Fig. 4.6

Solution

$$\omega L_p = \omega L_s = 5000 \times 5 \times 10^{-3} = 25\ \Omega \quad \text{and} \quad \omega M = 5\ \Omega$$

From equation (4.7)

$$R_{p(eff)} = 15 + \frac{25 \times 50}{50^2 + 25^2} = 15.4\ \Omega$$

$$X_{p(eff)} = 25 - \frac{25 \times 25}{50^2 + 25^2} = 24.8\ \Omega$$

Hence

$$I_p = 1/(15.4 + j24.8) \quad \text{and} \quad E_s = j5/(15.4 + j24.8)$$

$$I_s = E_s/Z_s = \frac{j5}{(15.4 + j24.8)(50 + j25)}$$

$$= \frac{5\underline{/90^\circ}}{29.19\underline{/58.2^\circ} \times 55.9\underline{/26.6^\circ}}$$

or $I = 3.06\underline{/5.2^\circ}$ mA (*Ans*)

Coefficient of Coupling

Whenever two coils are inductively coupled together, a current I_p flowing in the primary circuit will induce an e.m.f. into the secondary circuit. At the same time a self-induced voltage equal to $L\,dI_p/dt$ will also be induced into the primary circuit. The ratio of the two induced voltages is

$$V_s/V_p = M/L_p \tag{4.8}$$

If the coupling between the coils is perfect then

$$V_s = N_s \, d\Phi/dt \quad \text{and} \quad V_p = N_p \, d\Phi/dt$$

where N_p and N_s are the number of primary and secondary turns respectively, and Φ is the common flux set up by the primary current. The inductance of a winding is proportional to the *square* of the number of turns so that

$$L_p/L_s = (N_p/N_s)^2$$

Hence,

$$V_s/V_p = N_s/N_p = \sqrt{(L_s/L_p)}$$

and combining with equation (4.8) gives

$$M/L_p = \sqrt{(L_s/L_p)} \quad \text{or} \quad M = \sqrt{(L_pL_s)}$$

For all practical cases, however, 100% coupling between the windings is never achieved and this fact is taken into account by the introduction of a multiplying factor k known as the **coefficient of coupling**. Therefore,

$$M = k\sqrt{(L_pL_s)} \tag{4.9}$$

The value of k may be anywhere in between 0 (for zero coupling) and 1 (for perfect coupling).

The coupling coefficient may be expressed in terms of the Q-factors of the two windings. Thus

$$k = \frac{M}{\sqrt{(L_pL_s)}} = \frac{M}{\sqrt{\left(\frac{Q_pR_p}{\omega_p} \times \frac{Q_sR_s}{\omega_s}\right)}}$$

If $\omega_p = \omega_s = \omega$, then

$$k = \frac{\omega M}{\sqrt{(Q_pQ_sR_pR_s)}} \tag{4.9a}$$

Example 4.3

In the circuit of Fig. 4.7 the two windings are tuned to resonate at the same frequency. Calculate *a*) the resonant frequency, *b*) the value of C_s, *c*) the voltage appearing across C_s and *d*) the coefficient of coupling.

Fig. 4.7

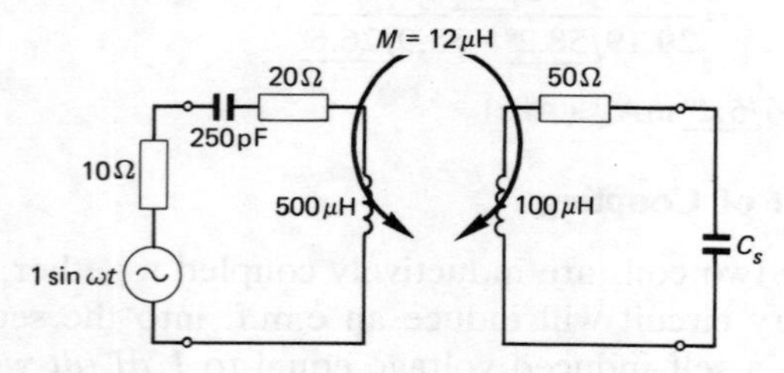

Solution

a) $f_0 = 1/2\pi\sqrt{(500 \times 10^{-6} \times 250 \times 10^{-12})} = 450.16$ kHz (*Ans*) *b*) $f_0 = 1/2\pi\sqrt{(C_pL_p)} = 1/2\pi\sqrt{(L_sC_s)}$ so that $C_pL_p = C_sL_s$

$$C_s = C_pL_p/L_s = 250 \times 500/100 = 1.25\ \text{nF} \quad (Ans)$$

c) Since both the primary and the secondary circuits are resonant, the effective primary impedance of the circuit is

$$Z_{p(eff)} = R_{p(eff)} = 30 + \tfrac{1}{50}[(2\pi \times 450.16 \times 10^3)^2 \times (12 \times 10^{-6})^2] = 53\ \Omega$$

$$I_p = E_p/Z_{p(eff)} = 10/53 = 0.189\ \text{A}$$

$$E_s = \omega MI_p = 2\pi \times 450.16 \times 10^3 \times 12 \times 10^{-6} \times 0.189 = 6.4\ \text{V}$$

$$I_s = E_s/Z_s = 6.4/50 = 0.128\ \text{A}$$

Therefore,

$$V_{Cs} = I_s/\omega C_s = 0.128/[2\pi \times 450.16 \times 10^3 \times 1.25 \times 10^{-9}] = 36.2\ \text{V} \quad (Ans)$$

Alternatively,

$$Q_s = \omega L_s/R_s = [2\pi \times 450.16 \times 10^3 \times 100 \times 10^{-6}]/50 = 5.66$$

and $V_{Cs} = QE_s = 5.66 \times 6.4 = 36.2\ \text{V}$ (*Ans*)

c) $k = \dfrac{M}{\sqrt{(L_pL_s)}} = \dfrac{12}{\sqrt{(500 \times 100)}} = 0.054$ (*Ans*)

Secondary-current/Frequency Characteristic

If the frequency of the voltage applied to the primary of two coupled circuits is varied either side of the resonant frequency, the shape of the **secondary-current/frequency characteristic** will depend upon the value of the coupling coefficient.

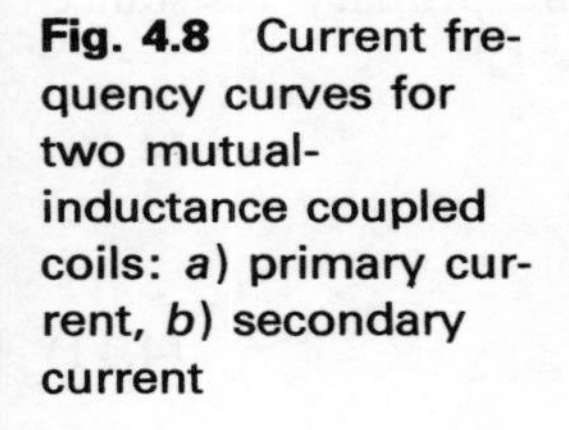

Fig. 4.8 Current frequency curves for two mutual-inductance coupled coils: *a*) primary current, *b*) secondary current

When the coupling coefficient is very small, the secondary current will also be small and the voltage induced into the primary winding by the secondary current is so minute that it has little effect upon the primary current. The primary-current/frequency characteristic is then determined solely by the selectivity of the primary circuit. This is shown by Fig. 4.8*a* and *b* by the curves marked *loose coupling*.

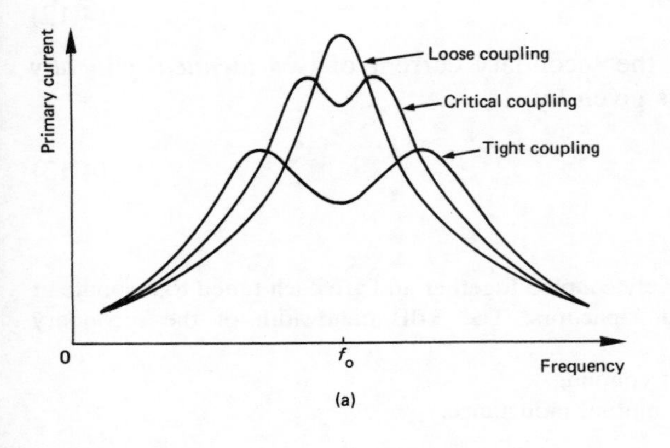

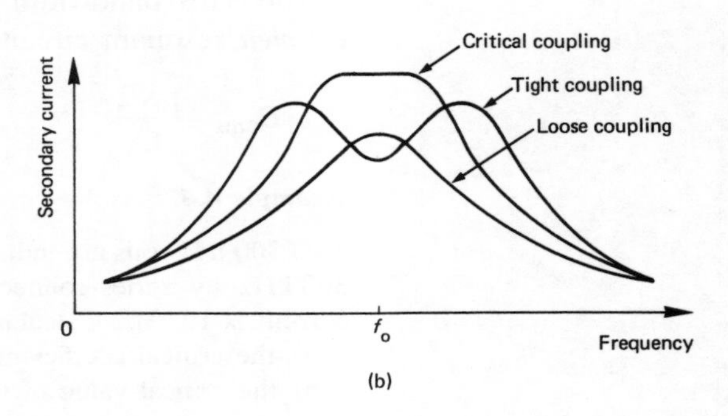

When the coupling between the windings is increased, the e.m.f. induced into the secondary winding by the primary current will also increase. This produces an increase in the secondary current, particularly in the neighbourhood of the resonant frequency. The e.m.f. induced into the primary circuit by the secondary current will be larger and so tends to oppose the flow of the primary current. The primary current at, and near, resonance is therefore reduced and the primary-current/frequency characteristic develops two peaks as shown by Fig. 4.8*a*.

If the coupling between the two windings is still further increased, the secondary current will rise with the result that the trough in the primary current response deepens. When the coupling coefficient has been increased to its *critical value* k_{crit} the secondary current has reached its maximum possible value and at this point its frequency characteristic exhibits a more or less flat top. This is shown in Fig. 4.8*b* by the curve marked as *critical coupling*.

If, now, the coupling coefficient is still further increased to become greater than the critical value k_{crit}, the secondary-current/frequency characteristic will also develop two peaks as shown by Fig. 4.8*b*. Any further increase in the coupling coefficient will not produce a corresponding increase in the secondary current but will only cause the twin peaks to move even further apart from one another. The two peaks occur at frequencies symmetrically spaced either side of the resonant frequency f_0 with a spacing equal to kf_0 Hertz.

When both the primary and the secondary circuits are resonant, the effective primary resistance $R_{p(eff)}$ is equal to $R_p + (\omega_0^2 M^2/R_s)$. If the coupling is critical, the coupled resistance is equal to the primary resistance proper. Thus

$$R_p = \omega_0^2 M^2/R_s \qquad (4.10)$$

Substituting equation (4.10) into (4.9a) gives

$$k_{crit} = \frac{\omega_0 M}{\sqrt{(Q_p Q_s \omega_0^2 M^2)}} = \frac{1}{\sqrt{(Q_p Q_s)}} \qquad (4.11)$$

If $Q_p = Q_s$, then

$$k_{crit} = 1/Q \qquad (4.12)$$

The 3 dB bandwidth of the secondary current of two *identical critically coupled* resonant circuits is given by

$$B_{3dB} = \frac{f_0\sqrt{2}}{Q} \qquad (4.13)$$

Example 4.4

Two 300 μH coils are inductively coupled together and are each tuned to resonate at 300 kHz by series-connected capacitors. The 3 dB bandwidth of the secondary current is 10 kHz. Calculate

a) the critical coefficient of coupling,

b) the critical value of the mutual inductance,

c) the capacitor value,

d) the separation between the secondary current peaks when the coupling coefficient is increased to 125% of its critical value.

Solution

a) From equation (4.13), $Q = \sqrt{2} \times 300/10 = 42.4$ and therefore

$$k_{crit} = 1/Q = 0.024 \quad (Ans)$$

b) $M = 0.024 \times 300 = 7.2\,\mu\text{H} \quad (Ans)$

c) $C_p = C_s = 1/4\pi^2 \times 300^2 \times 10^6 \times 300 \times 10^{-6} = 938\text{ pF} \quad (Ans)$

d) The coupling coefficient is $1.25 \times 0.024 = 0.03$ and therefore

Peak separation $= 0.03 \times 300 \times 10^3 = 9\text{ kHz} \quad (Ans)$

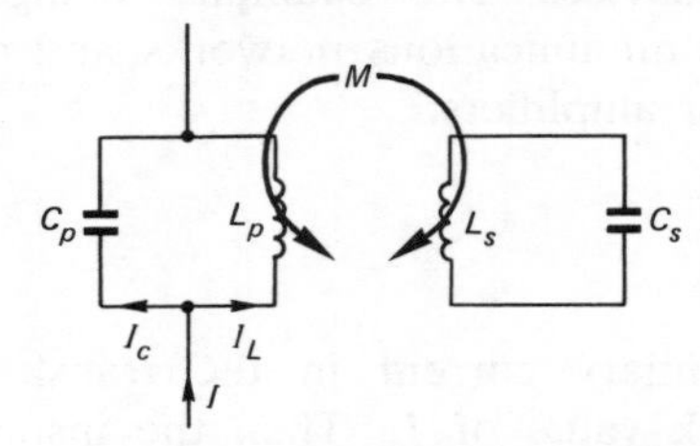

Fig. 4.9 Air-cored transformer with primary tuned by a parallel-connected capacitor

Very often the primary winding has its tuning capacitor connected in parallel as shown by Fig. 4.9. The current I_L flowing in the primary inductance will be equal to Q_pI. Alternatively, Thevenin's theorem can be used to convert the primary circuit into the equivalent series circuit. If $1/\omega C_p \ll$ (the source resistance), the series circuit previously discussed is obtained.

Probably the main application of air-cored transformers lies in radio receivers where they are often used in r.f. amplifiers [RSIII].

Iron-cored Transformers

The basic arrangement of an **iron-cored transformer** is shown in Fig. 4.10. Because of the eddy current losses discussed in Chapter 3 the iron core is constructed from a number of thin insulated laminations. Since the two windings are wound around a core made from a high-permeability material, the flux produced by a given current is much larger than that which would be produced using an air core. Because of this the coefficient of coupling is very much larger and it is usually very nearly unity.

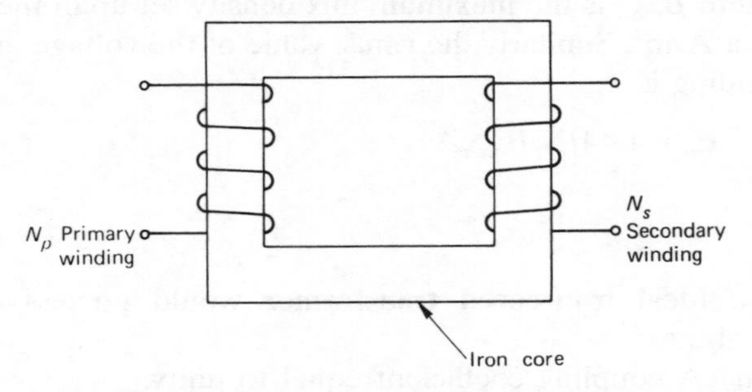

Fig. 4.10 Basic iron-cored transformer

The iron-cored transformer therefore possesses large values of self- and mutual inductance but these are obtained at the cost of introducing both eddy current and hysteresis losses. Another disadvantage, that will be discussed in Chapter 11, is that the relationship between the primary current and the flux it produces may be a non-linear one, particularly if the core material is at, or near, its saturated condition.

The main use of the iron-cored transformer is the stepping-up, or the stepping-down, of an alternating current or voltage. The voltages involved may be very high, as in electricity supply distribution, or may be from the standard a.c. mains supply voltage of 230 V to some much lower voltage needed to operate an electronic equipment such as, for example, a radio receiver. In such cases the transformer is operated as a constant-voltage and constant-frequency device. Transformers are also commonly employed as impedance-matching devices, two examples being line transformers in audio-frequency telecommunications networks and output transformers in audio-frequency power amplifiers.

The E.M.F. Equation

Suppose that the primary current in the transformer is of sinusoidal waveform with a peak value of I_p. Then the instantaneous value of the current is given by $i = I_p \sin \omega t$. Assuming the operation of the transformer to be linear the flux Φ set up in the core is

$$\Phi = \Phi_{max} \sin \omega t \text{ webers} \qquad (4.14)$$

Since the coefficient of coupling k is approximately equal to unity, the e.m.f. induced into the N_s secondary turns is

$$e_s = N_s \, d\Phi/dt \quad \text{or} \quad e_s = N_s \omega \Phi \cos \omega t \text{ volts}$$

The r.m.s. value of this voltage is

$$E_s = \frac{N_s \omega \Phi_{max}}{\sqrt{2}} = \frac{2\pi f N_s \Phi_{max}}{\sqrt{2}} = 4.44 f N_s \Phi_{max} \qquad (4.15)$$

This equation is often written in the form

$$E = 4.44 f N_s B_{max} A \qquad (4.16)$$

where B_{max} is the maximum flux density set up in the core of cross-sectional area A m^2. Similarly the r.m.s. value of the voltage induced into the primary winding is

$$E_p = 4.44 f N_p B_{max} A \qquad (4.17)$$

The Ideal Transformer

The **ideal iron-cored transformer** would possess each of the following attributes:

a) A coupling coefficient equal to unity.
b) Zero losses.
c) Infinite primary inductance.
d) Zero winding capacitances.

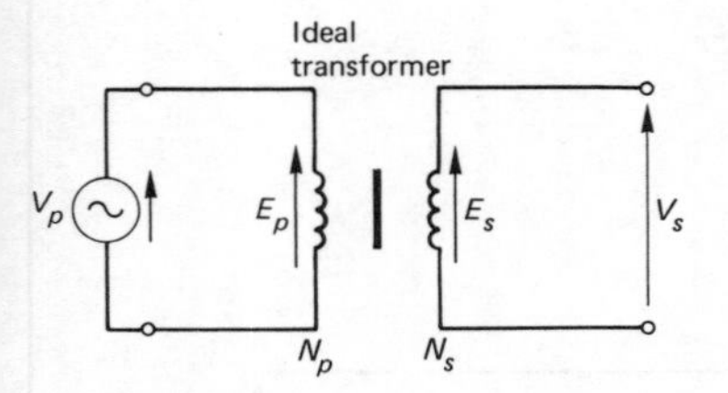

Fig. 4.11 Ideal transformer: secondary open-circuit

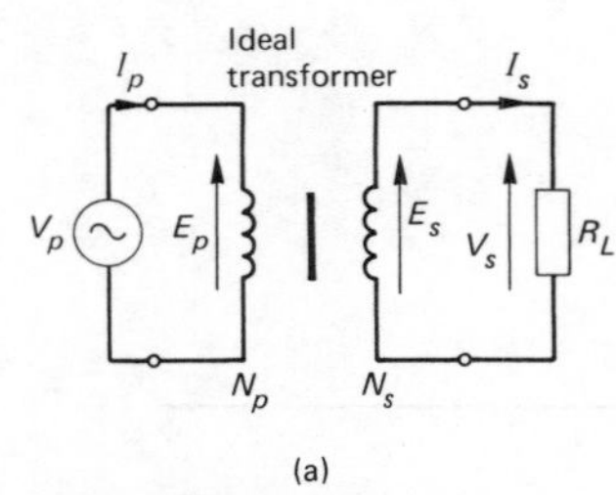

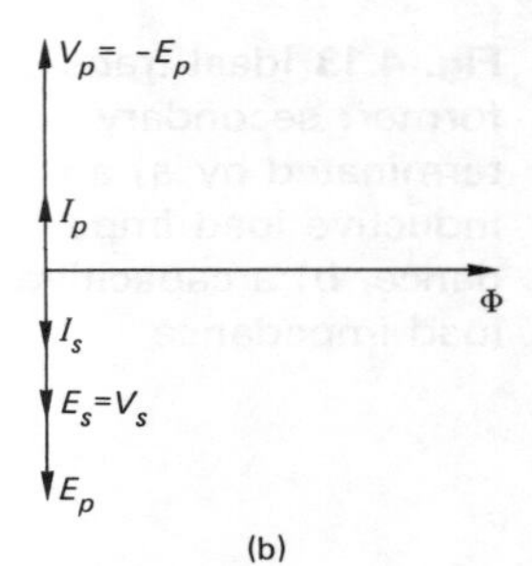

Fig. 4.12 Ideal transformer: secondary terminated by a resistance

Voltage, Current and Impedance Ratios

When a sinusoidal current I_p flows in the primary winding, the flux set up in the core links with both the windings and induces voltages

$$E_p = 4.44fN_p\Phi_{max} \quad \text{and} \quad E_s = 4.44fN_s\Phi_{max}$$

into them. The voltage ratio is therefore $E_p/E_s = N_p/N_s$. Further, since the windings are assumed to have zero resistance, the turns ratio is equal to the voltage ratio, i.e.

$$V_p/V_s = N_p/N_s \quad \text{(see Fig. 4.11)}$$

If the secondary terminals are closed in a resistance R_L, a secondary current I_s will flow and the transformer then will possess a current ratio. Fig. 4.12*a* shows the circuit and Fig. 4.12*b* gives the phasor diagram representing its currents and voltages. The flux Φ is taken as the reference phasor since it is common to both of the windings.

Since the primary inductance is assumed to be infinitely large, zero current is needed to magnetize the core. The applied voltage V_p leads the core flux Φ by 90° and the primary e.m.f. E_p is equal in magnitude to V_p but is of the opposite phase. The e.m.f. E_s induced into the secondary winding is in phase with E_p and its magnitude is equal to N_sE_p/N_p.

The secondary current I_s is equal to $E_s/R_L = V_s/R_L$ and it is in phase with E_s. In order to maintain the core flux Φ at a constant value, a primary current I_p must flow, 180° out of phase with I_s. The magnitude of I_p must be such that the m.m.f. it produces is equal to the m.m.f. produced by the secondary current. Therefore,

$$I_pN_p = I_sN_s \quad \text{or} \quad I_p/I_s = N_s/N_p$$

This means that the current ratio of an ideal transformer is the inverse of its turns and voltage ratios.

Further, the input resistance R_{in} of the transformer is

$$R_{in} = \frac{V_p}{I_p} = \frac{N_pV_s/N_s}{N_sI_s/N_p} = \frac{N_p^2}{N_s^2}R_L \quad \text{or} \quad \frac{R_{in}}{R_L} = \left(\frac{N_p}{N_s}\right)^2$$

Thus, the impedance ratio of an ideal transformer is the *square* of the turns ratio.

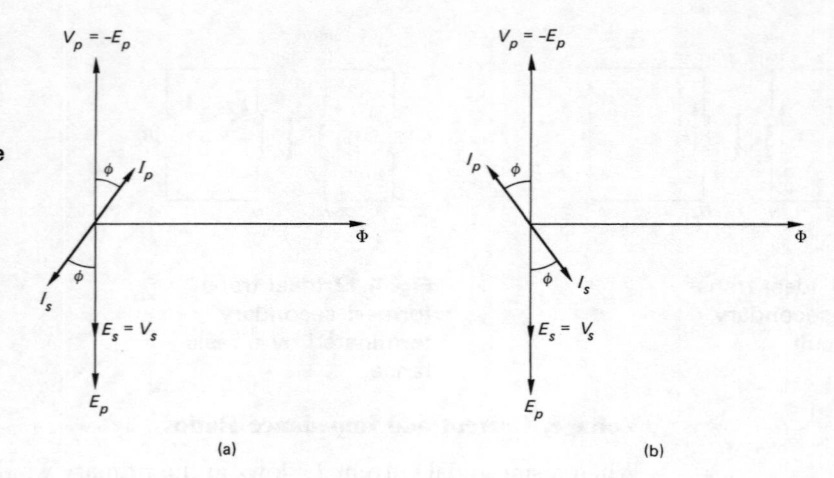

Fig. 4.13 Ideal transformer: secondary terminated by *a*) an inductive load impedance, *b*) a capacitive load impedance

When the load connected across the secondary terminals is either an inductive or a capacitive impedance, the primary and secondary currents will be out of phase with the primary and the secondary voltages respectively by an angle φ. This is shown by Fig. 4.13*a* and *b*.

Now $|Z_{in}| = |V_p|/|I_p|$ and $|Z_L| = |V_s|/|I_s|$ so that

$$|Z_{in}|/|Z_L| = (N_p/N_s)^2$$

Thus, the transformer action will only alter the *magnitude* of the load impedance but its phase angle will remain unchanged.

Example 4.5

An ideal transformer has a turns ratio of 5:1. Calculate the input impedance when the load connected across the secondary terminals is *a*) a 1000 Ω resistor, *b*) a 1 μF capacitor, and *c*) a 1000 Ω resistor in series with a 1 μF capacitor. The frequency is $5000/2\pi$ Hz.

Solution

a) $R_{in} = 25 \times 1000 = 25\ \text{k}\Omega$ (*Ans*)

b) $X_C = 1/j5000 \times 10^{-6} = -j200$. Therefore,

$$Z_{in} = -j200 \times 25 = -j5\ \text{k}\Omega \quad (Ans)$$

c) $Z_L = 1000 - j200\ \Omega = 1020\underline{/-11.3°}\ \Omega$. Therefore,

$$Z_{in} = 1020 \times 25\underline{/-11.3°}\ \Omega = 25.5\underline{/-11.3°}\ \text{k}\Omega \quad (Ans)$$

Practical Transformers

Any practical transformer will not satisfy the requirements for an ideal transformer listed on page 60.

Magnetizing Current

The **magnetizing current** I_m is the current that must flow in the primary winding in order to set up the required flux in the core because of the finite value of the primary inductance L_m. The magnetizing current I_m is purely inductive ($I_m = V_p/j\omega L_m$) and it therefore lags the primary voltage by 90°. Since I_m does not contribute to the secondary current, the primary inductance is drawn in the equivalent circuit of the transformer in parallel with the primary winding (Fig. 4.14*a*). I_m is considered to flow only in L_m. The windings are wound upon a core of a high permeability material so that L_m is of high value and the required magnetizing current is small.

Iron Losses

The practical transformer is subject to both **eddy current and hysteresis losses** in its core. The total loss is constant at a particular frequency and it is therefore represented in the equivalent circuit of Fig. 4.14*a* by a resistor R_c connected in parallel with the primary inductance. The total iron power losses are then equal to $I_c^2R_c$. The total no-load primary current I_0 is the phasor sum of the magnetizing current and the core-loss current. This is shown by the phasor diagram given in Fig. 4.14*b*. Usually I_c is much smaller than I_m.

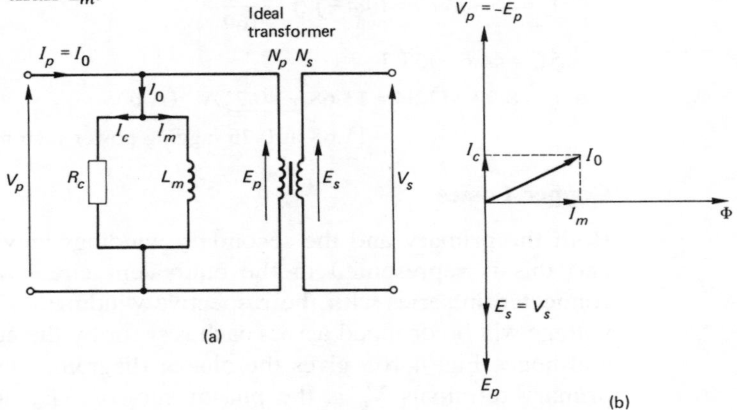

Fig. 4.14 Representation of magnetizing current and iron losses in *a*) equivalent circuit, *b*) phasor diagram of an iron-cored transformer: secondary open-circuited

When a load is connected across the secondary terminals of the transformer, a secondary current I_s must flow (Fig. 4.15*a*). As before, a primary current $I'_p = N_sI_s/N_p$ must flow to maintain the m.m.f. in the core. The total

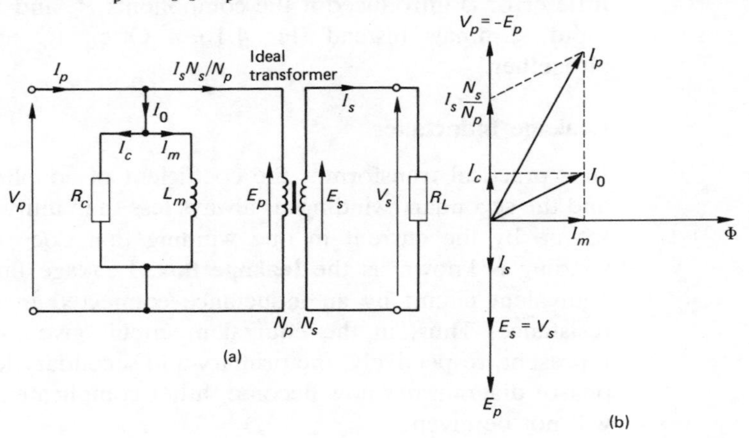

Fig. 4.15 As Fig. 4.14 but with the secondary terminals closed by a resistance

primary current I_p is then the phasor sum of I_0 and N_sI_s/N_p so that the phasor diagram must be modified to that shown by Fig. 4.15*b*.

Example 4.6

An iron-cored transformer has 800 primary turns and 160 secondary turns. The magnetizing current is 2 A and the core loss current is 0.4 A. When a secondary current flows the primary current is 60 A at 0.75 power factor lagging. Calculate the secondary current.

Solution The primary current is the phasor sum of the current I_0 and the current N_sI_s/N_p. I_0 is the phasor sum of the magnetizing current I_m and the core loss current I_c. Therefore, from Fig. 4.15*b*, taking V_p as the reference,

$$I_0 = (0.4 - j2)\text{ A}$$

and $I_p = (60 \times 0.75) - (j60 \times 0.66) = (45 - j39.7)\text{A}$

Hence,

$$I_p = 45 - j39.7 = (0.4 - j2) + \frac{800I_s}{160}$$

$$5I_s = 44.6 - j37.7$$

$$I_s = 8.92 - j7.54 = 11.68\underline{/-40.2^\circ}\text{ A} \quad (Ans)$$

$$= 11.68 \text{ at } 0.76 \text{ lagging power factor} \quad (Ans)$$

Copper Losses

Both the primary and the secondary windings inevitably possess resistance and this is represented in the equivalent circuit by resistors R_p and R_s connected in series with the respective windings as shown by Fig. 4.16*a*. A voltage will be dropped across each resistor by the current flowing through it and hence Fig. 4.16*b* gives the phasor diagram. The applied voltage to the primary terminals V_p is the phasor sum of $-E_p$ and the voltage dropped across the primary resistance R_p, i.e. I_pR_p. The voltage V_s appearing across the load resistor R_L is smaller than the secondary e.m.f. E_s because of the voltage drop I_sR_s across the secondary resistance R_s. Because $I_0 \ll I_sN_s/N_p$, little error is introduced if the components R_c and L_m are drawn across the input terminals instead (Fig. 4.16*c*). Often R_c and L_m can be omitted altogether.

Leakage Inductance

In a practical transformer the coefficient of coupling between the primary and the secondary windings is always less than unity. The fraction of the flux set up by the current in one winding that does *not* link with the other winding is known as the **leakage flux**. Leakage flux is represented in the equivalent circuit by an inductance connected in series with the winding resistance. Thus, in the equivalent circuit given in Fig. 4.17 L_p and L_s represent, respectively, the primary and secondary leakage inductances. The phasor diagram has now become rather complicated and confusing and so it will not be given.

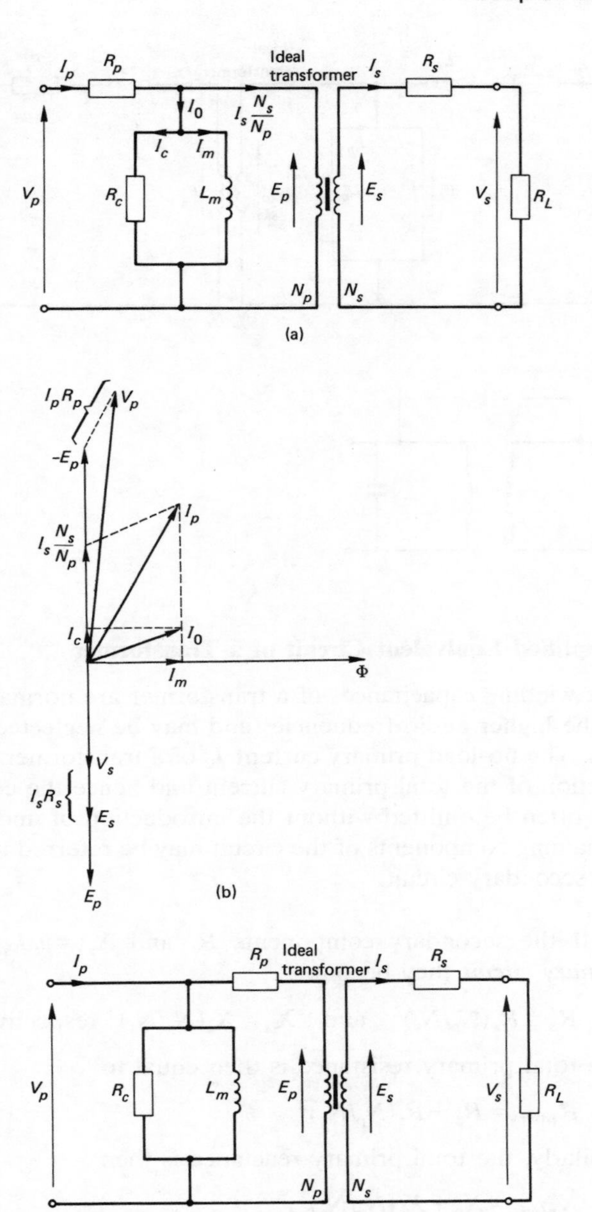

Fig. 4.16 Representation of the copper losses in an iron-cored transformer

Winding Capacitances

The total **winding capacitances** have components that originate from a number of sources such as the inter-turn, inter-layer, and inter-winding capacitances. These capacitances are usually represented in the equivalent circuit by capacitors drawn across the input and output terminals of the transformer (see Fig. 4.18).

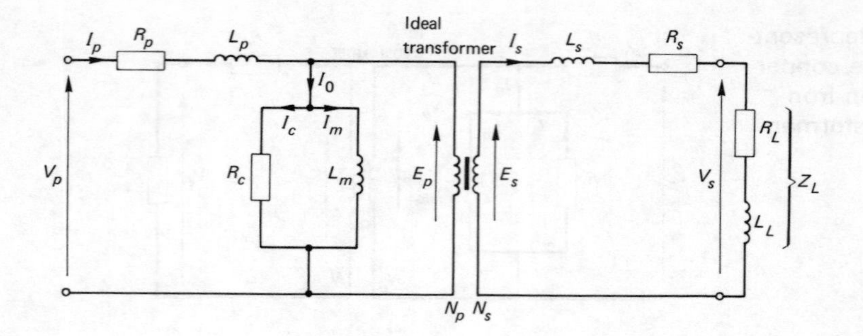

Fig. 4.17 Representation of leakage inductance in an iron-cored transformer

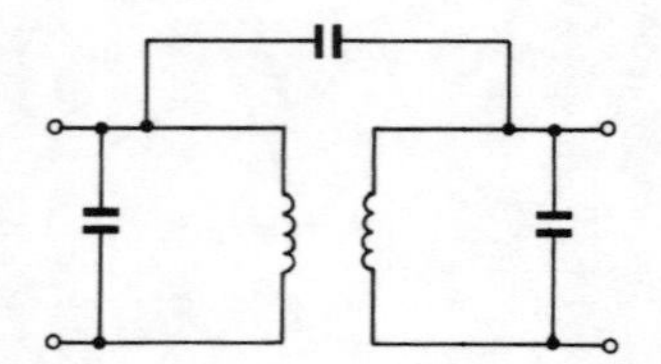

Fig. 4.18 Transformer capacitances

Simplified Equivalent Circuit of a Transformer

The winding capacitances of a transformer are normally only of importance at the higher audio-frequencies and may be neglected at all lower frequencies. The no-load primary current I_0 of a transformer is usually only a small fraction of the total primary current and hence the components R_c and L_m can often be omitted without the introduction of undue error. Further, the remaining components of the circuit may be referred to *either* the primary *or* the secondary circuit.

1 If the secondary components R_s and $X_s(=j\omega L_s)$ are referred to the *primary circuit* they become

$$R'_s = R_s(N_p/N_s)^2 \quad \text{and} \quad X'_s = X_s(N_p/N_s)^2 \text{ respectively}$$

The total primary resistance is then equal to

$$R_{p(eff)} = R_p + R_s(N_p/N_s)^2$$

Similarly, the total primary reactance is then

$$X_{p(eff)} = X_p + X_s(N_p/N_s)^2$$

The simplified equivalent circuit of the transformer is then given by Fig. 4.19. V'_s is the secondary load voltage referred to the primary, i.e.

$$V'_s = N_p V_s / N_s$$

2 Referring the primary components R_p and $X_p(=j\omega L_p)$ to the *secondary circuit* gives total secondary component values of

$$R_{s(eff)} = R_s + R_p(N_s/N_p)^2 \quad \text{and} \quad X_{s(eff)} = X_s + X_p(N_s/N_p)^2$$

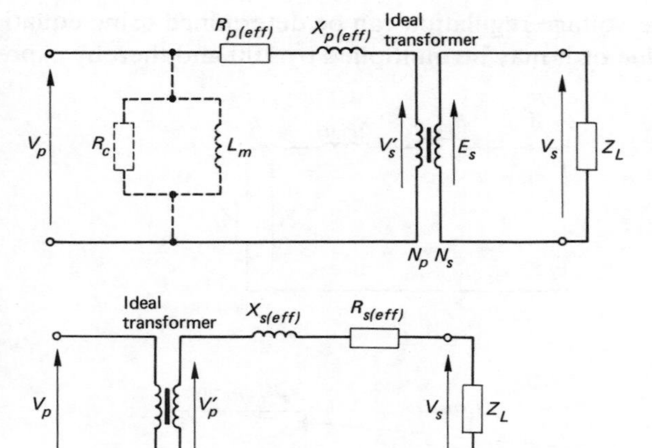

Fig. 4.19 Simplified transformer equivalent circuit: all components referred to the primary circuit

Fig. 4.20 Simplified transformer equivalent circuit: all components referred to the secondary circuit

The simplified equivalent circuit with all components referred to the secondary is shown in Fig. 4.20 in which $V_p' = V_p N_s / N_p$ and is the primary voltage referred to the secondary.

Example 4.7

A 40 kVA 4000/230 V transformer has the following data: primary resistance = 4 Ω, primary reactance = 15 Ω, secondary resistance = 0.05 Ω, secondary reactance = 0.1 Ω. Calculate the primary voltage needed to obtain the full-load current when the secondary terminals are short-circuited.

Solution Effective primary resistance

$$R_{p(eff)} = 4 + \left(\frac{4000}{230}\right)^2 \times 0.05 = 19.12\ \Omega$$

Effective primary reactance

$$X_{p(eff)} = 15 + \left(\frac{4000}{230}\right)^2 \times 0.1 = 45.25\ \Omega$$

$$\text{Full-load primary current} = \frac{40\,000}{4000} = 10\ \text{A}$$

Therefore,

$$V_p = 10(19.12 + j45.25) = 191.2 + j452.5\ \text{V}$$

and $|V_p| = 491.2$ V (*Ans*)

Voltage Regulation

The **voltage regulation** of a transformer is the change in the secondary voltage as the current taken by a load connected across the secondary terminals varies from zero to the maximum permitted current.

The voltage regulation of a transformer is defined by

$$\text{Regulation} = \frac{\text{no-load secondary voltage} - \text{full-load secondary voltage}}{\text{no-load secondary voltage}} \tag{4.18}$$

The voltage regulation can be determined using equation (4.18) as a *per-unit value* or it may be multiplied by 100 and thereby expressed as a percentage.

Fig. 4.21 Voltage regulation of a transformer

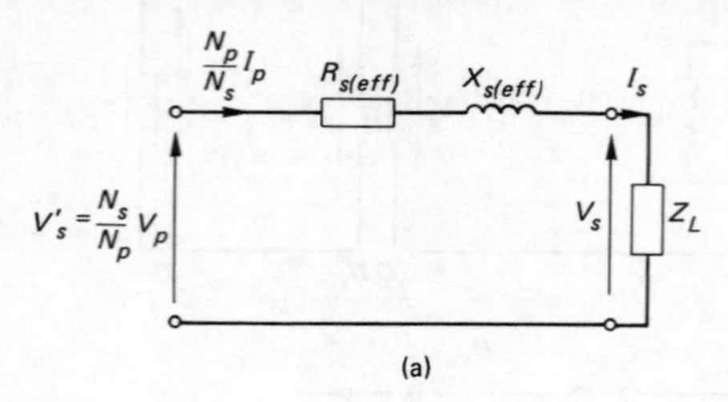

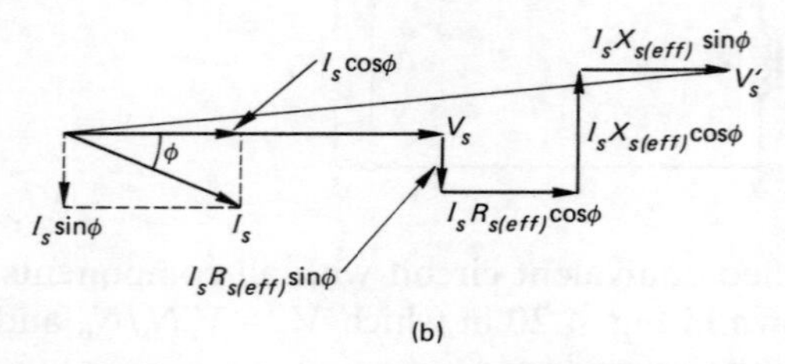

If the primary components R_p and X_p of the equivalent circuit are referred to the secondary (Fig. 4.21*a*), the phasor diagram of the transformer, assuming a lagging power factor, is given by Fig. 4.21*b*. The no-load current I_0 has been assumed to be negligibly small with respect to N_sI_s/N_p.

The voltage $V'_s = N_sV_p/N_p$ applied to the equivalent secondary circuit is the phasor sum of the secondary voltage V_s and the voltage dropped across $R_{s(eff)}$ and $X_{s(eff)}$. From Fig. 4.20*b*,

$$V'_s = \sqrt{[(V_s + I_sR_{s(eff)}\cos\varphi + I_sX_{s(eff)}\sin\varphi) + (I_sX_{s(eff)}\cos\varphi - I_sR_{s(eff)}\sin\varphi)]} \quad (4.19)$$

However, it can be seen from the phasor diagram that $I_sX_{s(eff)}\cos\varphi$ and $I_sR_{s(eff)}\sin\varphi$ tend to cancel out and generally little error is introduced if these terms are neglected. If this is done then

$$V'_s = V_s + I_sR_{s(eff)}\cos\varphi + I_sX_{s(eff)}\sin\varphi \quad (4.20)$$

The per unit *voltage regulation* of the transformer is then given by

$$\text{Regulation} = \frac{V'_s - V_s}{V'_s} = \frac{I_sR_{s(eff)}\cos\varphi + I_sX_{s(eff)}\sin\varphi}{V'_s} \quad (4.21)$$

If the load has a leading power factor then equation (4.21) must be altered to become equation (4.22):

$$\text{Regulation} = \frac{I_sR_{s(eff)}\cos\varphi - I_sX_{s(eff)}\sin\varphi}{V'_s} \quad (4.22)$$

Example 4.8

Calculate the percentage regulation of the transformer of Example 4.7 if the load power factor is *a*) unity, *b*) 0.85 lagging, and *c*) 0.85 leading.

Solution Referring all values to the secondary circuit,

$$R_{s(eff)} = 0.05 + \left(\frac{230}{4000}\right)^2 \times 4 = 0.063\ \Omega$$

$$X_{s(eff)} = 0.1 + \left(\frac{230}{4000}\right)^2 \times 15 = 0.15\ \Omega$$

a) $\%\text{ regulation} = \dfrac{(173.91 \times 0.063 \times 1) - 0}{230} \times 100 = 4.76\%$ (*Ans*)

b) $\%\text{ regulation} = \dfrac{(173.91 \times 0.063 \times 0.85) + (173.91 \times 0.15 \times 0.53)}{230} \times 100$

$= 10.06\%$ (*Ans*)

c) $\%\text{ regulation} = \dfrac{(173.91 \times 0.063 \times 0.85) - (173.91 \times 0.15 \times 0.53)}{230} \times 100$

$= -4.51\%$ (*Ans*)

The negative sign means that the terminal voltage increases with increase in the load current.

Efficiency of a Transformer

The efficiency of a transformer is always less than 100% because of the inevitable copper and iron losses. The core flux is, more or less, constant regardless of the load and so it is usual to assume that the iron losses are constant for all values of load current. Copper losses, on the other hand, are the result of I^2R dissipation in both of the windings and therefore increase in proportion to the squares of the primary and secondary currents.

The **efficiency** η of a transformer is given by

$$\eta = \frac{\text{output power}}{\text{input power}} \times 100\% \qquad (4.23)$$

The input power to a transformer is the sum of the output power and the total losses within the transformer and so equation (4.23) can be written as

$$\eta = \frac{I_s V_s \times \text{power factor}}{(I_s V_s \times \text{power factor}) + P_c + I_p^2 R_p + I_s^2 R_s} \times 100\% \qquad (4.24)$$

$$= \frac{I_s V_s \times \text{power factor}}{(I_s V_s \times \text{power factor}) + P_c + I_s^2 R_{s(eff)}} \times 100\% \qquad (4.25)$$

$$= \frac{I_s V_s \times \text{power factor}}{(I_s V_s \times \text{power factor}) + \text{iron loss} + \text{copper loss}} \times 100\% \qquad (4.26)$$

Example 4.9

A 1 kVA transformer has an iron loss of 25 W and a full load copper loss of 30 W. Calculate the efficiency of the transformer *a*) on full load and *b*) on half load if the load power factor is 0.85.

Solution

a) $\eta = \dfrac{1000 \times 0.85}{850 + 25 + 30} \times 100 = 93.92\%$ (*Ans*) *b*) On half load the copper losses will be reduced to

$(\frac{1}{2})^2 \times 30 = 7.5\text{ W}$

Therefore,

$$\eta = \frac{500 \times 0.85}{425 + 25 + 7.5} \times 100 = 92.9\% \quad (Ans)$$

To determine the necessary conditions for the maximum efficiency of a transformer, equation (4.25) must be differentiated with respect to I_s and the result equated to zero. Thus

$$\frac{d\eta}{dI_s} = \frac{V_s \times \text{p.f.}(I_s V_s \times \text{p.f.} + P_c + I_s^2 R_{s(eff)}) - (I_s V_s \times \text{p.f.})(V_s \times \text{p.f.} + 2I_s R_{s(eff)})}{(I_s V_s \times \text{p.f.}) + P_c + I_s^2 R_{s(eff)}} = 0$$

Hence,

$$(I_s V_s \times \text{p.f.}) + 2I_s^2 R_{s(eff)} = (I_s V_s \times \text{p.f.}) + P_c + I_s^2 R_{s(eff)}$$

or $\quad I_s^2 R_{s(eff)} = P_c$

This result means that a transformer will operate with its maximum possible efficiency when the load current is such that the copper losses are equal to the iron losses.

Transformer Tests

The testing of a transformer to ascertain its important parameters such as its voltage regulation and its efficiency is normally carried out by means of two tests. The first of these tests, known as the *open-circuit test*, measures the iron losses of the transformer. The other test, known as the *short-circuit test*, obtains the copper losses of the transformer.

1 *Open-circuit Test*

The circuit employed to carry out the open-circuit test is shown in Fig. 4.22. In order to keep the copper losses to the minimum possible value, the secondary terminals of the transformer are open-circuited. For this condition there are no copper losses in the secondary circuit and only a very small copper loss in the primary (which is due to I_0).

With the rated input voltage applied to the primary winding the indication of the wattmeter W_c will be very nearly equal to the iron losses. Further,

$$W_c = I_0 V_p \cos\varphi \quad \text{or} \quad \cos\varphi = W_c / V_p I_0$$

and finally

$$I_c = I_0 \cos\varphi \quad \text{and} \quad I_m = I_0 \sin\varphi$$

2 *Short-circuit Test*

The circuit used for the short-circuit test is shown by Fig. 4.23. Since the secondary winding is short-circuited it is possible to obtain the full-load current with only a low voltage applied across the primary winding. This

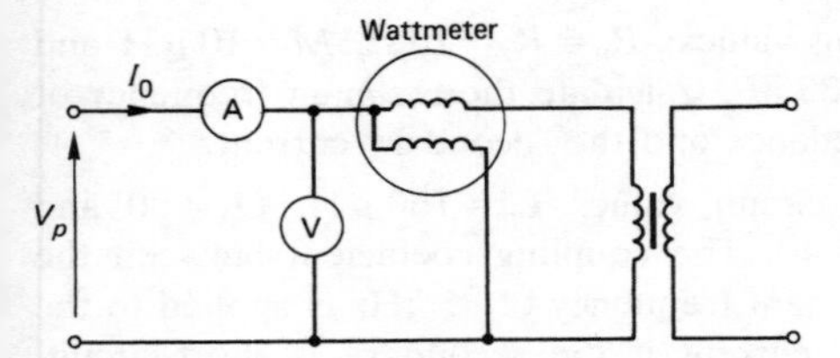

Fig. 4.22 Open-circuit test of a transformer

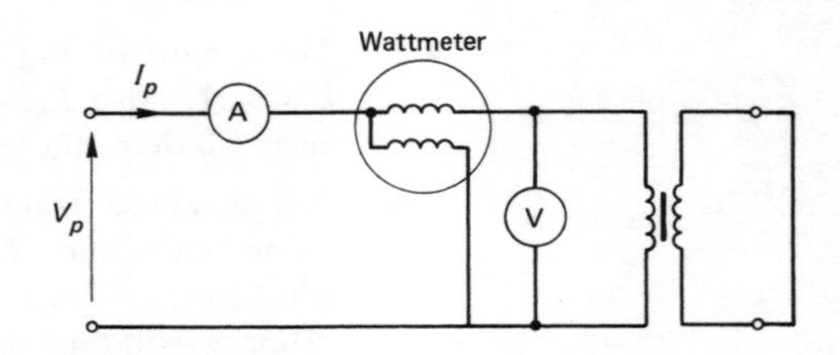

Fig. 4.23 Short-circuit test of a transformer

means that the iron losses are negligibly small and so the wattmeter indicates the full-load copper loss of the transformer.

Also the magnitude $|Z_{p(eff)}|$ of the effective primary impedance is equal to V_p/I_p, and the phase angle between V_p and I_p is obtained from $\cos\beta = W_c/V_pI_p$. Then,

the effective primary resistance $R_{p(eff)}$ is equal to $|Z_{p(eff)}|\cos\beta$

and the effective primary reactance $X_{p(eff)}$ is equal to $|Z_{p(eff)}|\sin\beta$.

Exercises 4

4.1 A 10 kVA 3200/400 V transformer has primary and secondary resistances of 2.5 Ω and 0.2 Ω respectively and a total reactance referred to the secondary of 0.3 Ω. Calculate the primary voltage needed to obtain the full-load secondary current when the secondary terminals are short-circuited.

4.2 A 100 kVA transformer has a nominal secondary voltage of 230 V. Its iron and full-load copper losses are 1000 W and 1500 W respectively. Calculate *a*) its full-load efficiency (assume unity power factor), *b*) its efficiency on half-load, *c*) the load at which the maximum efficiency occurs, and *d*) the maximum efficiency (assume unity power factor).

4.3 The primary and secondary windings of a 20 kVA 3200/230 V transformer have resistances of 5 Ω and 0.01 Ω respectively. The total reactance referred to the primary is 25 Ω. Calculate the voltage regulation of the transformer, *a*) for unity load power factor and *b*) for a load power factor of 0.82 lagging.

4.4 Both the secondary and the primary circuits of Fig. 4.24 have equal inductances and are tuned to resonance at a frequency of 500 000/2π Hz. When an e.m.f. E_p at this frequency is applied to the primary currents of 100 mA and 50 mA flow in the primary and secondary circuits respectively. Calculate *a*) the inductances L_p and L_s, *b*) the capacitances C_p and C_s, *c*) the mutual inductance M, *d*) the coupling coefficient, and *e*) the critical value of the coupling coefficient.

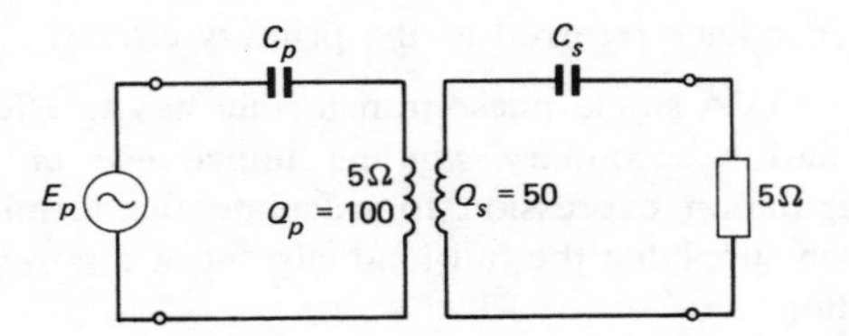

Fig. 4.24

4.5 The circuit of Fig. 4.5 has the following values: $R_p = R_s = 10\ \Omega$, $M = 10\ \mu H$ and $k = 0.01$. Also $L_p = L_s$ and $C_p = C_s = 2.533$ nF. Calculate the resonant frequency of each winding, the effective primary impedance and the secondary current.

4.6 An air-cored transformer has primary circuit values $L_p = 160\ \mu H$, $Q_p = 80$ and secondary values $L_s = 125\ \mu H$ and $Q_s = 40$. The coupling coefficient between the windings is 0.20. A 1 V signal at the resonant frequency of 85 kHz is applied to the primary winding. Calculate the primary current if the secondary is short-circuit. Calculate also the effective Q-factor of the primary winding.

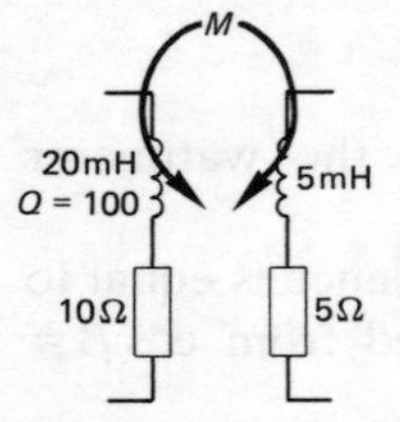

Fig. 4.25

4.7 For the circuit given in Fig. 4.25 calculate the values of the series-connected capacitors required to resonate each winding at the same frequency. If the coupling between the windings is critical calculate the effective primary resistance at the resonant frequency.

4.8 For the circuit of Fig. 4.5, $Q_p = 100$, $Q_s = 50$, $M = 10\ \mu H$ and $C_p = C_s = 1$ nF. If a voltage of 1 V at the resonant frequency of $10^6/2\pi$ Hz is applied to the primary calculate the voltage that appears across the secondary capacitor C_s.

4.9 Two identical tuned circuits each having $L = 300\ \mu H$, $R = 100\ \Omega$ and $C = 1.5$ nF are coupled together by a mutual inductance of 50 μH. An e.m.f. of 12 V at the resonant frequency is applied to the primary winding. If the secondary is short-circuited calculate the primary current, the coupling coefficient and the secondary current.

4.10 A 6600/400 V 10 kVA single-phase transformer has a load voltage of 385 V when supplying its full-load current at a power factor of 0.8 lagging. When the full-load current is supplied at a power factor of 0.62 leading, the load voltage is 401.6 V. Calculate the total resistance and reactance of the transformer when referred to the secondary.

4.11 A 6600/1100 V 10 kVA single-phase transformer has the following data:
primary resistance = 3.8 Ω, secondary resistance = 0.12 Ω
primary reactance = 8.8 Ω, secondary reactance = 0.26 Ω
Use the approximate regulation expression to calculate the percentage regulation at the low-voltage terminals when the transformer supplies the full load current at *a)* 0.86 lagging power factor and *b)* unity power factor.

4.12 A 6600/330 V 100 kVA single-phase transformer has its maximum efficiency at a 0.9 full load. When the secondary terminals are short-circuited 100 V are necessary to produce the full-load current in the short-circuit. Calculate *a)* the total resistance and reactance when referred to the secondary winding and *b)* the voltage across a load that takes the full-load current at a power factor of 0.8 lagging. Iron losses are 3000 W.

4.13 A 100 kVA 6600/330 V 50 Hz single-phase transformer is supplied at 6600 volts and 60 Hz and then has a hysteresis loss of 350 W and an eddy current loss of 150 W. Calculate *a)* the total iron loss when the transformer is supplied at 6600 V at 50 Hz. Assume the Steinmetz index to be 1.6. If the maximum efficiency of the transformer occurs at 0.9 full load, and 100 V is needed to circulate the full-load current when the secondary terminals are short-circuited, calculate the values of the resistance and reactance referred to the primary circuit.

4.14 A 2000/400 V 10 kVA single-phase transformer has an effective primary impedance of $(5 + j10)\ \Omega$ and a secondary winding impedance of $(0.25 + j1.2)\ \Omega$. Use the approximate regulation expression to calculate the terminal voltage on the low-voltage side when supplying the full-load current at a power factor of *a)* 0.8 lagging and *b)* 0.6 leading.

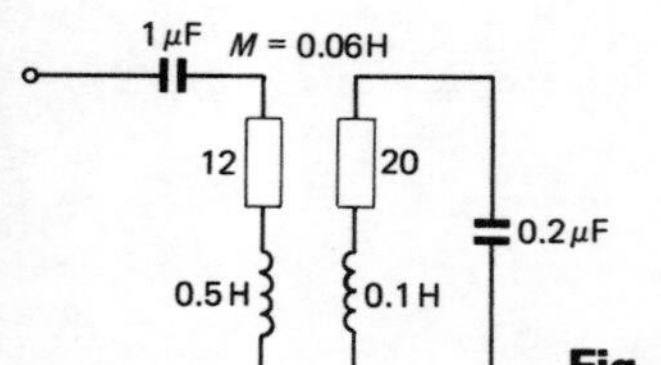

Secondary	V_p(V)	I_p(A)	*Input power* (W)
Open-circuit	220	0.7	72
Short circuit	8.2	2	8.0

Fig. 4.26

4.15 The circuit given in Fig. 4.19 could represent a 4 kVA 50 Hz single-phase transformer with a voltage ratio of 2000/400. When this transformer was tested using the open- and short-circuit tests, with the measurements taken on the low-voltage side, the data obtained is given in the table. Determine the component values for Fig. 4.19.

4.16 A 10 kVA 6600/330 V 50 Hz single-phase transformer has a primary resistance of 9.2 Ω and a secondary resistance of 0.1 Ω. The primary and secondary reactances are, respectively, 30 Ω and 0.3 Ω. Calculate *a*) the total equivalent resistance and reactance referred to the primary, *b*) the voltage regulation at a power factor of *a*) 0.8 lagging and *b*) 0.8 leading.

Short Exercises

4.17 A 20 kVA transformer has an iron loss of 1000 W. What value of copper loss will produce the maximum efficiency? If the power factor of the load is then 0.85 calculate the power dissipated in the load.

4.18 A 3200/400 V transformer has the following data: primary resistance = 5 Ω, primary reactance = 0.15 Ω, secondary resistance = 0.03 Ω, secondary reactance = 0.1 Ω. Determine the effective primary impedance.

4.19 Draw the phasor diagram of an iron-cored transformer with an inductive load taking into account the magnetizing current and the core losses.

4.20 An air-cored transformer has identical primary and secondary circuits inductively coupled together. The inductance of each winding is 458 μH and the 3 dB bandwidth of the transformer is 20 kHz. If the resonant frequency of the circuit is 1.2 MHz calculate, for critical coupling *a*) *k*, *b*) the mutual inductance *M*, *c*) the *Q*-factor of each winding and the overall *Q*-factor, and *d*) the tuning capacitances.

4.21 Calculate the input impedance of the circuit shown in Fig. 4.26 at a frequency of $5000/2\pi$ Hz.

4.22 A parallel-tuned circuit consists of a capacitor of negligible loss connected in parallel with an inductor of $Q = 80$ and $L = 50\ \mu$H. The circuit is at resonance at 1.2 MHz and takes a current of 6 mA from the source. Calculate the current flowing in the inductor. Also find the e.m.f. that is induced into an open-circuited coil that is inductively coupled by a mutual inductance of 2 μH.

4.23 In a short-circuit test carried out on a 6 kVA transformer an applied voltage of 24 V caused the full-load current of 14.5 A to flow in the primary circuit. The power then indicated by the wattmeter was 80 W. Calculate the per unit resistance and reactance of the transformer.

4.24 A 3300/240 V 25 kVA single-phase transformer has a maximum efficiency of 95% at 0.9 power factor. Calculate its iron losses.

5 Network Parameters

The Types of Parameter

When a voltage is applied to the input terminals of a four-terminal network, input and output currents will flow and an output voltage will be developed. Fig. 5.1 shows a four-terminal network with input and output currents I_1 and I_2 respectively and input and output voltages V_1 and V_2 respectively. It is assumed that the impedances within the network are *linear*, i.e. they obey Ohm' Law. The directions of current and voltage shown in the figure are conventional and may, in fact, be the other way around in any particular case.

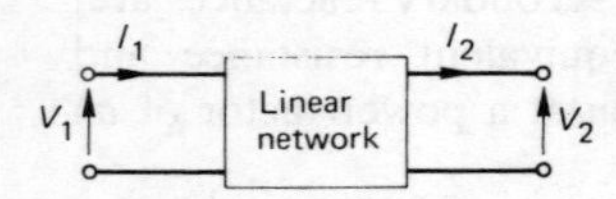

Fig. 5.1 Four-terminal network

Any two of the four variables can be taken as the *independent variables*, so that the other two automatically become the *dependent variables*. This leads to *six* different ways in which the terminal performance of the network can be described. However, not all of them are used in practice.

Suppose that the input current I_1 and the output voltage V_2 are taken to be the independent variables. Then the input voltage V_1 and the output current I_2 can be written down in terms of the output voltage, the input current and four circuit parameters. Thus

$$V_1 = h_{11}I_1 + h_{12}V_2 \tag{5.1}$$

$$I_2 = h_{21}I_1 + h_{22}V_2 \tag{5.2}$$

$$\begin{bmatrix} V_1 \\ I_2 \end{bmatrix} = \begin{bmatrix} h_{11} & h_{12} \\ h_{21} & h_{22} \end{bmatrix} \begin{bmatrix} I_1 \\ V_2 \end{bmatrix} \tag{5.3}$$

Equation (5.1) states that the input voltage V_1 is equal to a parameter h_{11} times the input current I_1 *plus* another parameter h_{12} times the output voltage V_2. The right-hand side of the equation must have the dimensions of a voltage and hence h_{11} must be an impedance while h_{12} must be a dimensionless quantity. Similarly the right-hand side of equation (5.2) must have the dimensions of a current; therefore h_{21} is dimensionless and h_{22} is an admittance.

If the output terminals of the network are short-circuited, $V_2 = 0$ and then

$$V_1 = h_{11}I_1 \quad \text{or} \quad h_{11} = V_1/I_1\,\Omega \quad (V_2 = 0) \tag{5.4}$$

$$\text{and} \quad I_2 = h_{21}I_2 \quad \text{or} \quad h_{21} = I_2/I_1 \quad (V_2 = 0) \tag{5.5}$$

Similarly, if $I_1 = 0$,

$$V_1 = h_{12}V_2 \quad \text{or} \quad h_{12} = V_1/V_2 \quad (I_1 = 0) \tag{5.6}$$

$$\text{and} \quad I_2 = h_{22}V_2 \quad \text{or} \quad h_{22} = I_2/V_2\,\text{S} \quad (I_1 = 0) \tag{5.7}$$

Because one of the parameters has the dimensions of impedance, another parameter has the dimensions of admittance, and the other two parameters are dimensionless, this set of parameters is known as the **hybrid parameters**. The h parameters are particularly appropriate for the representation of the a.c. performance of the bipolar transistor and they are widely employed for this purpose [EIV]. When the h parameters are used in transistor work, the four h parameters are generally re-labelled to indicate both the transistor configuration and the nature of the parameter. Thus, h_{11} becomes h_{ie}, where the suffix i indicates "input" and the suffix e indicates the common-emitter configuration. The other parameters are

forward current gain h_{fe}
output admittance h_{oe}
reverse voltage ratio h_{re}
h_{ie} is the input impedance of the transistor.

The other ways of importance in which the input and output variables of a 4-terminal network can be expressed are as follows.

1 *y parameters*

$$I_1 = y_{11}V_1 + y_{12}V_2 \tag{5.8}$$

$$I_2 = y_{21}V_1 + y_{22}V_2 \tag{5.9}$$

$$\begin{bmatrix} I_1 \\ I_2 \end{bmatrix} = \begin{bmatrix} y_{11} & y_{12} \\ y_{21} & y_{22} \end{bmatrix} \begin{bmatrix} V_1 \\ V_2 \end{bmatrix} \tag{5.10}$$

In this case, all four of the parameters have the dimensions of *admittance.* y_{11} and y_{21} can be determined by short-circuiting the output terminals to make $V_2 = 0$, while y_{12} and y_{22} can be found by setting V_1 to zero by short-circuiting the input terminals. Then,

$$y_{11} = I_1/V_1 \quad \text{with} \quad V_2 = 0$$

$$y_{12} = I_1/V_2 \quad \text{with} \quad V_1 = 0$$

$$y_{21} = I_2/V_1 \quad \text{with} \quad V_2 = 0$$

$$y_{22} = I_2/V_2 \quad \text{with} \quad V_1 = 0$$

2 *z parameters*

$$V_1 = z_{11}I_1 + z_{12}I_2 \tag{5.11}$$

$$V_2 = z_{21}I_1 + z_{22}I_2 \tag{5.12}$$

$$\begin{bmatrix} V_1 \\ V_2 \end{bmatrix} = \begin{bmatrix} z_{11} & z_{12} \\ z_{21} & z_{22} \end{bmatrix} \begin{bmatrix} I_1 \\ I_2 \end{bmatrix} \tag{5.13}$$

All four parameters have the dimensions of impedance and their values can be determined with first I_1 and then I_2 set to zero.

3 *General circuit, or transmission parameters*

$$V_1 = AV_2 + BI_2 \tag{5.14}$$

$$I_1 = CV_2 + DI_2 \tag{5.15}$$

$$\begin{bmatrix} V_1 \\ I_1 \end{bmatrix} = \begin{bmatrix} A & B \\ C & D \end{bmatrix} \begin{bmatrix} V_2 \\ I_2 \end{bmatrix} \tag{5.16}$$

In this case the parameters A and D are dimensionless, B has the dimensions of an impedance, and C is an admittance.

4 *g parameters*

$$I_1 = g_{11} V_1 + g_{12} I_2 \tag{5.17}$$

$$V_2 = g_{21} V_1 + g_{22} I_2 \tag{5.18}$$

$$\begin{bmatrix} I_1 \\ V_2 \end{bmatrix} = \begin{bmatrix} g_{11} & g_{12} \\ g_{21} & g_{22} \end{bmatrix} \begin{bmatrix} V_1 \\ I_2 \end{bmatrix} \tag{5.19}$$

The Choice of Parameters

The choice of which set of parameters to use in a particular case is determined by the type of network and the way in which it is interconnected with other networks. It has already been mentioned that the h parameters are widely employed in the analysis of bipolar transistor circuits at audio-frequencies. Some difficulty is experienced however in obtaining the values of the h parameters at high frequencies and for this reason (and others) high-frequency transistor analysis is generally carried out using y parameters.

The transmission or general circuit parameters are commonly used for the analysis of heavy-current circuits and for power frequency transmission lines. These parameters are also employed when two or more networks are

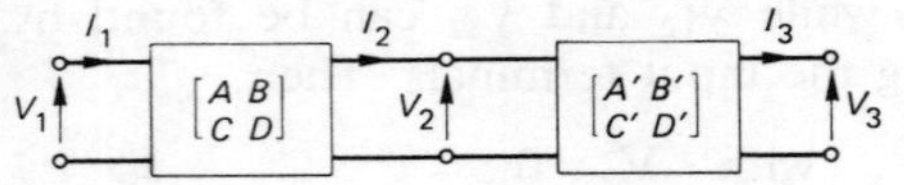

Fig. 5.2 Networks connected in cascade

I_1 I_2 I_3 V_1 $\begin{bmatrix} A & B \\ C & D \end{bmatrix}$ V_2 $\begin{bmatrix} A' & B' \\ C' & D' \end{bmatrix}$ V_3

connected in cascade since the overall performance of the cascade can be described by the product of their parameter matrices. Referring to Fig. 5.2,

$$\begin{bmatrix} V_1 \\ I_1 \end{bmatrix} = \begin{bmatrix} A & B \\ C & D \end{bmatrix} \begin{bmatrix} A' & B' \\ C' & D' \end{bmatrix} \begin{bmatrix} V_3 \\ I_3 \end{bmatrix} = \begin{bmatrix} AA'+BC' & AB'+BD' \\ CA'+DC' & CB'+DD' \end{bmatrix} \begin{bmatrix} V_3 \\ I_3 \end{bmatrix}$$

Similarly, a particular set of parameters is best suited to the overall behaviour of two networks connected in one of the other possible ways. Fig. 5.3a shows two networks connected in parallel; for this connection the y parameters are the most convenient because the overall matrix is equal to the sum of the individual y matrices of the two networks. Thus,

$$\begin{bmatrix} I_1 \\ I_2 \end{bmatrix} = \Big[[y_a] + [y_b] \Big] \begin{bmatrix} V_1 \\ V_2 \end{bmatrix} \tag{5.20}$$

When two networks are connected in series (Fig. 5.3b), the z parameters become the most convenient set and the overall matrix is

$$\begin{bmatrix} V_1 \\ V_2 \end{bmatrix} = \Big[[z_a] + [z_b] \Big] \begin{bmatrix} I_1 \\ I_2 \end{bmatrix} \tag{5.21}$$

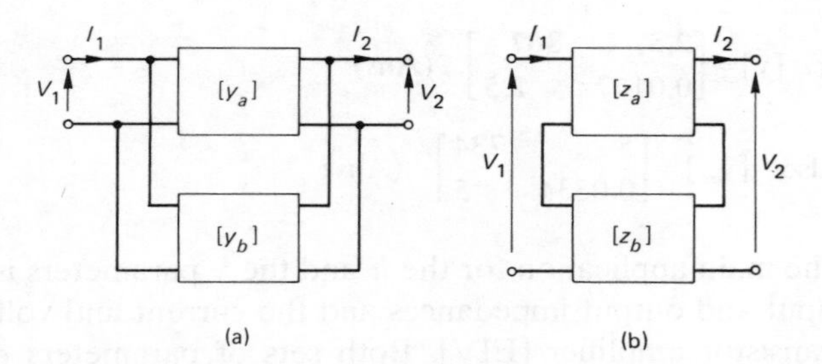

Fig. 5.3 Networks connected in *a*) parallel and *b*) series.

The other sets of parameters are not often used and they will not be further discussed in this book.

It is always possible to convert from one set of parameters to another set and the necessary relationships are given in Table 5.1.

Table 5.1 Relationships between parameters

From / To	$[z]$	$[y]$	$[h]$	$[g]$	$\begin{bmatrix} A & B \\ C & D \end{bmatrix}$
Note: $\lvert y\rvert = y_{11}y_{22} - y_{12}y_{21}$, $\lvert a\rvert = AD - BC$, $\lvert z\rvert = z_{11}z_{22} - z_{12}z_{21}$, $\lvert h\rvert = h_{11}h_{22} - h_{12}h_{21}$, $\lvert g\rvert = g_{11}g_{22} - g_{12}g_{21}$					
$[z]$	$\begin{bmatrix} z_{11} & z_{12} \\ z_{21} & z_{22} \end{bmatrix}$	$\frac{1}{\lvert y\rvert}\begin{bmatrix} y_{22} & -y_{12} \\ -y_{21} & y_{11} \end{bmatrix}$	$\frac{1}{h_{22}}\begin{bmatrix} \lvert h\rvert & h_{12} \\ -h_{21} & 1 \end{bmatrix}$	$\frac{1}{g_{11}}\begin{bmatrix} 1 & -g_{12} \\ g_{21} & \lvert g\rvert \end{bmatrix}$	$\frac{1}{C}\begin{vmatrix} A & \lvert a\rvert \\ 1 & D \end{vmatrix}$
$[y]$	$\frac{1}{\lvert z\rvert}\begin{bmatrix} z_{22} & -z_{12} \\ -z_{21} & z_{11} \end{bmatrix}$	$\begin{matrix} y_{11} & y_{12} \\ y_{21} & y_{22} \end{matrix}$	$\frac{1}{h_{11}}\begin{bmatrix} 1 & -h_{12} \\ h_{21} & \lvert h\rvert \end{bmatrix}$	$\frac{1}{g_{22}}\begin{bmatrix} \lvert g\rvert & g_{12} \\ -g_{21} & 1 \end{bmatrix}$	$\frac{1}{B}\begin{bmatrix} D & -\lvert a\rvert \\ -1 & A \end{bmatrix}$
$[h]$	$\frac{1}{z_{22}}\begin{bmatrix} \lvert z\rvert & z_{12} \\ -z_{21} & 1 \end{bmatrix}$	$\frac{1}{y_{11}}\begin{bmatrix} 1 & -y_{12} \\ y_{21} & \lvert y\rvert \end{bmatrix}$	$\begin{matrix} h_{11} & h_{12} \\ h_{21} & h_{22} \end{matrix}$	$\frac{1}{\lvert g\rvert}\begin{bmatrix} g_{22} & -g_{12} \\ -g_{21} & g_{11} \end{bmatrix}$	$\frac{1}{D}\begin{bmatrix} B & \lvert a\rvert \\ -1 & C \end{bmatrix}$
$[g]$	$\frac{1}{z_{11}}\begin{bmatrix} 1 & -z_{12} \\ z_{21} & \lvert z\rvert \end{bmatrix}$	$\frac{1}{y_{22}}\begin{bmatrix} \lvert y\rvert & y_{12} \\ -y_{21} & 1 \end{bmatrix}$	$\frac{1}{\lvert h\rvert}\begin{bmatrix} h_{22} & -h_{12} \\ -h_{21} & h_{11} \end{bmatrix}$	$\begin{matrix} g_{11} & g_{12} \\ g_{21} & g_{22} \end{matrix}$	$\frac{1}{A}\begin{bmatrix} C & -\lvert a\rvert \\ 1 & B \end{bmatrix}$
$\begin{bmatrix} A & B \\ C & D \end{bmatrix}$	$\frac{1}{z_{21}}\begin{bmatrix} z_{11} & \lvert z\rvert \\ 1 & z_{22} \end{bmatrix}$	$\frac{1}{y_{21}}\begin{bmatrix} -y_{22} & -1 \\ -\lvert y\rvert & -y_{11} \end{bmatrix}$	$\frac{1}{h_{21}}\begin{bmatrix} -\lvert h\rvert & -h_{11} \\ -h_{22} & -1 \end{bmatrix}$	$\frac{1}{g_{21}}\begin{bmatrix} 1 & g_{22} \\ g_{11} & \lvert g\rvert \end{bmatrix}$	$\begin{matrix} A & B \\ C & D \end{matrix}$

Example 5.1

A four-terminal network has the transmission matrix $\begin{bmatrix} 150 & 60 \\ 8 & 150 \end{bmatrix}$. Determine the y matrix for the network. What is the overall y matrix for two such networks connected in parallel?

Solution From Table 5.1

$$[y] = \frac{1}{60}\begin{bmatrix} 150 & -(150^2 - 480) \\ -1 & 150 \end{bmatrix}$$

or $[y]=\begin{bmatrix}2.5 & 367\\ 0.0167 & 2.5\end{bmatrix}$ (*Ans*)

Also $[y_{oo}]=\begin{bmatrix}5 & 734\\ 0.0334 & 5\end{bmatrix}$ (*Ans*)

The main application for the h and the y parameters is the calculation of the input and output impedances and the current and voltage gains of a bipolar transistor amplifier [EIV]. Both sets of parameters *can* be applied to the solution of network problems but such usage is not common. Even so, some examples of the use of y parameters in this way will be given later.

Transmission Parameters

The transmission, or general circuit, parameters are

$$V_1 = AV_2 + BI_2 \tag{5.14}$$

$$I_1 = CV_2 + DI_2 \tag{5.15}$$

$$\begin{bmatrix}V_1\\ I_1\end{bmatrix}=\begin{bmatrix}A & B\\ C & D\end{bmatrix}\begin{bmatrix}V_2\\ I_2\end{bmatrix} \tag{5.16}$$

If V_2 is made equal to zero,

$V_1 = BI_2$ or $B = V_1/I_2\ \Omega$

$I_1 = DI_2$ or $D = I_1/I_2$ (dimensionless)

If I_2 is made equal to zero,

$V_1 = AV_2$ or $A = V_1/V_2$ (dimensionless)

$I_1 = CV_2$ or $C = I_1/V_2$ S

For any *passive* network the relationship $AD - BC = 1$ is always true. Often, however, the values of the parameters of a particular network can be determined by comparing the equations of the network with equation (5.16).

1 *Series impedance Z*

From Fig. 5.4 suppose that a current I_2 flows. Then

$V_1 = V_2 + I_2Z$ and $I_1 = I_2$

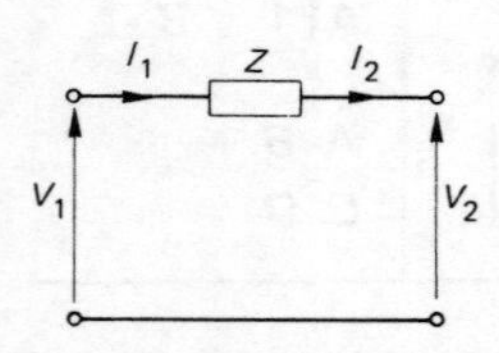

Fig. 5.4 Series impedance

Comparing with equation (5.16)

$$\begin{bmatrix}A & B\\ C & D\end{bmatrix}=\begin{bmatrix}1 & Z\\ 0 & 1\end{bmatrix}$$

Alternatively, if the output terminals of the network are first open-circuited and then short-circuited

$I_2 = I_1 = 0$ and $V_1 = V_2$

Hence

$V_1/V_2 = A = 1$ and $I_1/V_2 = C = 0$

Also, if $V_2 = 0$ $\quad V_1 = I_2Z \quad I_1 = I_2$

Hence

$$V_1/I_2 = B = Z \quad \text{and} \quad I_1/I_2 = D = 1$$

2 *Shunt admittance*
From Fig. 5.5, assuming I_2 exists,

$$V_1 = V_2 \quad \text{and} \quad I_1 = V_2 Y + I_2$$

Therefore, comparing with equation (5.16)

$$\begin{bmatrix} A & B \\ C & D \end{bmatrix} = \begin{bmatrix} 1 & 0 \\ Y & 1 \end{bmatrix}$$

Alternatively, when $V_2 = 0$, $V_1 = 0$ and $I_1 = I_2$. Hence,

$$V_1/I_2 = B = 0 \quad \text{and} \quad I_1/I_2 = D = 1$$

When $I_2 = 0$, $V_1 = V_2$ and $I_1 = V_2 Y$. Hence,

$$V_1/V_2 = A = 1 \quad \text{and} \quad I_1/V_2 = C = Y$$

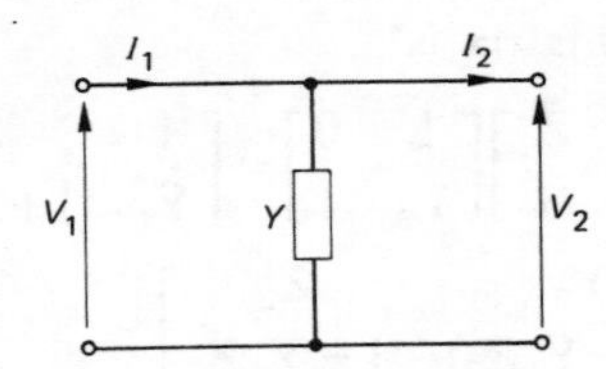

Fig. 5.5 Shunt admittance

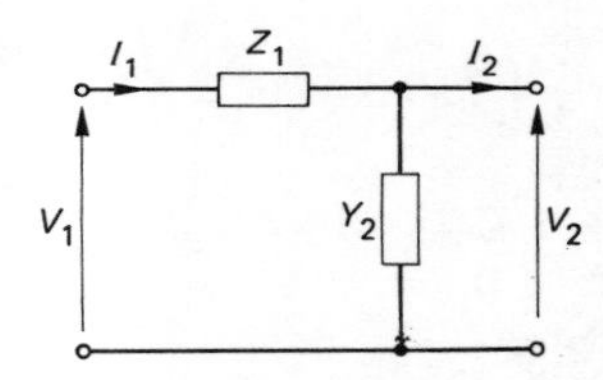

Fig. 5.6 L network

3 L *network*
From Fig. 5.6

$$V_1 = V_2 + I_1 Z_1 \quad \text{and} \quad I_1 = V_2 Y_2 + I_2$$

Hence,

$$V_1 = V_2 + (V_2 Y_2 + I_2) Z_1 = V_2(1 + Y_2 Z_1) + I_2 Z_1$$

Therefore,

$$\begin{bmatrix} A & B \\ C & D \end{bmatrix} = \begin{bmatrix} 1 + Y_2 Z_1 & Z_1 \\ Y_2 & 1 \end{bmatrix}$$

The L network can, alternatively, be regarded as the cascade connection of a series impedance Z_1 and a shunt admittance Y_2. In this case,

$$\begin{bmatrix} A & B \\ C & D \end{bmatrix} = \begin{bmatrix} 1 & Z_1 \\ 0 & 1 \end{bmatrix} \begin{bmatrix} 1 & 0 \\ Y_2 & 1 \end{bmatrix} = \begin{bmatrix} 1 + Z_1 Y_2 & Z_1 \\ Y_2 & 1 \end{bmatrix} \quad \text{(as before)}$$

If the L network is reversed so that the shunt admittance is across the input terminals of the network the transmission matrix becomes

$$\begin{bmatrix} A & B \\ C & D \end{bmatrix} = \begin{bmatrix} 1 & Z_1 \\ Y_2 & 1 + Z_1 Y_2 \end{bmatrix}$$

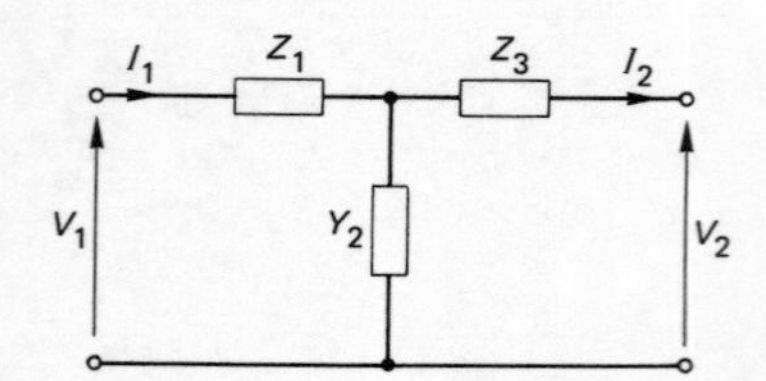

Fig. 5.7 T network

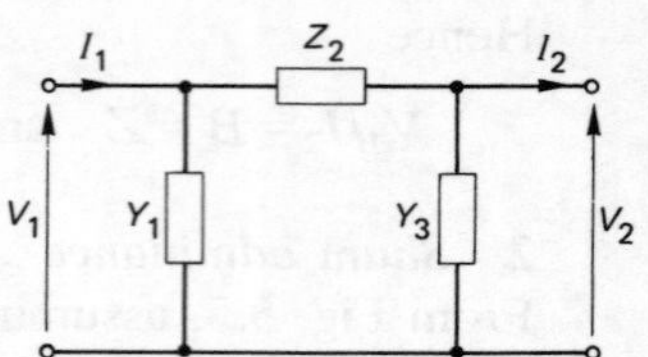

Fig. 5.8 π network

4 T *network*

Treating the T network (Fig. 5.7) as the cascade connection of a series impedance Z_1, a shunt admittance Y_2 and a series impedance Z_3 gives

$$\begin{bmatrix} A & B \\ C & D \end{bmatrix} = \begin{bmatrix} 1 & Z_1 \\ 0 & 1 \end{bmatrix}\begin{bmatrix} 1 & 0 \\ Y_2 & 1 \end{bmatrix}\begin{bmatrix} 1 & Z_3 \\ 0 & 1 \end{bmatrix} = \begin{bmatrix} 1+Z_1Y_2 & Z_1 \\ Y_2 & 1 \end{bmatrix}\begin{bmatrix} 1 & Z_3 \\ 0 & 1 \end{bmatrix}$$

$$= \begin{bmatrix} 1+Z_1Y_2 & Z_3(1+Z_1Y_2)+Z_1 \\ Y_2 & 1+Y_2Z_3 \end{bmatrix}$$

5 π *network*

The π network of Fig. 5.8 can be regarded as the cascade connection of a shunt admittance Y_1, a series impedance Z_2, and a shunt admittance Y_3, and hence its transmission matrix is

$$\begin{bmatrix} A & B \\ C & D \end{bmatrix} = \begin{bmatrix} 1 & 0 \\ Y_1 & 1 \end{bmatrix}\begin{bmatrix} 1 & Z_2 \\ 0 & 1 \end{bmatrix}\begin{bmatrix} 1 & 0 \\ Y_3 & 1 \end{bmatrix} = \begin{bmatrix} 1 & Z_2 \\ Y_1 & 1+Y_1Z_2 \end{bmatrix}\begin{bmatrix} 1 & 0 \\ Y_3 & 1 \end{bmatrix}$$

$$= \begin{bmatrix} 1+Z_2Y_3 & Z_2 \\ Y_1+Y_3(1+Y_1Z_2) & 1+Y_1Z_2 \end{bmatrix}$$

Example 5.2

Determine the transmission parameters of the network shown in Fig. 5.9.

Solution From equation (5.8),

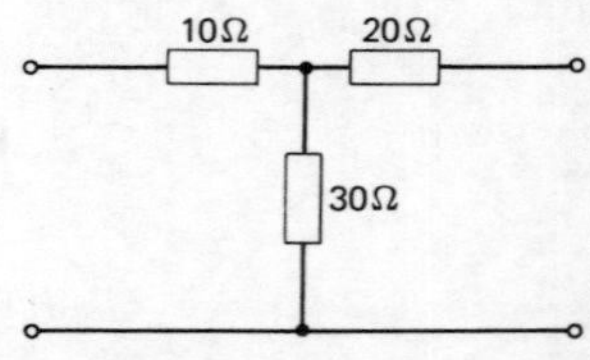

Fig. 5.9

1 $A = V_1/V_2 \quad (I_2 = 0)$

Since $V_2 = \dfrac{V_1 \times 30}{10+30} = 3V_1/4$

$A = 4/3 = 1.33$

2 $C = I_1/V_2 \quad (I_2 = 0)$

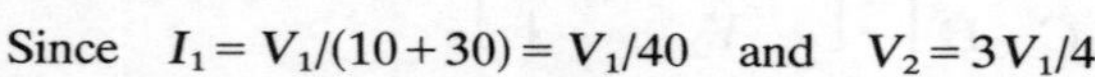

Since $I_1 = V_1/(10+30) = V_1/40$ and $V_2 = 3V_1/4$

$$C = \frac{V_1}{40} \times \frac{4}{3V_1} = 1/30\ \text{S} = 33.33 \times 10^{-3}\ \text{S}$$

3 $B = V_1/I_2 \quad (V_2 = 0)$

Since $I_1 = \dfrac{V_1}{10+\dfrac{20\times 30}{20+30}} = V_1/22$ and $I_2 = \dfrac{V_1}{22} \times \dfrac{30}{50}$

$$B = \frac{22 \times 50}{30} = 36.7\ \Omega$$

4 $D = I_1/I_2 \quad (V_2 = 0)$

Since $I_2 = I_1 \times \frac{30}{50}$ then $D = 5/3 = 1.67$

Thus the transmission matrix is

$$\begin{bmatrix} 1.33 & 36.7 \\ 33.33 \times 10^{-3} & 1.67 \end{bmatrix} \quad (Ans)$$

Note that

$$AD - BC = (1.33 \times 1.67) - (36.7 \times 33 \times 10^{-3}) = 1$$

The transmission parameters are generally used for the **determination of the output voltage and/or current** of a network in terms of its input voltage and/or current. The basic equations are repeated here:

$$V_1 = AV_2 + BI_2 \tag{5.14}$$

$$I_1 = CV_2 + DI_2 \tag{5.15}$$

From these equations, $I_2 = \frac{I_1 - CV_2}{D}$ and therefore

$$V_1 = AV_2 + \frac{B(I_1 - CV_2)}{D}$$

$$\text{or} \quad DV_1 - BI_1 = V_2(AD - BC) = V_2$$

Hence $V_2 = DV_1 - BI_1$ (5.22)

Similarly $I_2 = AI_1 - CV_1$ (5.23)

To obtain V_2 purely in terms of V_1 so that the voltage ratio of the network can be obtained, recourse must be made to the equation for V_1. This can be written as

$$V_1 = AV_2 + (BV_2/Z_L)$$

and therefore the voltage ratio is

$$\frac{V_2}{V_1} = \frac{V_2}{AV_2 + (BV_2/Z_L)} = \frac{Z_L}{AZ_L + B} \tag{5.24}$$

The input impedance of the network is the ratio $Z_{in} = V_1/I_1$. Therefore,

$$Z_{in} = \frac{AV_2 + BI_2}{CV_2 + DI_2} = \frac{AV_2 + (BV_2/Z_L)}{CV_2 + (DV_2/Z_L)}$$

$$= \frac{AZ_L + B}{CZ_L + D} \tag{5.25}$$

Example 5.3

A network has the following transmission parameters:

$$A = D = 1\underline{/60°} \qquad B = 200\underline{/70°}\ \Omega \qquad C = 1.5 \times 10^{-3}\underline{/-90}\ \text{S}$$

Calculate its input impedance when its output terminals are *a*) open-circuited and *b*) short-circuited.

Solution

a) When $Z_L = \infty$, the input impedance Z_{in} is given by $A/C\ \Omega$. Therefore

$$Z_{in} = \frac{1\angle 60°}{1.5\times 10^{-3}\angle -90°} = 666.7\angle 150°\ \Omega \quad (Ans)$$

b) When $Z_L = 0$, $Z_{in} = B/D\ \Omega$. Therefore,

$$Z_{in} = \frac{200\angle 70°}{1\angle 60°} = 200\angle 10°\ \Omega \quad (Ans)$$

Example 5.4

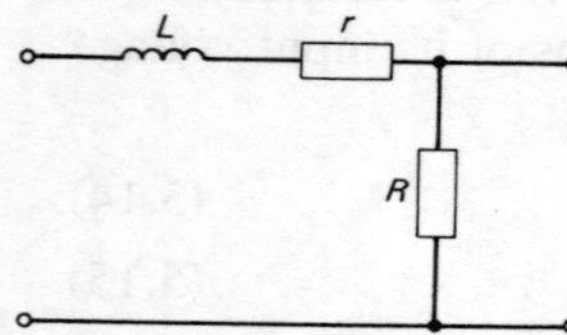

Fig. 5.10

Fig. 5.10 shows an L network whose transmission parameters are

$$\begin{bmatrix} 1+\frac{r+j\omega L}{R} & r+j\omega L \\ 1/R & 1 \end{bmatrix}$$

If $r = 20\ \Omega$ and $\omega L = 30\ \Omega$ calculate the value of R for two such networks connected in cascade to introduce an overall phase shift of 90°.

Solution When the output of the circuit is open-circuit equation (5.14) becomes $V_1 = AV_2$. This means that only the A parameter of the cascade is required. This is

$$\left(1+\frac{r+j\omega L}{R}\right)\left(1+\frac{r+j\omega L}{R}\right)+\frac{(r+j\omega L)}{R}$$

$$= \frac{R^2+r^2-\omega^2L^2+j2r\omega L+3rR}{R^2}+j3\omega L/R$$

For the network to introduce a 90° phase shift the real part of this equation must be equal to zero. Therefore,

$$R^2+3rR+r^2-\omega^2L^2=0$$

$$R^2+60R+400-900=0$$

$$R = \frac{-60\pm\sqrt{(3600+2000)}}{2} = 7.42\ \Omega \quad (Ans)$$

(ignoring the negative result).

Example 5.5

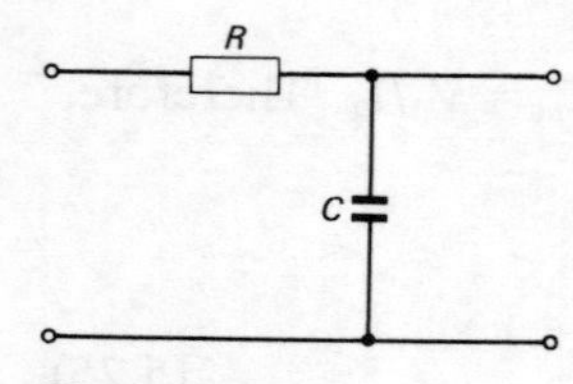

Fig. 5.11

Three of the circuits shown in Fig. 5.11 are connected in cascade. Derive an expression for the frequency at which the combination will introduce a phase shift of 180° when terminated in an open-circuit.

Solution For a single network $Z_1 = R$ and $Y_2 = j\omega C$. Hence the transmission matrix is

$$\begin{bmatrix} 1+j\omega CR & R \\ j\omega C & 1 \end{bmatrix}$$

For three such networks in cascade the overall matrix is

$$\begin{bmatrix} 1+j\omega CR & R \\ j\omega C & 1 \end{bmatrix}^3$$

but as in the previous example only the A term is needed. This is

$$(1+j\omega CR)^3+j\omega CR(1+j\omega CR)+j\omega CR(1+j\omega CR)+j\omega CR$$
$$=1-5\omega^2C^2R^2+6j\omega CR-j\omega^3C^3R^3$$

For V_2 to be 180° out of phase with V_1, the j terms must sum to zero. Therefore $6\omega_0 CR=\omega_0^3C^3R^3$ or $f_0=\sqrt{6}/2\pi CR$ (*Ans*)

Star-Delta Transform

The **star-delta transform** has already been discussed in Chapter 2. The transmission parameters provide an alternative method of deriving the necessary equations.

Consider the T and π networks shown in Fig. 5.12 (Figs. 2.9*c* and *d* reproduced). The transmission matrices are

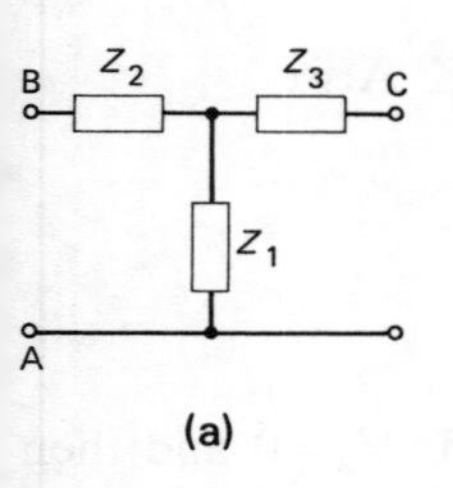

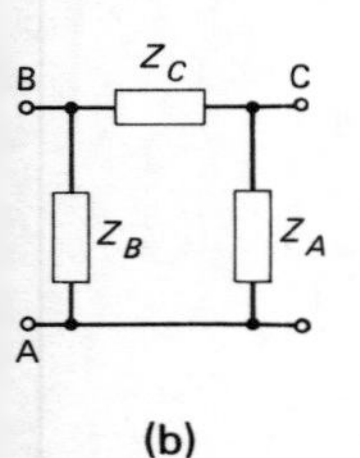

Fig. 5.12

$$\begin{bmatrix} 1+(Z_2/Z_1) & Z_3[1+(Z_2/Z_1)]+Z_2 \\ 1/Z_1 & 1+(Z_3/Z_1) \end{bmatrix}$$

and
$$\begin{bmatrix} 1+(Z_C/Z_A) & Z_C \\ (1/Z_B)+\dfrac{1}{Z_A}[1+(Z_C/Z_B)] & 1+(Z_C/Z_B) \end{bmatrix}$$

Hence $$Z_C=\frac{Z_1Z_3+Z_2Z_3+Z_1Z_2}{Z_1} \tag{5.26}$$

$Z_2/Z_1=Z_C/Z_A$ or

$$Z_A=Z_CZ_1/Z_2=\frac{Z_1Z_3+Z_2Z_3+Z_1Z_2}{Z_2} \tag{5.27}$$

and $Z_3/Z_1=Z_C/Z_B$ or

$$Z_B=Z_CZ_1/Z_3=\frac{Z_1Z_3+Z_2Z_3+Z_1Z_2}{Z_3} \tag{5.28}$$

Similarly,

$$1/Z_1=\frac{1}{Z_B}+\frac{1}{Z_A}\left(1+\frac{Z_C}{Z_B}\right)$$
$$=\frac{1}{Z_B}+\frac{Z_C}{Z_AZ_B}+\frac{1}{Z_A}=\frac{Z_A+Z_B+Z_C}{Z_AZ_B}$$

or $$Z_1=\frac{Z_AZ_B}{Z_A+Z_B+Z_C} \tag{5.29}$$

$Z_3/Z_1=Z_C/Z_B$ or

$$Z_3=Z_1Z_C/Z_B=\frac{Z_AZ_C}{Z_A+Z_B+Z_C} \tag{5.30}$$

Lastly $Z_2/Z_1=Z_C/Z_A$ or

$$Z_2=Z_1Z_C/Z_A=\frac{Z_BZ_C}{Z_A+Z_B+Z_C} \tag{5.31}$$

y Parameters

When networks which are connected in parallel are to be analyzed it is more convenient to use y parameters. The methods to be used to determine the values of the y parameters of a particular network are very similar to those already described. For this reason, only one example is given and then the results for the more commonly used networks are quoted in Table 5.2.

Consider the series impedance of Fig. 5.4 again.

Since $I_2 = I_1$ and $V_1 = V_2 + I_2 Z$ then

$$I_1 = (V_1/Z) - (V_2/Z)$$

Comparing with equation (5.8), $y_{11} = 1/Z$ and $y_{12} = -1/Z$. Also

$$I_2 = (V_1/Z) - (V_2/Z)$$

so that, comparing with equation (5.9),

$$y_{21} = 1/Z \quad \text{and} \quad y_{22} = -1/Z$$

Alternatively, with the output terminals short-circuited, $V_2 = 0$ and then $V_1 = I_1 Z_1$ so that

$$I_1/V_1 = y_{11} = 1/Z$$

Also, $V_1 = I_2 Z$ (since $I_2 = I_1$) and so

$$I_2/V_1 = y_{21} = 1/Z$$

With the input terminals short-circuited, $V_1 = 0$ and

$$I_1 = -y_{12} V_2 = -V_2/Z \quad \text{and} \quad y_{12} = 1/Z$$

Finally $I_2 = -V_2/Z$ or $I_2/V_2 = y_{22} = -1/Z$

The y parameter matrices for each of the other circuits considered earlier are given in Table 5.2.

Exercises 5

5.1 Determine the transmission parameters of the network shown in Fig. 5.13.

5.2 Determine the overall transmission matrix of the two cascaded networks shown in Fig. 5.14.

5.3 A network has the following transmission parameters:

$$A = 1\angle 30^\circ \qquad B = 150\angle 50^\circ\ \Omega \qquad C = 0.02\angle -6.8^\circ\ \text{S} \qquad D = 3\angle 0^\circ$$

Calculate the input impedance of the network when its output terminals are *a*) open-circuited and *b*) short-circuited.

5.4 A network has the transmission parameters $A = D = 14.14\angle 0^\circ$, $B = 50\angle 60^\circ\ \Omega$ and $C = 4\angle -60^\circ$ S. Calculate *a*) the ratio V_{out}/V_{in} and *b*) the input impedance of the network when a 600 Ω resistor is connected across the output terminals.

5.5 Determine *a*) the y parameters and *b*) the *h* parameters of the network shown in Fig. 5.15.

5.6 A network has the following *h* parameters: $h_{11} = 1000\ \Omega$, $h_{12} = h_{21} = 0.5$ and $h_{22} = 0.015$. Determine its transmission parameters.

Table 5.2

Figure no.	*y matrix*
5.5	$\begin{bmatrix} \infty & \infty \\ \infty & \infty \end{bmatrix}$
5.6	$\begin{bmatrix} 1/Z_1 & -1/Z_1 \\ -1/Z_1 & -\left(\dfrac{1+Z_1Y_2}{Z_1}\right) \end{bmatrix}$
5.6 (reversed)	$\begin{bmatrix} \dfrac{1+Z_1Y_2}{Z_1} & -1/Z_1 \\ 1/Z_1 & -1/Z_1 \end{bmatrix}$
5.7	$\begin{bmatrix} \dfrac{1+Z_3Y_2}{Z_1+Z_3+Z_1Y_2Z_3} & \dfrac{-1}{Z_1+Z_3+Z_1Y_2Z_3} \\ \dfrac{1}{Z_1+Z_3+Z_1Y_2Z_3} & \dfrac{-(1+Z_3Y_2)}{Z_1+Z_3+Z_1Y_2Z_3} \end{bmatrix}$
5.8	$\begin{bmatrix} \dfrac{1+Z_1Y_2}{Z_1} & -Y_2 \\ -Y_2 & \dfrac{1+Y_2Z_3}{Z_3} \end{bmatrix}$

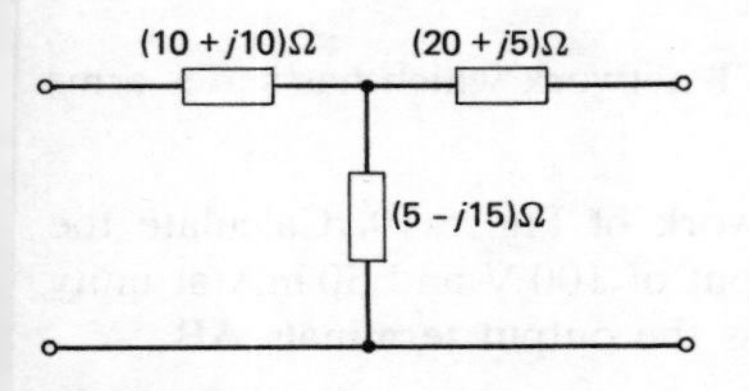

Fig. 5.13

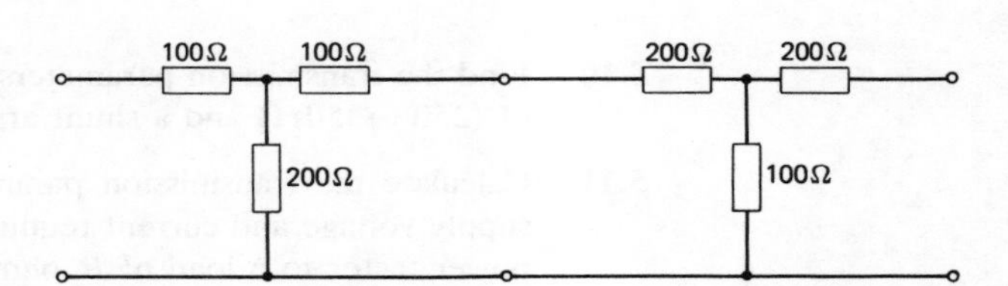

Fig. 5.14

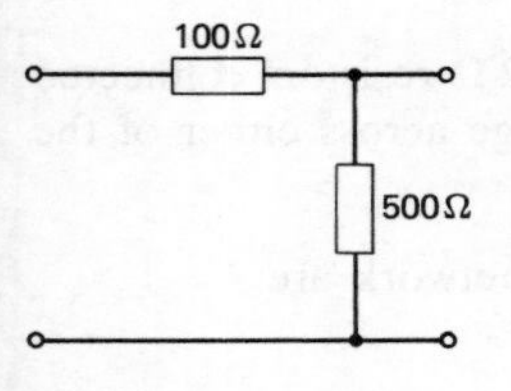

Fig. 5.15

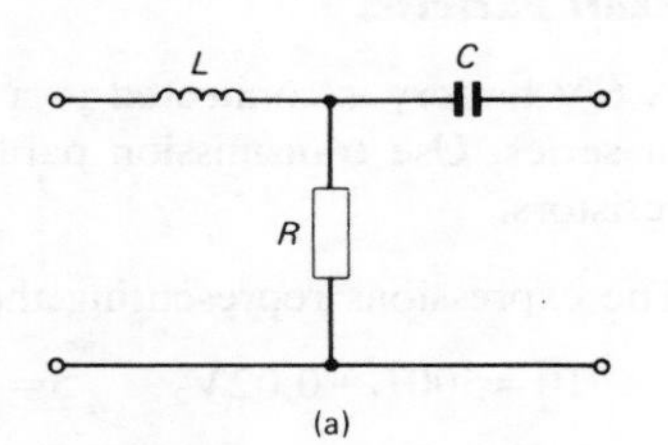

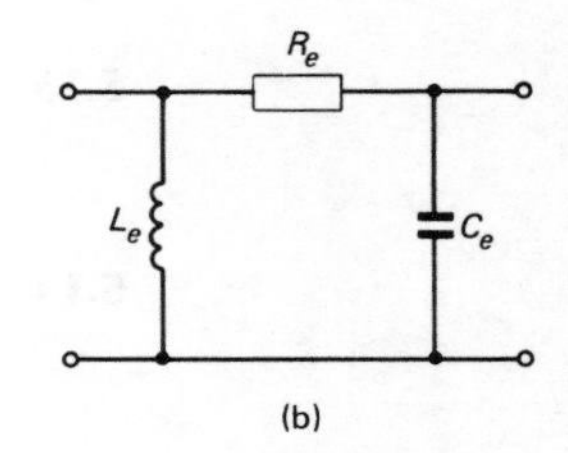

Fig. 5.16

5.7 Write down the transmission matrices for the networks shown in Fig. 5.16*a* and *b* and hence determine the necessary relationship between the components of the two circuits to be equivalent to one another.

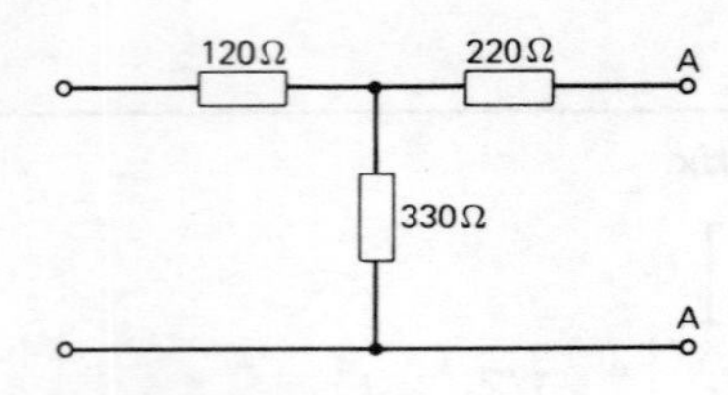

Fig. 5.17

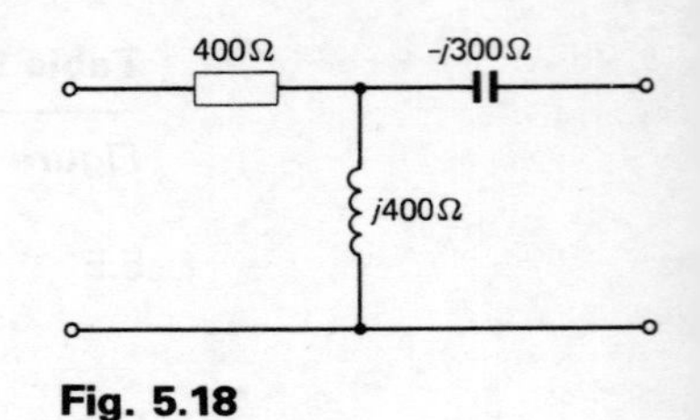

Fig. 5.18

5.8 For the network in Fig. 5.17 determine *a*) the transmission parameters and *b*) the voltage ratio V_2/V_1 when a 1200 Ω resistor is connected across the terminals AA.

5.9 Calculate the transmission parameters of the circuit given in Fig. 5.18. If, when the circuit is terminated in a 500 Ω resistor, the output voltage is 100 V calculate the input voltage.

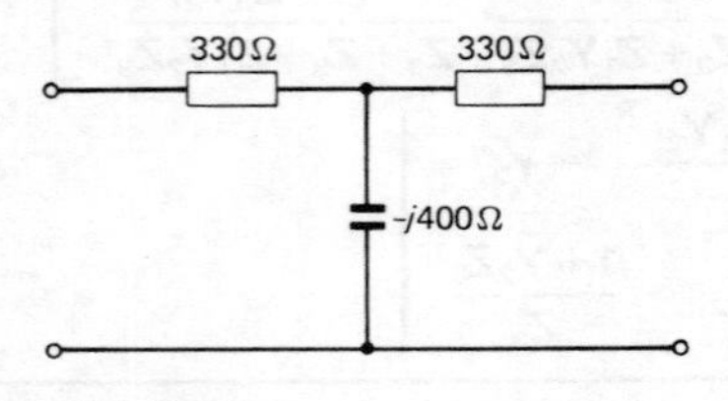

Fig. 5.19

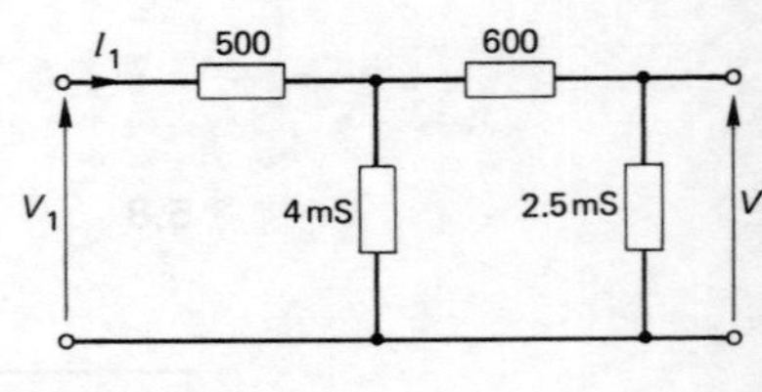

Fig. 5.20

5.10 Find the transmission parameters of a symmetrical T network which has series arms of $(250+j350)$ Ω and a shunt arm of $(-j180)$ Ω.

5.11 Calculate the transmission parameters of the network of Fig. 5.19. Calculate the supply voltage and current required to give an output of 100 V and 50 mA at unity power factor to a load of *R* ohms connected across the output terminals AB.

5.12 Calculate the transmission parameters of the circuit shown in Fig. 5.20. If the output voltage is $25\angle 0°$ volts and the output current is $16\angle 0°$ mA calculate the input voltage.

Short Exercises

5.13 A 6 V battery is connected to a circuit that consists of two 12 Ω resistors connected in series. Use transmission parameters to calculate the voltage across either of the resistors.

5.14 The expressions representing the behaviour of a 4-terminal network are

$$10=500I_1+0.02V_2 \qquad 5=50I_1+0.04V_2$$

What kind of parameters have been used and for what purpose are they best suited?

5.15 A network has the impedance matrix $\begin{bmatrix}100 & 10\\ 20 & 40\end{bmatrix}$ and is connected in series with another network. The overall impedance matrix is then $\begin{bmatrix}150 & 50\\ 30 & 80\end{bmatrix}$. Write down the impedance matrix of the second network.

5.16 One set of parameters has not been given in this chapter, i.e. those for which I_2 and V_2 are the independent variables. Write down equations for this set of parameters, using coefficients E, F, G and H and deduce the dimensions of each parameter.

5.17 A passive network has the following transmission parameters: $A = D = 100$, $B = 500\ \Omega$. Calculate the value of C.

5.18 A 4-terminal network has the h parameter matrix

$$\begin{bmatrix} 1000 & 1\times10^{-6} \\ 120 & 4\times10^{-4} \end{bmatrix}$$

Determine the y matrix for the network.

5.19 A network has the transmission matrix

$$\begin{bmatrix} 1.5 & 40 \\ 0.0313 & 1.5 \end{bmatrix}$$

Two such networks are connected in cascade. Determine the transmission matrix of the combination.

5.20 A symmetrical T network has each series impedance equal to $50\underline{/60^\circ}\ \Omega$ and a shunt admittance of $1.2\times10^{-3}\underline{/90^\circ}$ S. Calculate the transmission parameters of the network.

6 Energy Transfer

Energy is the capability to do work. Any machine that possesses some energy is able to perform work, converting energy from one form to another. Conversely, work must have been done in order to provide the machine with energy in the first place. A machine will accept input energy in one form and convert some of it into output energy in some other form. Some of the output energy will not be useful since it will be provided in the form of heat. The part of the input energy that does not appear at the output is stored within the machine. This means that work done and energy are two aspects of the same thing and they are both measured in terms of the same unit; namely the *Joule*.

1 joule is the work done when a force of 1 newton acts over a distance of 1 metre in the direction of the force.

There are several different forms of energy; those considered in this chapter are electrical, mechanical, chemical, heat, light and sound.

Principle of Energy Conversion

The principle of energy conversion states that energy can be neither created nor destroyed, but only changed from one form to another. This means that all the input energy to a machine either must appear at the output of the machine or must be stored within it.

Several examples of energy conversion are well known:

A In both primary and secondary cell batteries, chemical energy is changed into electrical energy.
B In power stations, chemical energy is obtained from the burning of coal and oil and is changed first into heat, then into mechanical energy, and, finally, by means of a generator into electrical energy.
C In a telephone transmitter sound energy is changed into electrical energy, whilst in the receiver the opposite process takes place; that is, electrical energy is converted into sound energy.
D In an electric light bulb, electrical energy is changed into both light and heat energy.
E In an electric fire, electrical energy is changed into heat energy with some light energy also.
F Electric motors and generators of many kinds convert electrical energy into mechanical energy or vice versa.

It is well known, in some of these cases at least, that not all of the input energy is converted into the required form of output energy. Take, for example, D and E. The electric light bulb produces quite a lot of unwanted heat energy as well as the wanted light energy, while an electric fire generally produces some light energy as well as heat.

This is always true; some of the input energy is always used for the production of energy in one, or more, unwanted forms. As a result the efficiency of any energy conversion device or transducer always falls short of 100%. The losses which occur in various types of electric motor will be mentioned later (p. 98) in this chapter.

Electrical Energy

Work is done in an electrical circuit when a number of electrons move from one point to another. The **work done** is QV where Q is the total charge in coulombs that is transferred and V is the potential difference between the two points. Since $Q=It$, where I is the current in amperes and t is the time in seconds, the **energy** is given by

$$W=QV=IVt \text{ joules} \tag{6.1}$$

$$=I^2Rt=V^2t/R \text{ joules} \tag{6.2}$$

Power is the rate of doing work and its unit is the watt. 1 watt is the power dissipated when energy is used at the rate of 1 J/s, i.e. energy = power × time.

Energy in an Electric Field

If a CAPACITANCE of $C\,\mu$F is charged to have a voltage of V volts across its terminals, the charge Q stored is

$$Q=CV \text{ coulombs} \tag{6.3}$$

If the voltage across the capacitor is then increased to $V+dV$ volts in a time of dt seconds the current taken must be equal to

$$I=C\,dV/dt$$

The energy supplied to the capacitor is

$$VI\,dt=VC\frac{dV}{dt}\times dt=VC\,dV \text{ joules}$$

The **total energy** W supplied as the capacitance is charged from zero volts to V volts terminal voltage is

$$W=C\int_0^V V\,dV=\tfrac{1}{2}CV^2 \text{ joules} \tag{6.4}$$

or $$W=\tfrac{1}{2}QV \text{ joules} \tag{6.5}$$

Example 6.1

A 1 μF capacitor is charged from a 24 V d.c. supply and is then disconnected from the supply. Calculate the energy stored in the capacitor.

Two uncharged capacitors, one of 2 μF and one of 5 μF, are now connected in series with one another and then the series combination is connected in parallel with the 1 μF capacitor. Calculate the energy stored in each capacitor and the total energy stored. Assume no loss of charge occurs at any time.

Solution

$$W = \tfrac{1}{2} \times 1 \times 10^{-6} \times 24^2 = 288 \times 10^{-6}\text{ J} \quad (\textit{Ans})$$

When the 2 μF and 5 μF capacitors are connected in series the total capacitance is

$$1 + [5 \times 2/(5+2)] = 2.43\ \mu\text{F}$$

The voltage V across the combination (Fig. 6.1) is

$$V = Q/C = \frac{24 \times 1 \times 10^{-6}}{2.43 \times 10^{-6}} = 9.88\text{ V}$$

The energy stored in the 1 μF capacitor is

$$\tfrac{1}{2} \times 1 \times 10^{-6} \times 9.88^2 = 48.8 \times 10^{-6}\text{ J} \quad (\textit{Ans})$$

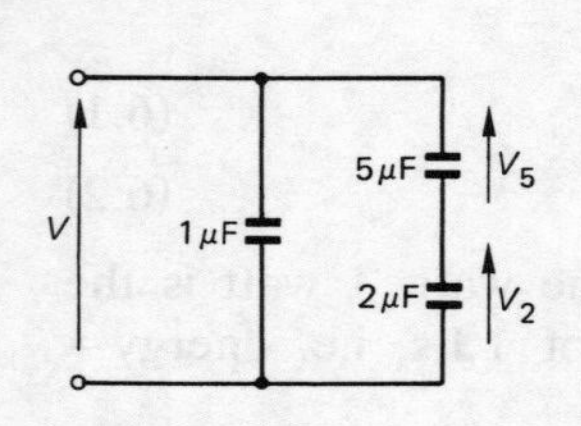

Fig. 6.1

The voltage V_5 across the 5 μF capacitor $= 9.88 \times 2/7 = 2.82$ V. Hence, the energy stored is

$$\tfrac{1}{2} \times 5 \times 10^{-6} \times 2.82^2 = 19.9 \times 10^{-6}\text{ J} \quad (\textit{Ans})$$

The voltage V_2 across the 2 μF capacitor $= 9.88 \times 5/7 = 7.06$ V. So, the energy stored in this capacitor is

$$\tfrac{1}{2} \times 2 \times 10^{-6} \times 7.06^2 = 49.8 \times 10^{-6}\text{ J} \quad (\textit{Ans})$$

The total energy stored is

$$(48.8 + 19.9 + 49.8) \times 10^{-6} = 118.5 \times 10^{-6}\text{ J} \quad (\textit{Ans})$$

The total energy stored is less than that which was stored in the 1 μF capacitor originally. This is because some energy has been dissipated in the form of heat (I^2R) when current flowed to re-distribute the charge when the new capacitors were joined into the circuit.

Example 6.2

A 2 μF capacitor is connected in series with a 100 kΩ resistor and a 120 V d.c. supply. Calculate the rate at which energy is being stored in the capacitor when its terminal voltage is 40 V.

Solution When $V_C = 40$ volts the voltage across the resistor must be 80 volts and so the current flowing is $80/100 = 0.8$ mA. At this instant therefore, the power supplied to the circuit is

$$120 \times 0.8 \times 10^{-3} = 96\text{ mW}$$

and the power dissipated in the resistor is

$$80 \times 0.8 \times 10^{-3} = 64\text{ mW}$$

Therefore, the rate at which energy is stored in the capacitor is

$$96 - 64 = 32\text{ mW} \quad (\textit{Ans})$$

Force between Two Charged Plates

Consider two metal plates of area A square metres separated from one another by a dielectric of permittivity $\varepsilon = \varepsilon_0\varepsilon_r$ and thickness d metres. The capacitance of the system is $C = \varepsilon A/d$ and so the stored energy is

$$\frac{1}{2}\frac{\varepsilon A}{d}V^2 \text{ joules}$$

The energy stored in the dielectric per cubic metre is

$$W = \frac{\varepsilon A V^2}{2dAd} = \frac{\varepsilon V^2}{2d^2} \text{ J/m}^3 \qquad (6.6)$$

Suppose the plates are disconnected from both the supply voltage and the external circuit so that charge can neither be supplied nor lost. If, then, one of the plates is fixed in position and the other plate is moved away through a distance of dx metres, the stored charge will not change. The capacitance C of the plates will be reduced and so the voltage V between the plates must increase. Since C is inversely proportional to d, and V is inversely proportional to C, the ratio V/d will remain *constant.* Consequently the energy stored per cubic metre is *unchanged.* This means that the energy stored in the extra volume of dielectric must have been supplied by the work done in moving the plate. Therefore,

$$F\,dx = \frac{\varepsilon V^2}{2d^2} \times A\,dx$$

So the **force F between two charged plates** is

$$F = \frac{\varepsilon A V^2}{2d^2} \text{ joules} \qquad (6.7)$$

$$\text{or} \quad F = \frac{QV}{2d} = W/d \text{ joules} \qquad (6.8)$$

Energy in a Magnetic Field

When the current flowing in an INDUCTANCE of L henrys increases in value, the e.m.f. induced into the inductance is equal to $-L\,di/dt$ volts. The energy stored in the magnetic field of the inductance in a time of dt seconds will be

$$iv\,dt = i\,dtL\frac{di}{dt} = Li\,di \text{ joules}$$

The **total energy** absorbed when the current increases from zero to I amperes is

$$W = L\int_0^I i\,di = \tfrac{1}{2}LI^2 \text{ joules} \qquad (6.9)$$

Example 6.3

A coil has an inductance of 2 H and an effective resistance of 10 Ω and it is connected across a 12 V d.c. supply. Calculate the rate at which energy is being stored in the magnetic field when the current is 0.4 A. Also find the maximum energy stored.

Solution When the current is 0.4 A the voltage dropped across the series resistor is $0.4 \times 10 = 4$ V. The input power is $12 \times 0.4 = 4.8$ W and the power dissipated is $4 \times 0.4 = 1.6$ W. Therefore, the rate at which energy is being stored is

$4.8 - 1.6 = 3.2$ W (*Ans*)

Maximum current = final current = 12/10 = 1.2 A. Therefore, maximum energy stored is

$\frac{1}{2} \times 1.2^2 \times 2 = 1.44$ joules (*Ans*)

When energy calculations are to be performed upon a magnetic circuit, a knowledge of the various **magnetic circuit equations** is required. These should have been studied at an earlier stage and so this chapter will only quote the relevant formulae. (l is the length of the magnetic path in metres.)

A Magnetomotive force (m.m.f.) = NI ampere-turns (AT) (6.10)
B Magnetizing force $H = NI/l$ ampere-turns/metre (AT/m) (6.11)
C Magnetic flux Φ = m.m.f./reluctance webers (Wb) (6.12)
D Flux density B = flux/area = Φ/A tesla (T) (6.13)

E Absolute permeability $(\mu = \mu_r\mu_0) = \dfrac{\text{flux density}}{\text{magnetizing force}}$

$$= B/H \text{ henry/metre (H/m)} \tag{6.14}$$

F $$\text{Relative permeability } \mu_r = \frac{\text{flux density in material}}{\text{flux density in air}} \tag{6.15}$$

(The flux density must be caused by the same magnetizing force.)

G μ_0 = permeability of air (strictly a vacuum) = $4\pi \times 10^{-7}$ H/m (6.16)
H Reluctance S = m.m.f./flux = $NI/\Phi = Hl/BA = l/\mu A$ AT/Wb (6.17)
I Inductance = flux linkages per ampere = $N\Phi/I$ (6.18)

$$= \frac{N}{I} \cdot \frac{NI}{S} = N^2/S \tag{6.19}$$

The energy stored in an inductance is given by equation (6.9),

$W = \frac{1}{2}LI^2$ joules

Hence

$$W = \tfrac{1}{2}I^2\frac{N^2}{S} = \tfrac{1}{2}I^2N^2\frac{\mu A}{l} = \frac{1}{2}\left(\frac{NI}{l}\right)^2\mu Al = \tfrac{1}{2}\mu H^2Al \text{ joules}$$

Therefore, the energy stored per cubic metre is

$$W = \tfrac{1}{2}\mu H^2 = \tfrac{1}{2}BH = \frac{B^2}{2\mu} \text{ J/m}^3 \tag{6.20a}$$

When this expression is used to determine the force of attraction between two magnetic surfaces it is the energy stored in the field within the air gap that is important and hence equation (6.20*a*) becomes

$$W = \frac{B^2}{2\mu_0} \text{ J/m}^3 \tag{6.20b}$$

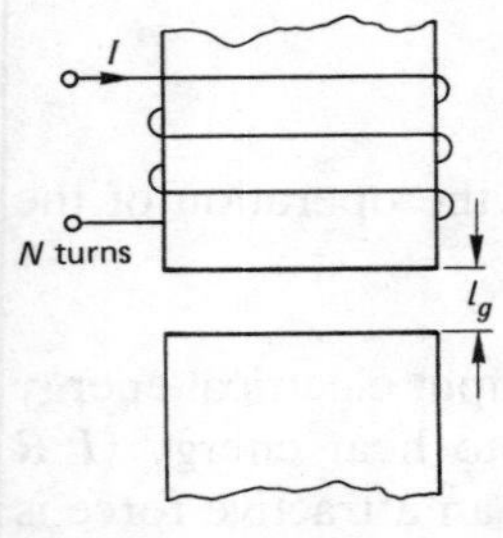

Fig. 6.2

1 Consider Fig. 6.2 which shows two parallel flat surfaces, made from a magnetic material, and separated from one another by an air gap of length l_g metres. The upper surface has a coil of N turns wound around it that carries a current of I amperes. Suppose a force of F newtons is exerted upon the lower surface to move it away from the upper surface through a distance of dl_g. The reluctance of the air gap will then increase and, if the flux density in the air gap is to be maintained constant at its original value, the current I must also increase by the appropriate amount. The flux linkages will then be unchanged and so there will be zero induced e.m.f. in the winding. This means that zero electrical energy will have been supplied to the circuit. Therefore the extra energy stored in the air gap must have been derived from the work done by the applied force. Hence,

$$F\,dl_g = \frac{B^2}{2\mu_0} \cdot A\,dl_g$$

$$F = \frac{B^2 A}{2\mu_0} \text{ newtons} \tag{6.21}$$

Example 6.4

Data for the relay shown in Fig. 6.3 is
mean length of magnetic circuit = 32 cm
total length of air gaps = 3.5 mm
mean cross-sectional area of magnetic circuit = 0.6 cm^2
number of turns = 8000 coil current = 45 mA $\mu_r = 700$
Calculate the force exerted on the armature.

Fig. 6.3 basic relay

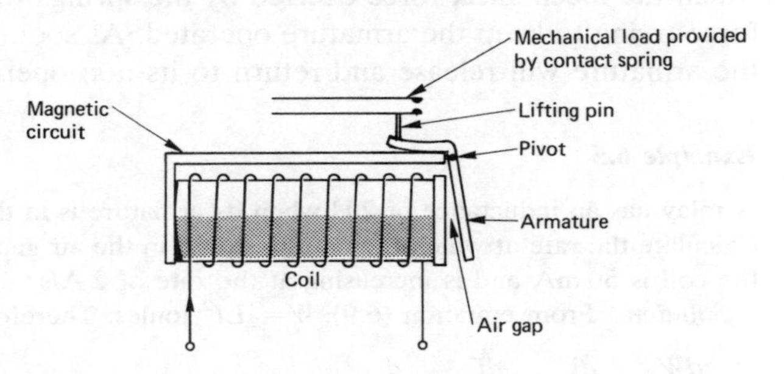

Solution Total reluctance is

$$S = \frac{0.32}{700 \times 4\pi \times 10^{-7} \times 60 \times 10^{-6}} + \frac{3.5 \times 10^{-3}}{4\pi \times 10^{-7} \times 60 \times 10^{-6}}$$

$$= 52.48 \times 10^6 \text{ AT/Wb}$$

$$\text{Flux } \Phi = \frac{8000 \times 45 \times 10^{-3}}{52.48 \times 10^{6}} = 6.86\ \mu\text{Wb}$$

$$\text{Flux density } B = \frac{6.86 \times 10^{-6}}{60 \times 10^{-6}} = 0.114\ \text{T}$$

Therefore, from equation (6.21) the force F exerted on the armature is

$$F = \frac{0.114^2 \times 60 \times 10^{-6}}{2 \times 4\pi \times 10^{-7}} = 0.31\ \text{N} \quad (Ans)$$

2 The various energy changes that take place during the operation of the relay (see Example 6.4) are as follows:

A When a current first flows in the coil winding, the input electrical energy is converted mainly into magnetic energy but also into heat energy (I^2R dissipation in the winding resistance). During this time an attractive force is exerted upon the armature but its magnitude is not large enough for it to be able to overcome the mechanical forces (mainly exerted by the contact springs) opposing any movement of the armature.

B When the magnetic flux density in the air gap has built up sufficiently, the attractive force exerted upon the armature becomes large enough to balance the opposing mechanical forces and the armature starts to move. As the armature moves, mechanical energy is needed to lift the contact springs and to supply kinetic energy to the moving parts. This energy is supplied by the electrical circuit.

C When the armature is brought to a halt at the end of its travel, all of the kinetic energy it possesses is dissipated.

D Once the armature has fully operated, electrical energy must still be supplied to the relay to keep the armature in its operated position. This energy must provide a sufficiently great attractive force to balance the restoring force exerted by the contact springs.

E When the current in the coil is reduced, the point will be reached at which the mechanical force exerted by the springs will exceed the magnetic force trying to keep the armature operated. As soon as this point is reached the armature will release and return to its non-operated position.

Example 6.5

A relay has an inductance of 2 H when its armature is in the non-operated position. Calculate the rate at which the energy stored in the air gap changes if the current in the coil is 50 mA and is increasing at the rate of 2 A/s.

Solution From equation (6.9), $W = \frac{1}{2}LI^2$ joules. Therefore,

$$\frac{dW}{dt} = Li\frac{di}{dt} + \tfrac{1}{2}i^2\frac{dL}{dt} = Li\frac{di}{dt}$$

since L is a constant. Therefore,

$$\frac{dW}{dt} = 2 \times 50 \times 10^{-3} \times 2 = 0.2\ \text{W} \quad (Ans)$$

Forces in Inductive Circuits

If the physical dimensions of a current-carrying conductor are changed, the number of flux linkages will alter. Since inductance is the number of flux linkages per ampere, the inductance of the circuit will also be changed. In turn, the energy stored in the magnetic field will vary and so some work will have been done on the circuit. The change in the number of flux linkages will induce an e.m.f. $N\,d\Phi/dt$ into the circuit. If the current flowing in the circuit is maintained at its original value, electrical energy must be supplied either *to* or *by* the circuit to satisfy equation (6.22).

$$\text{Energy supplied} = \text{work done} + \text{change in stored energy} \tag{6.22}$$

Suppose that a force of F newtons is exerted upon the N current-carrying conductors shown in Fig. 6.4 with the result that the conductors are moved through a distance of dx metres. Then the work done is $F\,dx$ joules. If the change in the position of the conductors reduces the number of flux linkages, the e.m.f. V which is induced will act in such a direction that the current is kept constant at its original value, This means that energy is supplied *by* the electrical source to the circuit and is equal to $IV\,dt$ joules. Now

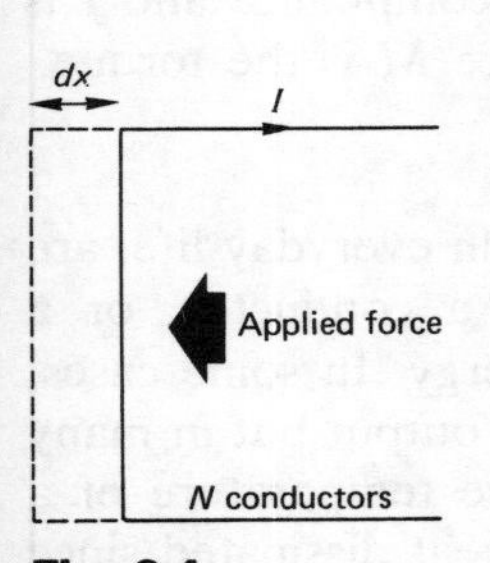

Fig. 6.4

$$IV\,dt = IN\frac{d\Phi}{dt}\,dt = Id(N\Phi) = Id(LI) = I^2\,dL \text{ joules}$$

Since the current is maintained constant at I amperes, the change in the stored magnetic energy will be $\frac{1}{2}I^2\,dL$, where dL is the change in the inductance of the circuit. The increase in the electrical input energy is equal to the sum of the work done and the change in the stored magnetic energy. Therefore,

$$I^2\,dL = F\,dx + \tfrac{1}{2}I^2\,dL$$

$$F = \tfrac{1}{2}I^2\frac{dL}{dx} \text{ newtons} \tag{6.23}$$

If the change in the dimensions of the current-carrying conductors is such that the number of flux linkages is increased, energy will be delivered *by* the circuit *to* the electrical source. In this case

$$IV\,dt - \tfrac{1}{2}LI^2 = F\,dx$$

which leads to equation (6.23) again.

If the system of conductors is rotational and is subject to an applied torque T, the corresponding equation is

$$T = \tfrac{1}{2}I^2\frac{dL}{d\theta} \tag{6.24}$$

where $d\theta$ is the angular distance through which the conductors move.

Non-electrical Energy

Mechanical Energy

Energy in a mechanical system exists by virtue of either the *position* or the *velocity* of a body. If a body is positioned above the Earth it will possess *potential energy*. A body of mass m travelling with a velocity v m/s possesses a kinetic energy of $\frac{1}{2}mv^2$ joules.

If a body is moved through a distance of x metres by the application of a force of F newtons, the work done is Fx joules. In the case of a body suspended above the Earth the applied force is the force of gravity. For a rotating body with an angular velocity of ω radians per second the kinetic energy is given by $\frac{1}{2}J\omega^2$ joules, where J is the moment of inertia.

A spring possesses potential energy whenever it is deformed. For a linear spring the potential energy is $W=\frac{1}{2}kf^2$ where k is the compliance and f is the restoring force. For a rotary spring $W=\frac{1}{2}kM^2$ where M is the torque.

Heat, Sound and Light Energy

Three other forms of energy, which are commonly met in everyday life, are heat, sound and light. Whenever a current flows in a conductor, or a component, electrical energy is converted into heat energy. In some cases, such as an electric fire, the heat produced is the wanted output but in many other cases it is an embarrassment. The increase in the temperature of a conductor, or a component, is not proportional to the heat dissipated since some of the generated heat is lost by convection, conduction or radiation, or a combination of them. When an increase in the temperature is undesirable, as in a transistor for example, the component is often mounted on a "heat sink" to assist in the removal of the unwanted heat as rapidly as possible.

For electrical energy to be converted into sound or light energy, a *transducer* is required. Some possibilities have been mentioned earlier in this chapter. Others are to be found in the fields of radio, television and electronics: for example, photo-diodes and light-emitting diodes.

Electrical–Mechanical Energy Transfer

Electric machines, of many kinds, are widely used for the conversion of either electrical energy into mechanical energy (e.g. a motor or a relay) or mechanical energy into electrical energy (a generator). In either case the machine must be supplied with energy from some external source. Some of this input energy is lost, mainly in the form of heat energy, but most of it is converted into another desired form.

Thus, in the case of an electric motor, the electrical input energy is used to drive a rotating shaft which, when coupled to an external mechanical load, will produce output mechanical energy. Conversely, a generator is provided with input mechanical energy, in the form of a rotating shaft, and this is converted into electrical energy. In both cases the available output energy is always smaller than the input energy so that the conversion efficiency is always less than 100%.

The ENERGY BALANCE EQUATION for an electrical machine is

$$\begin{matrix}\text{Electrical input energy}\\+\\\text{mechanical input energy}\end{matrix}=\begin{matrix}\text{stored magnetic}\\\text{field energy}\end{matrix}+\begin{matrix}\text{stored}\\\text{mechanical}\\\text{energy}\end{matrix}+\begin{matrix}\text{energy}\\\text{dissipation}\end{matrix} \qquad (6.24)$$

The energy dissipation is nearly all heat energy.

If the machine is operated as a motor its output will consist of mechanical energy and hence this will be *negative* in equation (6.24). Conversely, the output energy of a generator is electrical and then the electrical *input* energy in equation (6.24) is *negative.*

The **energy balance equation** can be expressed more succinctly using symbols, as shown by equation (6.25).

$$W_E + W_M = W_F + W_S + W_C \tag{6.25}$$

where W_E is the electrical input energy
W_M is the input mechanical energy
W_F is the stored magnetic energy
W_S is the stored mechanical energy
W_C is the total energy dissipation.

The energy balance diagram corresponding to equation (6.25) is shown in Fig. 6.5.

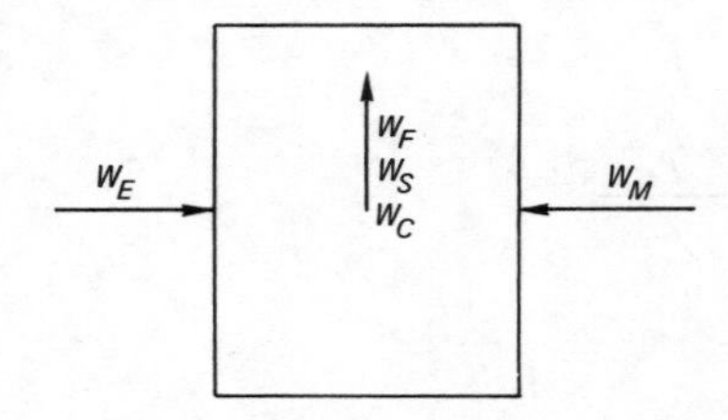

Fig. 6.5 Energy balance diagram

The energy balance concept is still applicable if either *changes* of energy or *rates of change* of energy are considered. Equations (6.26) and (6.27) respectively apply.

$$\delta W_E + \delta W_M = \delta W_F + \delta W_S + \delta W_C \tag{6.26}$$

$$\frac{\delta W_E}{\delta t} + \frac{\delta W_M}{\delta t} = \frac{\delta W_F}{\delta t} + \frac{\delta W_S}{\delta t} + \frac{\delta W_C}{\delta t} \tag{6.27a}$$

$$P_E + P_M = \frac{\delta W_F}{\delta t} + \frac{\delta W_S}{\delta t} + \frac{\delta W_C}{\delta t} \tag{6.27b}$$

Equation (6.27*b*) is generally known as the **power balance equation**. It states that the rate of change of stored energy in a machine is equal to the total power input.

When the machine first starts to run, its electrical and mechanical losses are unequal and the machine is able to increase its speed. At some particular speed the electrical and mechanical losses become equal to one another and then the machine operates under *steady-state conditions.* Before steady state conditions prevail the question as to which of the two types of loss is the greater depends upon the type of machine. For a generator, the mechanical losses are greater than the electrical losses until the steady running speed is reached. For a motor, on the other hand, the electrical losses are the greater until loss balance is achieved. Once steady-state conditions exist, the currents and velocities at various points in the machine are constant and there are then *no* changes in either electrical or mechanical energy storage to consider.

The energy balance and power balance equations are difficult to solve since they include *transient terms*, i.e. terms that exist only while the machine is either speeding-up or is slowing-down. When a machine is in one of its steady-state conditions (either stationary or running at a constant speed) there are *no* changes in the stored energy and so

$$\frac{\delta W_F}{\delta t} = \frac{\delta W_S}{\delta t} = 0$$

Then equation (6.27*b*) can be written in the form

$$P_E + P_M = \frac{\delta W_C}{\delta t} = P_C \tag{6.28}$$

The energy balance diagram is shown by Fig. 6.6.

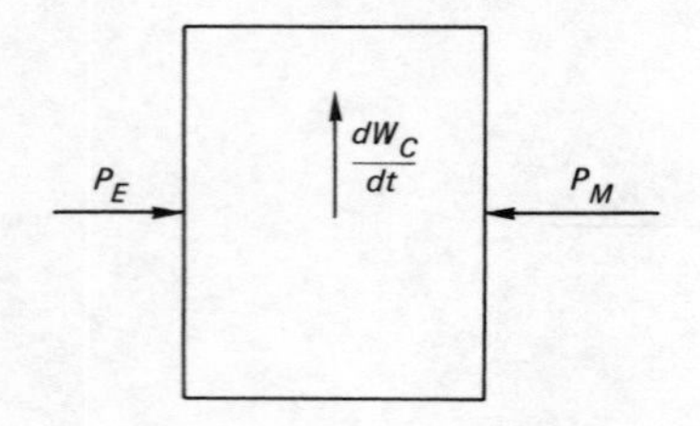

Fig. 6.6 Simplified energy balance diagram

Example 6.6

A shunt-wound machine is rated at 240 V, 3 kW and 2500 r.p.m. An input power of 200 W is needed to provide the field current. When the machine is operated as a motor with no load, a further input power of 150 W is needed to supply the armature losses. Assuming the power losses are constant calculate the efficiency of the machine when it is used as a generator producing the rated output power. The armature has a resistance of 1.2 Ω.

Solution At full load the armature current is 3000/240 = 12.5 A and so the armature loss is $12.5^2 \times 1.2 = 187.5$ W. The energy balance equation becomes

$$-3000 + 200 + W_M = 150 + 187.5 + 200$$

$$W_M = 3000 + 150 + 187.5 = 3337.5 \text{ W}$$

and the efficiency is

$$\frac{3000}{3337.5 + 200} \times 100 = 84.8\% \quad (Ans)$$

Power Losses in Electrical Motors

In the calculation of the efficiency of an electric motor the various sources of loss, both electrical and mechanical, must be identified. The losses can be illustrated by a **power flow diagram**.

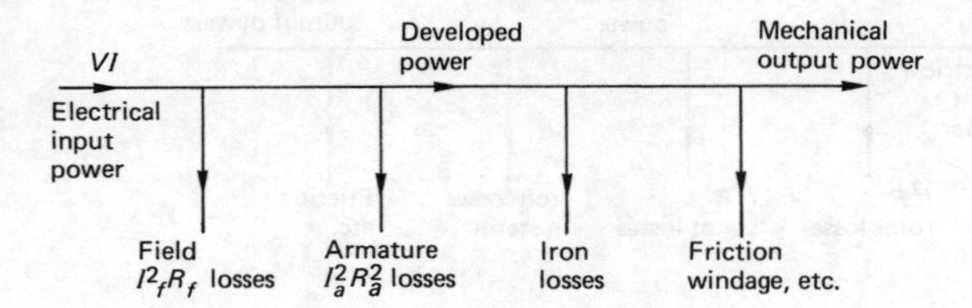

Fig. 6.7 Power flow diagram of a d.c. shunt machine

1 *D.C. Motor* The power flow diagram of a d.c. motor is shown in Fig. 6.7. Some of the losses shown vary with the load on the motor and hence with the input current *I*, and some of them are relatively constant. In a shunt-wound motor the field current, and hence the field loss, is constant. The armature loss varies with the load. The friction loss varies with the speed of the machine but not with the current taken from the supply. The iron losses are approximately constant with load.

The efficiency η of a d.c. machine is 100 times the ratio of the output power to the input power. Thus,

$$\eta = \frac{VI - I_a^2 R_a - I_f^2 R_f - F}{VI} \times 100\% \tag{6.29}$$

where F represents the sum of the iron, friction and windage, etc. losses.

If it assumed that the field current is much smaller than the armature current so that $I_a \simeq I$, then η can be written as

$$\eta = \frac{V - IR_a - \frac{1}{I}(I_f^2 R_f + F)}{V} \times 100\% \tag{6.30}$$

The maximum value of the motor's efficiency can be found by differentiating equation (6.30) with respect to the input current I and then equating the result to zero. Thus:

$$\frac{d\left[V - IR_a - \frac{1}{I}(I_f^2 R_f + F)\right]}{dI} = 0$$

$$-R_a + \frac{1}{I^2}(I_f^2 R_f + F) = 0$$

$$I^2 R_a = I_f^2 R_f + F$$

Thus, the efficiency of the motor is at its maximum possible value when the load (i.e. the input current) is such that the variable losses are equal to the fixed losses.

In the case of a series-wound motor the field losses will also vary with the load.

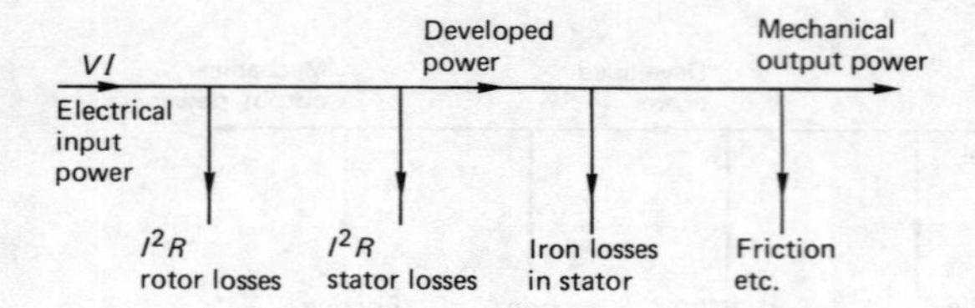

Fig. 6.8 Power flow diagram of an induction motor

2 *Cage Rotor Induction Motor* The power flow diagram for a cage rotor induction motor is shown in Fig. 6.8. Some of the electrical input power is lost in the form of heat energy dissipated in the effective resistances of the rotor and stator windings and some (small) iron losses. The remainder of the input energy is converted into mechanical form and then some of this is, in turn, lost because of friction, etc. The fixed losses are the stator core loss and friction/windage losses. The variable losses consist of the I^2R losses in both the rotor and the stator and the core loss in the rotor. However, the stator core loss is usually very small.

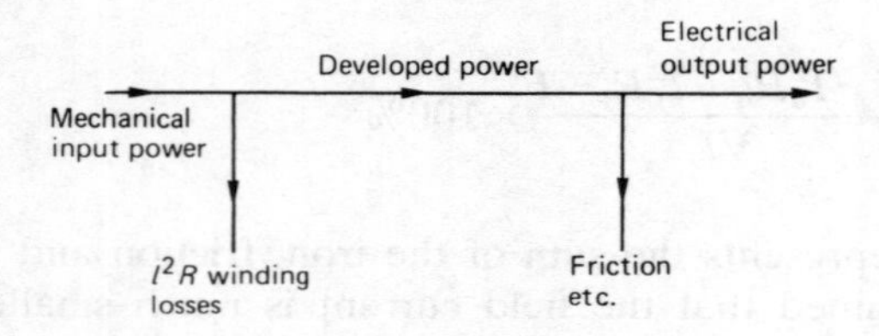

Fig. 6.9 Power flow diagram of an a.c. motor

3 *A.C. Motor* Fig. 6.9 shows the power flow diagram for an a.c. motor. For a motor of this type the fixed losses are *a)* the core loss, *b)* friction and windage losses. All the other losses vary with the load on the armature.

Exercises 6

6.1 Fig. 6.10 shows a conductor carrying a current *I* amps that passes axially through the centre of an iron ring. The ring consists of two halves held together by a magnetic force of attraction. The ring has a mean diameter of 45 cm and a cross-sectional area of 50 cm^2. Calculate the force that must be exerted to pull the two halves of the ring apart when the current is 160 A. (Take $\mu_r = 400$.)

6.2 A conductor has a diameter of 1.6 cm and it is bent into the shape shown in Fig. 6.11. The inductance of the wire is then

$$L = \mu_0 R(\log 10R/0.015)$$

Calculate the average force exerted on the wire when the current flowing in it is 125 A peak.

6.3 The basic construction of a relay is shown in Fig. 6.3. The coil has 1000 turns and the air gap has a length of 0.6 cm and a cross-sectional area of 1.4 cm^2. Calculate the current needed to operate the relay if the force exerted on the armature by the contact springs is 4.8 N.

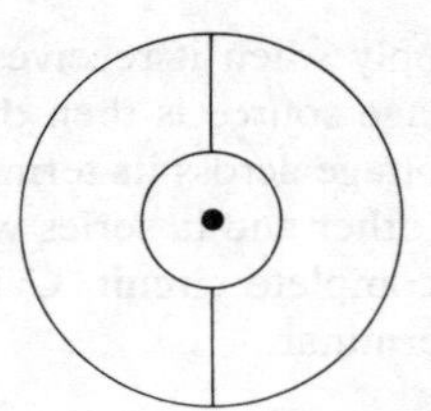

Fig. 6.10

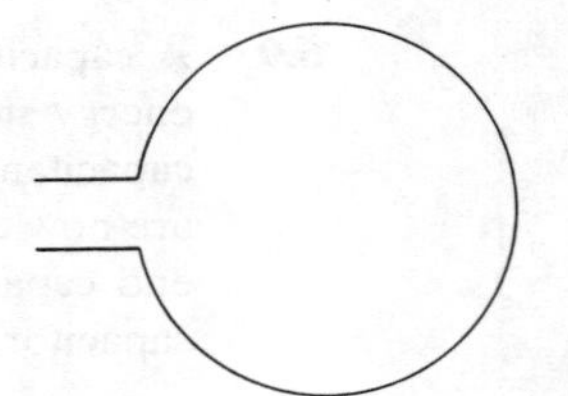

Fig. 6.11

6.4 A d.c. machine is rated at 240 V, 2000 W at 2000 r.p.m. The power dissipated in the field circuit is 100 W. When the machine is operated as a motor the armature losses are 200 W. Calculate the efficiency of the machine when it is used as a generator. State any assumptions made but take armature resistance as 1 Ω.

6.5 A capacitor is charged with 20 mC and it is then disconnected from the voltage supply. The energy stored in the capacitor is then 2.9 J. Calculate *a*) the capacitance and *b*) the voltage across the capacitor. This capacitor is then connected across a 10 μF capacitor. Calculate the voltage across each capacitor.

6.6 Fig. 6.12 shows an electromagnetic device. When current flows in the coil the magnetic force developed attracts the armature upwards. The length of the air gap varies from 2 cm when the device is non-operated to 0.76 cm when it is operated. There are 1500 turns and the current passed through the coil is 4 A. Calculate *a*) the flux density in the air gap, *b*) the inductance of the coil when the plunger is (i) un-operated (ii) operated, *c*) the work done as the armature moves from its non-operated to its operated positions.

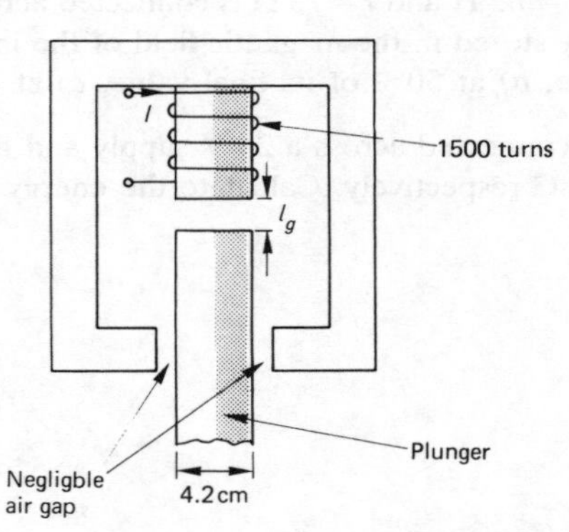

Fig. 6.12

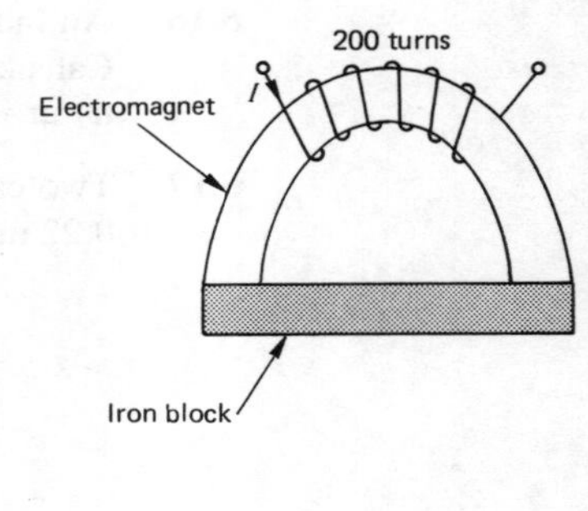

Fig. 6.13

6.7 Fig. 6.13 represents an electromagnet holding up a block of iron. The mean length of the magnetic circuit is 0.76 m and its cross-sectional area is 4 cm^2. Calculate the force exerted on the iron block when a current of 20 A flows through the coil. (Take $\mu_r = 700$).)

6.8 List twelve examples of energy transfer. Give three of each of *a*) electrical–mechanical, *b*) electrical–sound, *c*) electrical–heat, and *d*) electrical–light energy conversion.

6.9 A capacitor is connected to a d.c. supply when it receives a charge of 1.8 mC. The energy stored is then 1.56 J. The voltage source is then disconnected. Calculate the capacitance of the capacitor and the voltage across its terminals. Two 2 μF capacitors are now connected in series with each other and in series with the first capacitor. The end capacitors are linked to form a complete circuit. Calculate the charge in each capacitor and the voltage across its terminals,

Short Exercises

6.10 Calculate the energy stored in an air gap 1.5 mm wide and 8 cm^2 area when the magnetic flux density in the gap is 0.85 T.

6.11 A magnetic system has a m.m.f. of 200 AT and a total reluctance of $S = 200 \times 10^3$ AT/Wb. Calculate the energy stored in the air gap.

6.12 A magnetic circuit has 0.35 joules of energy stored in its air gap when the m.m.f. is 300 AT. If there are 800 turns calculate the inductance of the circuit.

6.13 The energy stored in an air gap is 4 J. What will be the energy stored if the length of the air gap is doubled? Assume that the flux density remains constant.

6.14 A relay has an inductance of 3 H. Calculate the rate of change of the energy stored in the air gap as the current increases from 20 mA at the rate of 15 A/s. Assume the inductance to remain constant.

6.15 A 10 μF capacitor is charged to 50 V and it is then disconnected from the supply. A 2 μF, initially discharged, capacitor is then connected across the 10 μF capacitor. Calculate *a*) the voltage across the combination and *b*) the energy stored.

6.16 An inductance of $L = 5.5$ H and $r = 75\ \Omega$ is connected across a d.c. voltage of 150 V. Calculate the energy stored in the magnetic field of the inductor when the current is *a*) at its initial value, *b*) at 50% of its final value, *c*) at its final value.

6.17 Two capacitors are connected across a 24 V supply and then have stored charges of 0.22 mC and 0.24 mC respectively. Calculate the energy stored in each capacitor.

7 Electric and Magnetic Fields

A field is defined as a region of space in which a certain physical state prevails and in which certain actions take place. Thus an *electric field* is a region in the vicinity of an electric charge in which a force will be exerted upon any other electric charge. The electric charge is supposed, for convenience, to consist of a number of lines of electric flux that can be drawn or *mapped*.

Whenever a current flows in a conductor, a *magnetic field* will be set up around that conductor. The magnetic field is the region in which a force may be exerted upon any other conductor and it is also supposed to consist of a number of lines of flux.

Both electric and magnetic fields can be represented by an equation of the form

$$\text{Flux density} = \text{a constant} \times \text{potential gradient} \qquad (7.1)$$

Several other fields can also be accurately represented by an equation of the same form; amongst these are included current conduction, fluidic dynamic and thermal conduction fields. Because of the analogy between the various fields it is possible to simulate the behaviour of one field by a consideration of the analogous electric field.

The Electric Field

When an electric charge is brought into the vicinity of another electric charge, a force of either attraction or repulsion (depending on the signs of the charges) is experienced. The space throughout which the force exists is said to be an **electric field**.

The strength of the force exerted is measured in terms of the **electric field strength** E of the field. The field strength at any point is the force exerted on a unit charge at that point and it is measured in terms of volts/metre.

The **electric potential** V at any point in the field is the work done, in joules, in moving unit charge from a point of zero potential to that point. Within an electric field there will be always a number of points at the same potential and these points are said to be *equi-potentials*. If a line is drawn to join up a number of equi-potentials, an equi-potential line or surface is formed.

The *potential difference* between two points d metres apart is the work done in moving unit charge between these two points. If this potential difference is V volts then the electric field strength between the two points is

$$E = -V/d \text{ volts/metre}$$

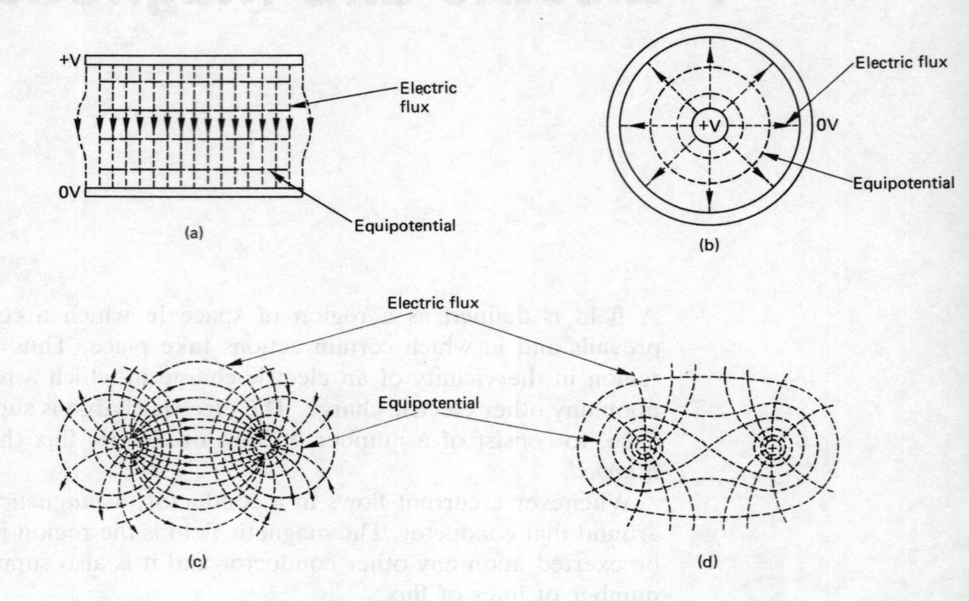

Fig. 7.1 Electric fields of *a*) two parallel metal plates, *b*) two concentric metal cylinders, and *c*), *d*) a twin line

The **electric flux** Ψ is the quantity of charge that is moved across a given area in a dielectric. The flux moved across a surface which encloses a charge of q coulombs is equal to that charge, i.e. $\Psi = q$.

The *electric flux density* D is the flux per metre. At any point in the field the flux density is equal to the product of the electric field strength and the permittivity of the dielectric, i.e.

$$D = \varepsilon E \text{ C/m}^2$$

Figs. 7.1*a*, *b*, *c* and *d* show respectively the electric fields in a parallel-plate capacitor, two concentric cylinders, and a twin line. In each case the lines of electric flux are really three-dimensional and leave a positive charge and terminate on a less positive or negative charge. In Fig. 7.1*d* it is assumed that both positive charges are of equal magnitude and so the flux lines are only shown leaving a charge. The equi-potential lines are everywhere at right angles to the lines of flux; in the case of the parallel-plate capacitor this means that the equi-potentials are parallel to the plates. The flux lines are sometimes known as *streamlines* or *flowlines.*

The *capacitance* of a system enclosing an electric field relates the charge stored to the potential difference:

$$C = Q/V \qquad (7.2)$$

For a parallel-plate capacitor the electric flux is equal to the charge Q stored and the flux density D is equal to Q/A C/m, where A is the area of the plates. In turn, the electric field strength is $E = D/\varepsilon = Q/\varepsilon A$ and the voltage across the plates is $V = Ed = Qd/\varepsilon A$, where d is the plate separation. Therefore,

$$C = Q/V = \varepsilon A/d \text{ F} \qquad (7.3)$$

Energy Stored in an Electric Field

It was shown in Chapter 6 that the energy stored in a capacitor is given by

$$W=\tfrac{1}{2}CV^2=\frac{1}{2}\frac{\varepsilon A}{d}\cdot E^2d^2$$

and the **energy** stored per cubic metre of dielectric is

$$W=\tfrac{1}{2}DE \text{ J/m}^3 \tag{7.4}$$

Example 7.1

A rectangular slab of a dielectric material is 2 cm thick and has a voltage of 600 V applied across it. The energy stored is then 2 μJ. Calculate the flux density if the area of the slab is 24 cm^2.

Solution The volume of the dielectric is

$$2\times10^{-2}\times24\times10^{-4}=48\times10^{-6} \text{ m}^3$$

and so the energy stored per cubic metre is

$$2\times10^{-6}/48\times10^{-6}=41.67 \text{ mJ/m}^3$$

Therefore,

$$D=\frac{0.0417\times4\times10^{-2}}{600}=2.78\times10^{-6} \text{ C/m}^2 \quad (Ans)$$

Magnetic Fields

It is well known that a magnet is able to exert an attractive force upon bodies made of certain materials. The region of space within which this force is experienced is said to be a region in which the **magnetic field** of the magnet exists. The field is supposed to consist of a number of lines of magnetic flux. The direction of the flux lines at any point in the field is chosen to indicate the direction in which the force exerted acts at that point. Fig. 7.2, for example, shows the magnetic field of a bar magnet; the direction of each of the flux lines is from North to South.

When a current flows in a conductor a magnetic field is set up around that conductor. The direction of the field is determined by the direction in which the current flows in the conductor. When the current flows into the paper (Fig. 7.3*a*), the magnetic field is in the clockwise direction. Conversely (Fig.

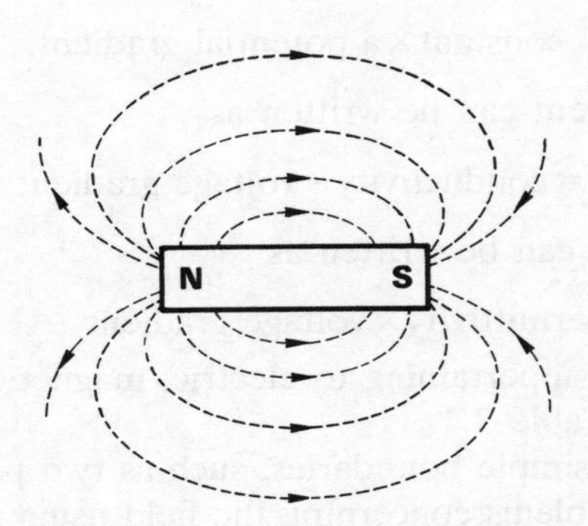

Fig. 7.2 Magnetic field of a bar magnet

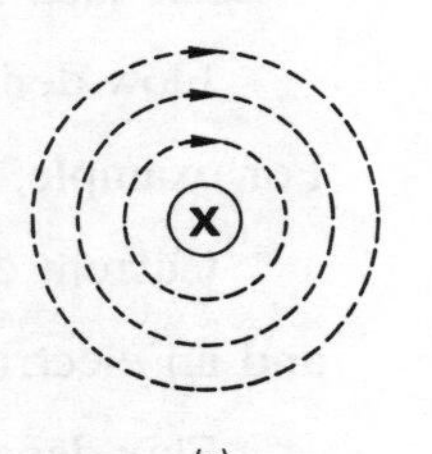

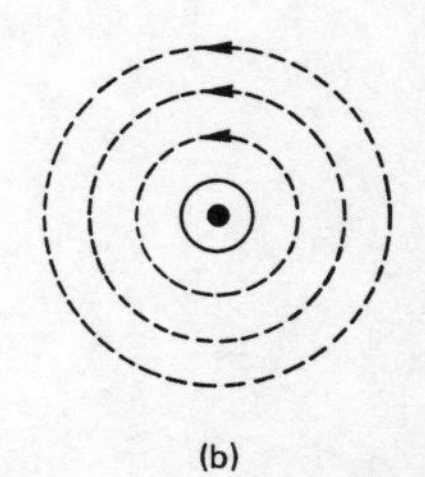

Fig. 7.3 Magnetic field set up around a current-carrying conductor

7.3*b*), when the current flows out of the paper an anticlockwise field is set up.

When the current first starts to flow in the conductor, a back e.m.f. is produced. Work must be done to overcome this e.m.f. so that the current can increase and the magnetic field can build up. Some of the energy supplied is used to provide the inevitable I^2Rt heat energy developed in the resistance of the conductor. The remainder of the energy is stored in the magnetic field. When the current ceases to flow, the magnetic field collapses back into the conductor and the energy stored in the field is returned to the source.

The maximum **energy** stored in an inductance L was shown on page 91 to be equal to $\frac{1}{2}LI^2$ joules. But $L = N\Phi/I$ and $H = NI/l$ and therefore

$$W = \tfrac{1}{2}\Phi I = \tfrac{1}{2}BA \cdot Hl \quad (\text{since } N = 1)$$

$$W = \tfrac{1}{2}BH = \tfrac{1}{2}\mu H^2 \text{ J/m}^3 \tag{7.5}$$

Example 7.2

Calculate the energy stored in a magnetic material of relative permeability 1000 when the magnetizing force is 2000 AT/m.

Solution From equation (7.5),

$$W = \tfrac{1}{2} \times 10^3 \times 4\pi \times 10^{-7} \times 4 \times 10^6 = 2513.3 \text{ J/m}^3 \quad (Ans)$$

Field Plotting

Many of the problems encountered in electrical engineering involve the determination of the potentials existing at various points in a field. The field in question might be an electric field or a magnetic field but, equally likely, the problem may be concerned with the flow of current in a resistive field or the determination of the temperature at various points in a thermal field, or perhaps the flow of a liquid in a cooling system.

Several different kinds of field can be described mathematically by the use of Laplace's equation, i.e.

$$\frac{\partial^2\theta}{\partial x^2} + \frac{\partial^2\theta}{\partial y^2} + \frac{\partial^2\theta}{\partial z^2} = 0 \tag{7.6}$$

where θ is the scalar value of a variable in the field.

Each such field is essentially of the form

$$\text{Flow density} = \text{a constant} \times \text{a potential gradient} \tag{7.7}$$

For example, a current can be written as

$$\text{Current density} = \text{conductivity} \times \text{voltage gradient} \tag{7.8}$$

and an electric field can be written as

$$\text{Flux density} = \text{permittivity} \times \text{voltage gradient} \tag{7.9}$$

The relationships appertaining to electric, magnetic, current and thermal fields are given in Table 7.1.

When a field has simple boundaries, such as two parallel planes, it is not difficult to solve problems concerning the field using mathematics. This will be demonstrated later in this chapter. In most other cases however, the problem is usually too difficult to be tackled in this manner and an approximate method of solution must be adopted.

For a paper solution, a *mapping* method can be adopted which is capable of producing reasonably simple and accurate solutions although often at the expense of considerable effort.

Experimental methods are also available in which an electric current is set up to simulate the field to be investigated. Such simulation is possible because of the similarities between the mathematical equations describing the electric field and the other fields mentioned earlier. Two experimental methods are often used: one, known as the electrolytic tank, uses a conducting liquid medium to simulate the field under investigation; the other method uses a special resistance paper.

Mapping Method

For a field to be represented by a **mapping method** a *flat* plane must, of course, be used and so the term $\partial^2\theta/\partial z^2$ in Laplace's equation becomes equal to zero. The plane of the field is divided into a number of "squares" formed by equi-potentials and streamlines. The two sets of lines must be drawn at right angles to one another at each point of intersection. Unless the field is uniform the two sets of lines will *not* be straight and then only approximate squares can be drawn. For this reason the "squares" are known as *curvilinear squares.*

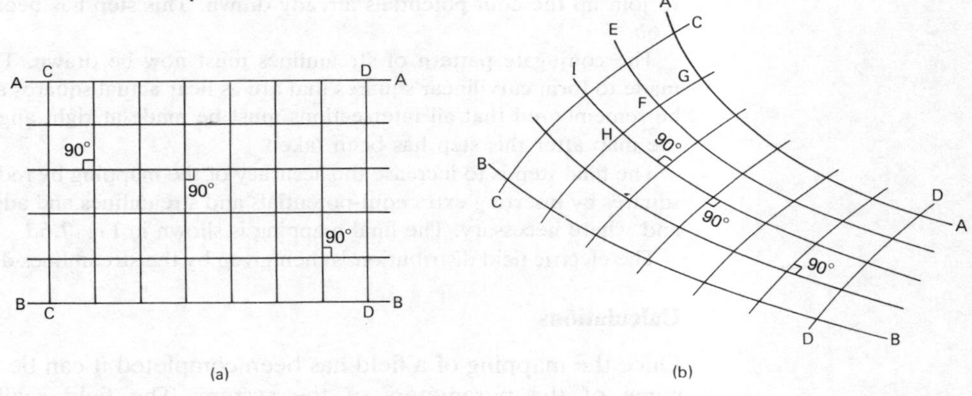

Fig. 7.4 Mapping a field *a*) linear, *b*) non-linear

An example of the method is shown by Fig. 7.4 in which the area enclosed by the equi-potentials AA and BB and the lines CC and DD has been divided into a number of actual squares in Fig. 7.4*a* and curvilinear squares in Fig. 7.4*b*. No problem exists for the linear field of Fig. 7.4*a* but greater care is needed for the non-linear field. Once the pattern of equi-potentials has been drawn, e.g. lines AA, BB and all those in between, the conjugate pattern remains to be drawn. This must be done by *a*) making sure that all lines cross the equi-potentials at right angles and *b*) forming approximate (curvilinear) squares such as AEFG and FEIH for which AE/AG$\simeq$ EI/EF. Several attempts may be needed to obtain a reasonable map in which the majority of the "squares" are very nearly actual squares.

Table 7.1

Field	*Potential difference*	*Potential gradient*	*Flow density*
Current	Voltage V	$E = V/d$	$J = \sigma E$
Electric	Voltage V	$E = V/d$	$D = \varepsilon E$
Magnetic	Ampere-turns NI	$H = NI/l$	$B = \mu H$
Thermal	Degrees θ	degrees/m θ/d	$\theta/R_{th} \cdot A$

Greater accuracy can always be obtained by drawing smaller squares but obviously this will entail more work. Thus a reasonable compromise between accuracy and time must be settled upon.

Example 7.3

Fig. 7.5 shows two metal plates with the upper plate at a potential 100 V more positive than the lower plate. Map the electric field between the plates.

Solution The first step is to assume that the electric field is uniform on both the left-hand and the right-hand sides of the system where the two plates are parallel to one another. Other equi-potentials can then be drawn at regular intervals between the plates; this has been done in Fig. 7.6*a* for the equi-potentials at 75 V, 50 V and 25 V.

The next step is to draw smooth curves, following the curvature of the upper plate, to join up the equi-potentials already drawn. This step has been carried out in Fig. 7.6*b*.

The conjugate pattern of streamlines must now be drawn. The attempt must be made to form curvilinear squares that are as near actual squares as possible. It should be remembered that all intersections must be made at right angles. Fig. 7.6*c* shows the map after this step has been taken.

The final step is to increase the accuracy of the mapping by reducing the size of the squares by inserting extra equi-potentials and streamlines and adjusting the pattern if and where necessary. The final mapping is shown in Fig. 7.6*d*.

The electric field distribution is then given by the streamlines drawn in Fig. 7.6*d*.

Calculations

Once the mapping of a field has been completed it can be used to determine some of the parameters of the system. The field really exists in three dimensions so that each curvilinear square in the map is actually a cube. If an electric field is considered then each cube will have a capacitance of ε per metre. If the mapping of the field contains y rows of squares with x squares in each row then the capacitance of the system is

$$C = \varepsilon \times x/y \text{ per metre}$$

Example 7.4

The mapping of the electric field between two concentric cylinders contains 4 rows of squares with 26 squares in each row. Calculate the capacitance per metre of the cylinders if the relative permittivity of the dielectric is 4. Take $\varepsilon_0 = 8.85 \times 10^{-12}$ F/m.

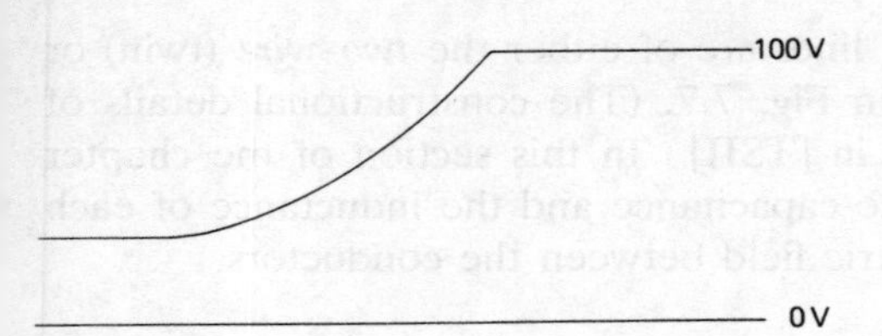

Fig. 7.5

Fig. 7.6

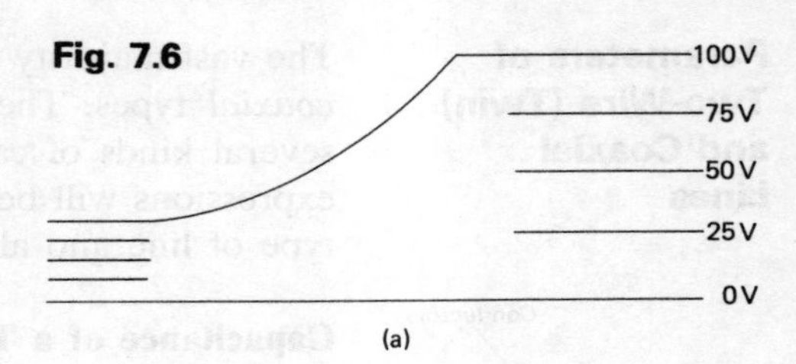

(a)

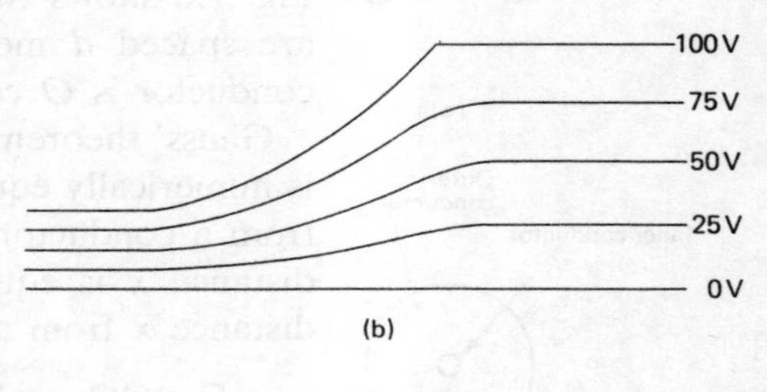

(b)

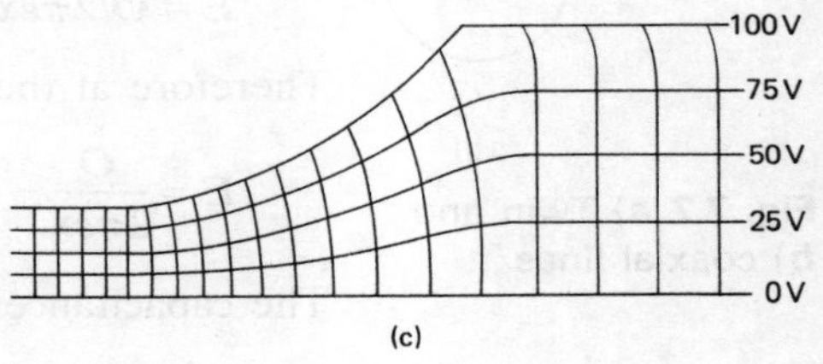

(c)

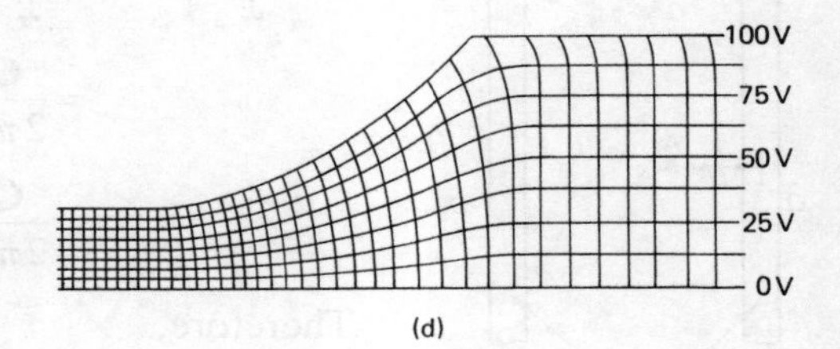

(d)

Solution

$$C = \frac{4 \times 8.85 \times 10^{-12} \times 26}{4} = 0.23 \times 10^{-9} \text{ F/m} \quad (Ans)$$

The results for the other types of field mentioned earlier are analogous to those already obtained. Thus

In a current field conductance $G = \sigma x/y$

where σ is the conductivity.

In a magnetic field reluctance $S = \mu x/y$

In a thermal field thermal conductance $= C_{th} x/y$

where C_{th} is the thermal conductivity.

Parameters of Two-Wire (Twin) and Coaxial Lines

The vast majority of transmission lines are of either the *two-wire* (twin) or *coaxial* types. These are shown in Fig. 7.7. (The constructional details of several kinds of cables are given in [TSII].) In this section of the chapter expressions will be derived for the capacitance and the inductance of each type of line and also for the electric field between the conductors.

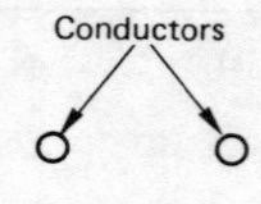

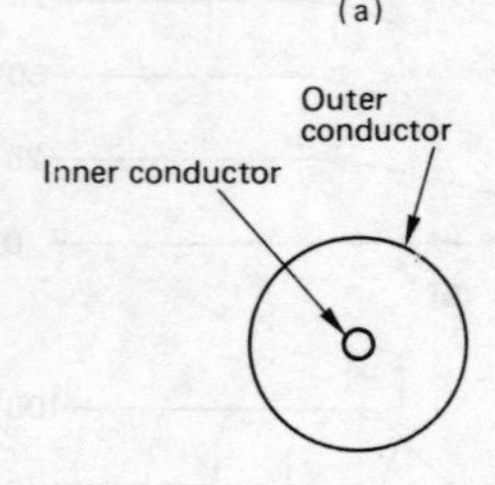

Fig. 7.7 *a*) Twin and *b*) coaxial lines

Capacitance of a Two-Wire Line

Fig. 7.8 shows two long parallel conductors each of radius r metres which are spaced d metres apart. Suppose the charge per unit length on each conductor is Q coulombs and the permittivity of the dielectric is $\varepsilon = \varepsilon_r \varepsilon_0$.

Gauss' theorem states that the outward normal flux from a closed surface is numerically equal to the charge enclosed by that surface. At a distance x from a conductor the surface is of area $2\pi x$ and hence the flux density at distance x is equal to $Q/2\pi x$. Since $D = \varepsilon E$ the electric field strength at distance x from a conductor is given by equation (7.10)

$$E = Q/2\pi\varepsilon x \text{ V/m} \tag{7.10}$$

Therefore at the point distance x from the left-hand conductor,

$$E = \frac{Q}{2\pi\varepsilon x} - \frac{(-Q)}{2\pi\varepsilon(d-x)} = \frac{Q}{2\pi\varepsilon}\left[\frac{1}{x} + \frac{1}{d-x}\right]$$

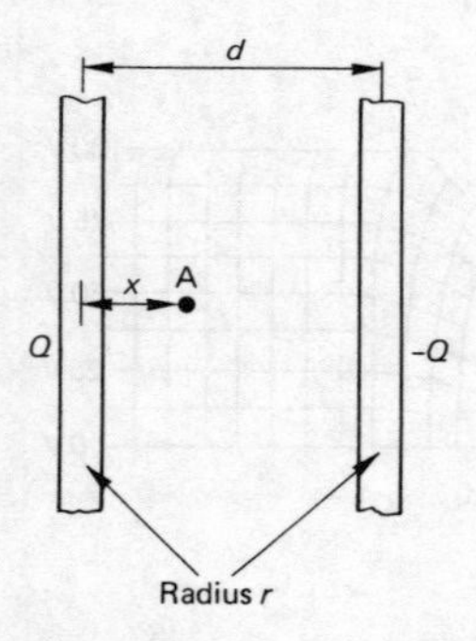

Fig. 7.8 Two parallel conductors

The capacitance C of the line is $C = Q/V = Q/\int E\,dx$.

$$\int_r^{d-r} E\,dx = \int_r^{d-r} \frac{Q}{2\pi\varepsilon}\left[\frac{1}{x} + \frac{1}{d-x}\right]$$

$$= \frac{Q}{2\pi\varepsilon}[\log_e x - \log_e(d-x)]_r^{d-r}$$

$$= \frac{Q}{2\pi\varepsilon}\left[\log_e\left(\frac{d-r}{r}\right) - \log_e\left(\frac{r}{d-r}\right)\right] = \frac{Q}{\pi\varepsilon}\log_e\left(\frac{d-r}{r}\right)$$

Therefore,

$$C = Q/V = \frac{\pi\varepsilon}{\log_e(d-r)/r} \text{ F/m} \tag{7.11}$$

The electric field between the two conductors is given by equation (7.10), i.e.

$$E = \frac{Q}{2\pi\varepsilon x} = \frac{V\pi\varepsilon}{\log_e[(d-r)/r]} \times \frac{1}{2\pi\varepsilon x}$$

$$E = \frac{V}{2x\log_e(d-r)/r} \tag{7.12}$$

Capacitance of a Coaxial Cable

Fig. 7.9 shows two concentric cylinders or a coaxial cable. The inner conductor has a radius of r while the inner radius of the outer conductor is

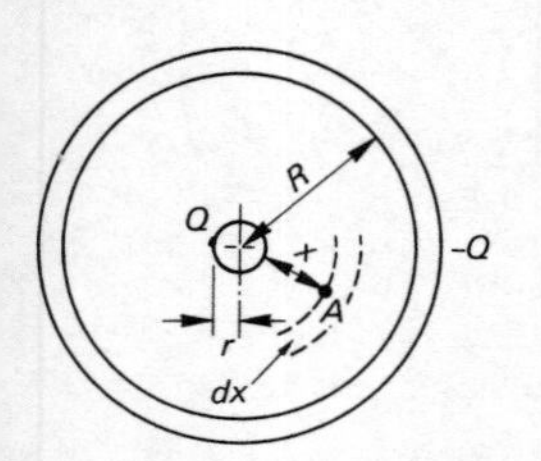

Fig. 7.9 Two concentric cylindrical conductors

R. The electric field strength at a point distance x from the centre of the inner conductor is

$$E = \frac{Q}{2\pi\varepsilon x}\ \text{V/m} \tag{7.13}$$

The voltage between the conductors is

$$V = \int_r^R E\,dx = \frac{Q}{2\pi\varepsilon}\int_r^R \frac{1}{x}\,dx = \frac{Q}{2\pi\varepsilon}[\log_e x]_r^R$$

$$V = \frac{Q}{2\pi\varepsilon}\log_e(R/r)\ \text{V}$$

$$\text{Capacitance } C = Q/V = \frac{2\pi\varepsilon}{\log_e(R/r)}\ \text{F/m} \tag{7.14}$$

Electric Field between a Pair of Coaxial Conductors

$$E = \frac{Q}{2\pi\varepsilon x} = \frac{2\pi\varepsilon V}{\log_e(R/r)} \times \frac{1}{2\pi\varepsilon x}$$

$$E = \frac{V}{x\log_e(R/r)}\ \text{V/m} \tag{7.15}$$

The electric field strength varies inversely with the distance x from the inner conductor and it has its maximum value at the surface of the inner conductor, i.e. when $x = r$. Thus,

$$E_{max} = \frac{V}{r\log_e(R/r)} \tag{7.16}$$

Transposing equation (7.16) gives

$$R = re^{+V/rE_{max}} \tag{7.17}$$

The best cable has the minimum outer diameter for given values of applied voltage and permissible maximum electric stress. To determine the minimum radius for the outer conductor differentiate equation (7.17) with respect to r and equate the result to zero. Then,

$$\frac{dR}{dr} = 0 = e^{V/rE_{max}} - re^{V/rE_{max}} \cdot V/r^2E_{max}$$

$$r = V/E_{max}$$

Substituting this result into equation (7.17), the minimum value for the diameter of the outer conductor is

$$R = re^{VE_{max}/VE_{max}} = re^1 = 2.7182r \tag{7.18}$$

Example 7.5

A coaxial cable has an inner diameter of 0.26 cm and an outer diameter of 0.95 cm. Calculate its capacitance and the maximum potential gradient when the applied voltage is 250 V, given that $\varepsilon_r = 1$.

Solution From equation (7.14),

$$C = \frac{2\pi \times 8.85 \times 10^{-12}}{\log_e(0.95/0.26)} = 43 \text{ pF/m} \quad (Ans)$$

From equation (7.16),

$$E_{max} = \frac{250}{2.6 \times 10^{-3} \log_e(0.95/0.26)} = 74.21 \text{ kV/m} \quad (Ans)$$

Example 7.6

A concentric cylindrical capacitor has an outer conductor whose inner diameter is 60 mm. Calculate *a*) the diameter of the inner conductor to give minimum electric stress at the surface of the inner conductor and *b*) the maximum voltage that can be applied to the capacitor if the breakdown voltage of the dielectric is 6 kV/mm.

Solution

a) From equation (7.18),

$$r = R_{min}/e = 60/2.718 = 22.08 \text{ mm} \quad (Ans)$$

b) From equation (7.16)

$$6000 = V_{max}/22.08 \log_e e$$

$$V_{max} = 6000 \times 22.08 \times 1 = 132.48 \text{ kV} \quad (Ans)$$

Inductance of a Coaxial Pair

Suppose the outer conductor shown in Fig. 7.9 carries a current I amperes in one direction while the inner conductor carries the same current in the opposite direction. The magnetic field strength at the point marked as A due to the current flowing in the inner conductor is $H = I/2\pi x$ AT/m. The flux passing through the small element dx is (since $BA = \mu HA = \mu I dx$ since $N = 1$).

$$d\Phi = \left[\frac{\mu_0 I}{2\pi x} + \frac{\mu_0 I}{2\pi(R-x)}\right] dx \text{ per metre length}$$

The total flux linking the two conductors is

$$\Phi = \frac{\mu_0 I}{2\pi} \int_r^R \left(\frac{1}{x} + \frac{1}{R-x}\right) dx = \frac{\mu_0 I}{2\pi} \log_e(R/r)$$

and so $$L = \Phi/I = \frac{\mu_0 \log_e(R/r)}{2\pi} \qquad (7.19)$$

There is a further component to the inductance because of the flux linkages inside the two conductors themselves. This extra component, which is usually small, is equal to $\mu_0/8\pi$.

Inductance of a Two-Wire Line

Fig. 7.10 shows a twin line whose conductors are of radius r separated by a spacing of d metres. Suppose the left-hand conductor carries a current of I amperes in the direction into the paper, while the right-hand conductor carries the same current out of the paper.

The magnetic field strength H at distance x from the left-hand conductor is $H=I/2\pi x$ AT/m. The flux passing through the small distance dx is

$$\Phi=\left[\frac{\mu_0 I}{2\pi x}+\frac{\mu_0 I}{2\pi(d-x)}\right]dx \text{ per metre length}$$

The total flux linking the two conductors is

$$\int_r^{d-r}\Phi\,dx=\frac{\mu_0 I}{2\pi}\int_r^{d-r}[1/x+1/(d-x)]\,dx$$

$$=\frac{\mu_0 I}{\pi}\log_e[(d-r)/r]$$

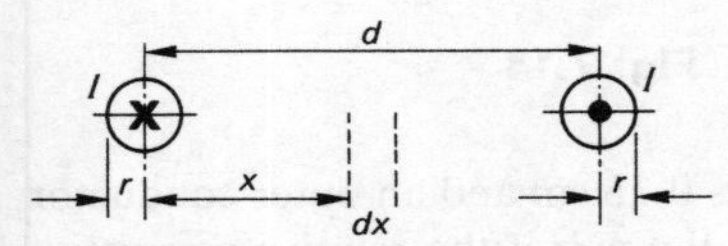

Fig. 7.10 Two parallel conductors

Now $L=N\Phi/I=\Phi/I$ so that

$$L=\frac{\mu_0}{\pi}\log_e(d-r)/r \text{ H/m} \qquad (7.20)$$

As for the coaxial line there is an extra component to the inductance equal to $\mu_0/4\pi$ because of flux linkages inside the conductors.

Exercises 7

7.1 A capacitor consists of two 400 mm long concentric cylinders. The capacitance of the component is 60 pF and the maximum electric field strength is 3.5 MV/m when the applied voltage is 235 kV r.m.s. Calculate the inner and outer radii of the two conductors.

7.2 Show that the voltage at any point distance x from the centre of two coaxial cylinders is given by

$$V=\frac{V\log_e(R/x)}{\log_e(R/r)}$$

where R and r are the radii of the outer and the inner cylinders respectively.

7.3 Map the electric field in the system shown in Fig. 7.11.

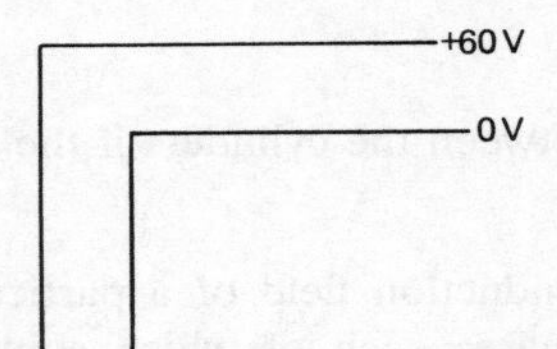

Fig. 7.11

7.4 A dielectric of relative permittivity 5 is 5 cm thick and has an area of 20 cm². The applied voltage is 1000 V. Calculate the energy stored in the dielectric. Calculate also the capacitance of a parallel plate capacitor made using this dielectric.

7.5 Discuss the analogy between the following fields: *a*) the current flowing between two electrodes inserted into an electrolyte, *b*) the magnetic field between two magnetized surfaces, *c*) the flow of heat energy through a wall.

7.6 Calculate the capacitance and the inductance per metre of a two-wire line that consists of two 1.5 cm diameter conductors which are spaced 15 cm apart. Assume the dielectric to be air.

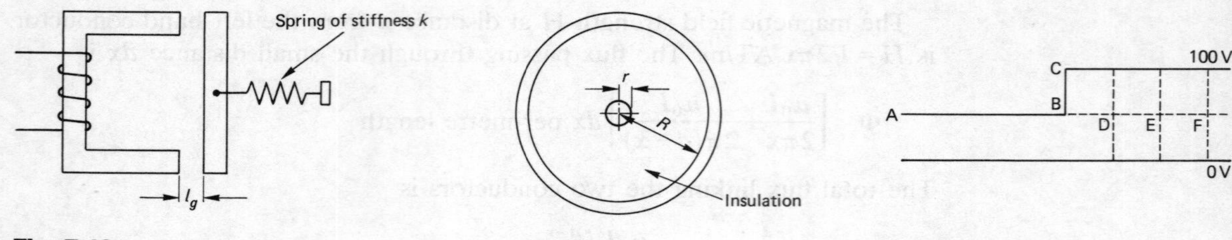

Fig. 7.12 Fig. 7.13 Fig. 7.14

7.7 A coaxial cable pair has an inner conductor of radius 0.15 cm and an outer conductor of inner radius 0.5 cm. Calculate the capacitance of the pair if the relative permittivity of the dielectric is 2.5.

7.8 Fig. 7.12 shows an actuator with a spring-loaded armature. When the device is non-activated the length of the air gap is l_0. Show that when a current I flows in the coil to operate the device and reduce the air gap to l_1 the electrical input energy needed is given by

$$W_E = \frac{1}{2}k(l_0^2 - l_1^2)$$

7.9 Write down the expression for the capacitance per metre of two concentric cylinders. Use the analogy between electric and conduction fields to calculate the insulation resistance between the two conductors shown in Fig. 7.13 if the radius of the inner is 6 mm, the inner radius of the outer is 22 mm, and the resistivity of the insulating material is 9×10^{12} Ωm.

Short Exercises

7.10 Explain what is meant by *a)* a streamline, *b)* an equi-potential in an electrical conduction field. Redraw Fig. 7.1*d* and insert two equi-potential surfaces into your figure.

7.11 The voltage at any point at radius x from the centre of the inner of two concentric cylinders is given by

$$V_x = V \log_e\left(\frac{R/x}{R/r}\right)$$

Calculate the voltage midway between the cylinders if the applied voltage is 100 kV and $R/r = 3$.

7.12 The mapping of the current conduction field of a particular system was achieved using 5 rows of curvilinear squares each of which contained 12 squares. If the conductivity of the material is 60×10^6 siemen/metre calculate the current that would flow when 120 mV are applied to the system.

7.13 Sketch the electric field distributions in *a)* a parallel-plate capacitor and *b)* a concentric cylinder capacitor. Explain why the equi-potentials are parallel to the plates and at right angles to the lines of electric flux.

7.14 What is meant by curvilinear squares? Describe the context in which they are employed and say why actual squares are not always used.

7.15 For the system shown in Fig. 7.14 map the electric field and estimate the potentials at D, E and F.

8 Network Theory

Introduction

A 4-terminal, or two-port, network is one that has two input and two output terminals. Very often one input and one output terminal are common. The four basic versions of networks are known as the T, the π, the L and the *lattice* networks and they are shown, respectively, in Figs. 8.1*a, b, c* and *d.* When a T or a π network is *symmetrical* it is customary to label the components to make the total series impedance equal to Z_1 and the total shunt impedance equal to Z_2. This practice is followed in this chapter (see Fig. 8.7).

Fig. 8.1 Showing *a*) T, *b*) π, *c*) *L*, and *d*) lattice networks

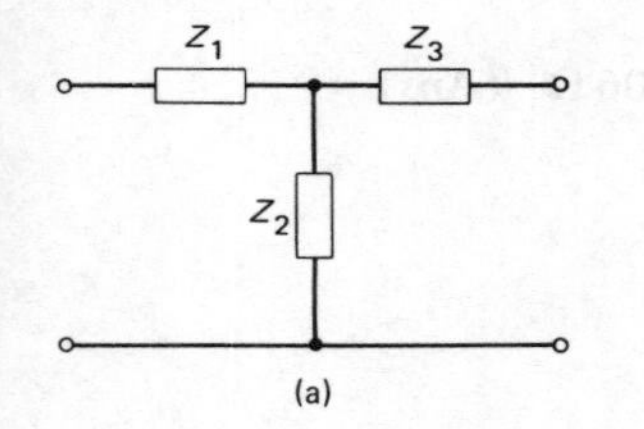

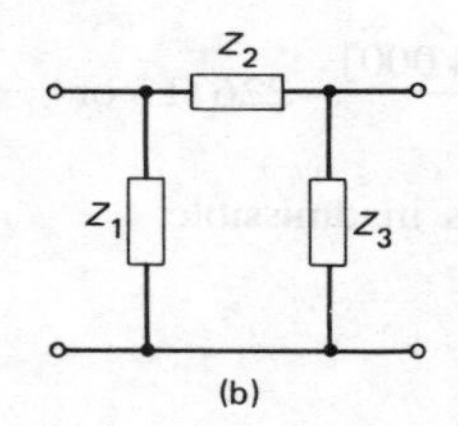

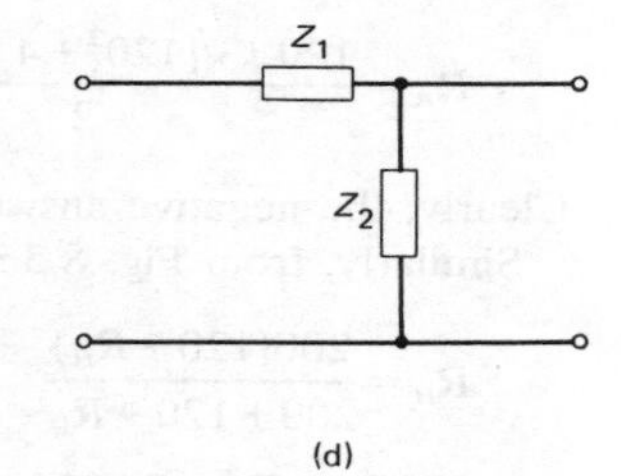

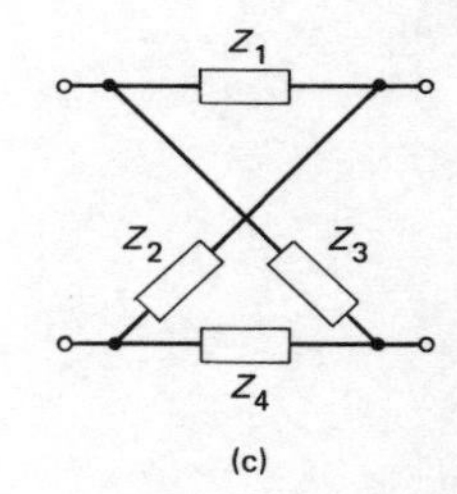

Iterative, Image and Characteristic Impedances

Iterative Impedance

The **iterative impedances** of a network are the two impedances such that, if one of them is connected across one pair of terminals, the *same* value of impedance is measured at the other pair of terminals. There is only one iterative impedance for each direction of transmission. Thus, referring to Fig. 8.2, when an impedance Z_A is connected across the terminals BB, the input impedance at the terminals AA is also Z_A. Conversely, when an impedance Z_B is connected across the terminals AA the impedance measured at terminals BB is also Z_B.

The method to be used for the determination of the values of the iterative impedances Z_A and Z_B of a particular network will be demonstrated by an example.

Fig. 8.2 Iterative impedances

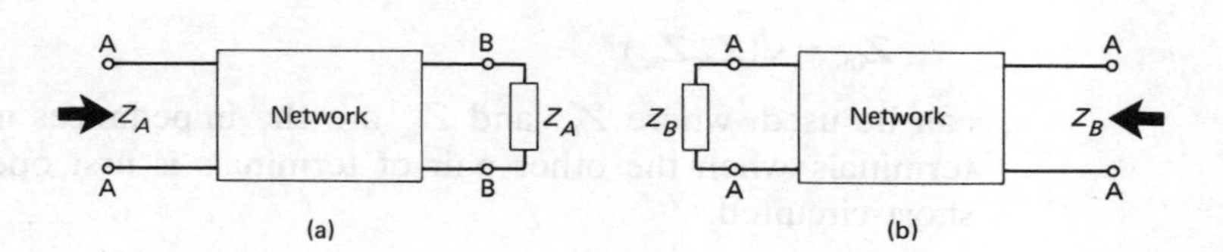

Example 8.1

Calculate the iterative impedances of the network shown in Fig. 8.3*a*.

Fig. 8.3

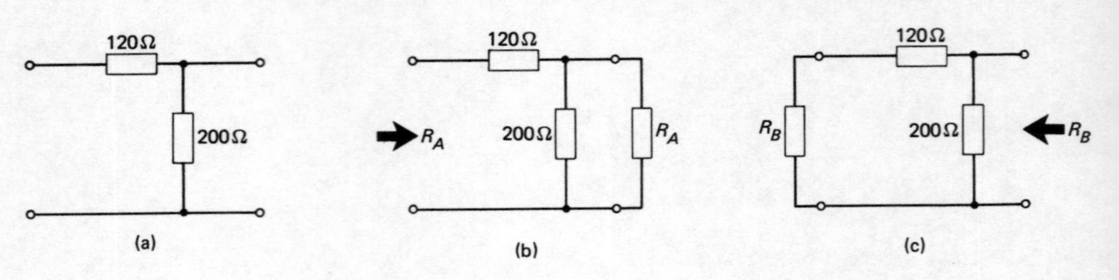

Solution From the definition of an iterative impedance, when a resistor R_A is connected across the output terminals of the network the input impedance will also be R_A. Therefore, from Fig. 8.3*b*,

$$R_A = 120 + \frac{200R_A}{200+R_A} = \frac{24\,000+120R_A+200R_A}{200+R_A}$$

Hence $R_A^2 - 120R_A - 24\,000 = 0$ or

$$R_A = \frac{120 \pm \sqrt{[120^2 + 4\times 24\,000]}}{2} = 226\ \Omega \quad \text{or} \quad -106\ \Omega \quad (Ans)$$

Clearly, the negative answer is inadmissible.

Similarly, from Fig. 8.3*c*,

$$R_B = \frac{200(120+R_B)}{200+120+R_B}$$

$$320R_B + R_B^2 = 24\,000 + 200R_B$$

or $R_B^2 + 120R_B - 24\,000 = 0$

Solving $R_B = 106\ \Omega$ or $-226\ \Omega$ (*Ans*)

Again, the negative answer is inadmissible. It will have been noted that the negative result obtained for R_A is actually the value of R_B and vice versa. This means, of course, that for a resistive network only one iterative impedance calculation need be carried out.

Image Impedance

The two **image impedances** of a network are such that, when one of them is connected across one pair of terminals, the other is measured at the other pair of terminals and vice versa. The concept is illustrated by Fig. 8.4.

The two image impedances of a network can be calculated using the method employed for the determination of the iterative impedances. Alternatively, and somewhat more simply, the expression

$$Z_{im} = \sqrt{(Z_{oc}Z_{sc})} \tag{8.1}$$

can be used, where Z_{oc} and Z_{sc} are the impedances measured at a pair of terminals when the other pair of terminals is first open-circuited and then short-circuited.

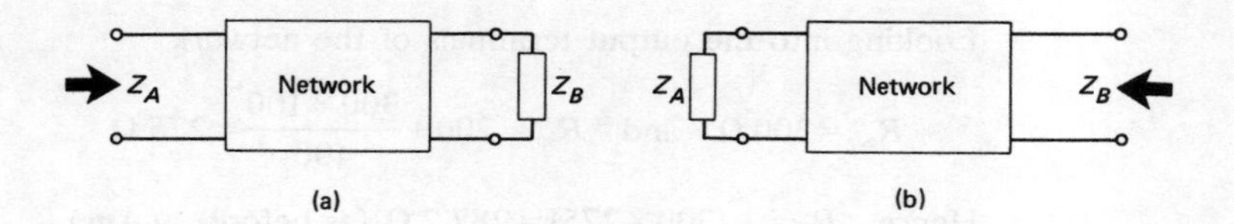

Fig. 8.4 Image impedances

Example 8.2

Calculate the image impedances of the network shown in Fig. 8.5*a*.

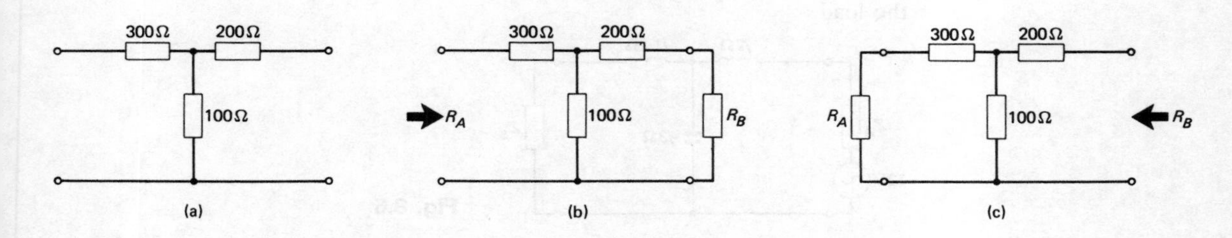

Fig. 8.5

Solution From Fig. 8.5*b*,

$$R_A = 300 + \frac{100(200 + R_B)}{100 + 200 + R_B}$$

$$300R_A + R_A R_B = (9 \times 10^4) + 300R_B + (2 \times 10^4) + 100R_B$$

$$300R_A + R_A R_B = (11 \times 10^4) + 400R_B \tag{8.2}$$

From Fig. 8.5*c*,

$$R_B = 200 + \frac{100(300 + R_A)}{100 + 300 + R_A}$$

$$400R_B + R_A R_B = (8 \times 10^4) + 200R_A + (3 \times 10^4) + 100R_A$$

$$-300R_A + R_A R_B = (11 \times 10^4) - 400R_B \tag{8.3}$$

Adding equations (8.2) and (8.3),

$$2R_A R_B = 22 \times 10^4 \qquad R_A = 11 \times 10^4 / R_B$$

Substituting into equation (8.2)

$$\frac{300 \times 11 \times 10^4}{R_B} + (11 \times 10^4) = (11 \times 10^4) + 400R_B$$

$$R_B^2 = 3 \times 11 \times 10^4/4 \quad \text{and} \quad R_B = 287.2\ \Omega \quad (Ans)$$

Therefore $R_A = 11 \times 10^4/287.2 = 383\ \Omega$ (*Ans*)

Using the simpler method, i.e. using equation (8.1) and looking into the input terminals of the network,

$$R_{oc} = 300 + 100 = 400\ \Omega$$

Also $R_{sc} = 300 + \frac{200 \times 100}{200 + 100} = 366.7\ \Omega$

Hence $R_A = \sqrt{(400 \times 366.7)} = 383\ \Omega$ (as before) (*Ans*)

Looking into the output terminals of the network

$$R_{oc} = 300\ \Omega \quad \text{and} \quad R_{sc} = 200 + \frac{300 \times 100}{400} = 275\ \Omega$$

Hence $R_B = \sqrt{(300 \times 275)} = 287.2\ \Omega$ (as before) (*Ans*)

Exercise 8.3

For the circuit shown in Fig. 8.6 calculate *a*) the image impedances, *b*) the input and output currents, and *c*) the input power to the network and the power dissipated in the load.

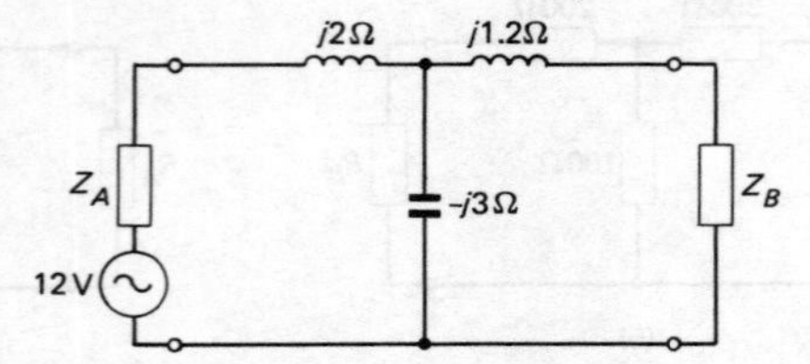

Fig. 8.6

Solution

a) $Z_A = \sqrt{(Z_{oc}Z_{sc})} = \sqrt{\left[(j2 - j3)\left(j2 + \frac{-j3 \times j1.2}{j1.2 - j3}\right)\right]} = 2\ \Omega$ (*Ans*)

$$Z_B = \sqrt{\left[(j1.2 - j3)\left(j1.2 + \frac{-j3 \times j2}{j2 - j3}\right)\right]} = 3.6\ \Omega \quad (Ans)$$

b) $V_{in} = 12/2 = 6$ V. Therefore $I_{in} = 6/2 = 3$ A (*Ans*)

$$I_2 = 3 \times \frac{-j3}{3.6 + j1.2 - j3} = 1 - j2\ \text{A} \quad (Ans)$$

c) $P_{in} = 6^2/2 = 18$ W (*Ans*)

$$P_{out} = |1 - j2|^2 \times 3.6 = 2.2361^2 \times 3.6 = 18\ \text{W} \quad (Ans)$$

Note that there is no loss in the network because it contains no resistances.

Characteristic Impedance

When a network is symmetrical, its iterative impedances and its image impedances are equal to one another and have a common value for both directions of transmission. The common value of impedance is known as the **characteristic impedance** of the network.

An expression for the characteristic impedance can be derived using the same methods as for the iterative and image impedances. Figs. 8.7*a* and *b*

Fig. 8.7 Symmetrical T and π networks

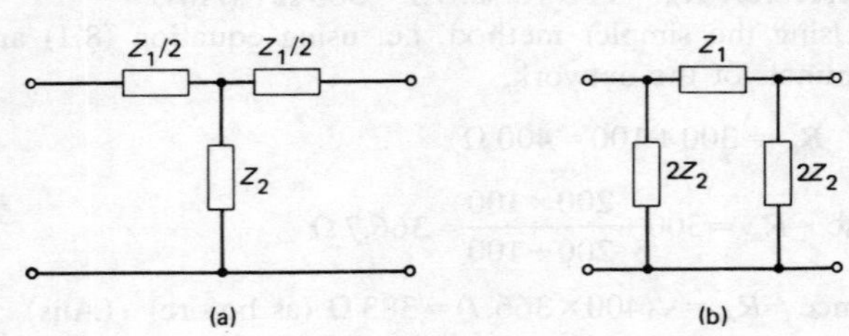

show, respectively, a symmetrical T and a symmetrical π network. From Fig. 8.7*a*,

$$Z_{0T} = \frac{Z_1}{2} + \frac{Z_2\left(\frac{Z_1}{2} + Z_{0T}\right)}{\frac{Z_1}{2} + Z_2 + Z_{0T}}$$

$$\frac{Z_{0T}Z_1}{2} + Z_{0T}Z_2 + Z_{0T}^2 = \frac{Z_1^2}{4} + \frac{Z_1Z_2}{2} + \frac{Z_1Z_{0T}}{2} + \frac{Z_1Z_2}{2} + Z_{0T}Z_2$$

$$Z_{0T}^2 = \frac{Z_1^2}{4} + Z_1Z_2$$

$$Z_{0T} = \sqrt{\left(\frac{Z_1^2}{4} + Z_1Z_2\right)} \tag{8.4}$$

Alternatively,

$$Z_{oc} = \frac{Z_1}{2} + Z_2 \quad \text{and} \quad Z_{sc} = \frac{Z_1}{2} + \frac{Z_1Z_2/2}{\frac{Z_1}{2} + Z_2}$$

$$Z_{sc} = \frac{\frac{Z_1^2}{4} + Z_1Z_2}{\frac{Z_1}{2} + Z_2}$$

Therefore

$$Z_{0T} = \sqrt{\left(\frac{Z_1^2}{4} + Z_1Z_2\right)} \quad \text{as before}$$

For the symmetrical π network of Fig. 8.7*b*,

$$Z_{oc} = \frac{2Z_2(Z_1 + 2Z_2)}{4Z_2 + Z_1} \quad \text{and} \quad Z_{sc} = \frac{2Z_2Z_1}{2Z_2 + Z_1}$$

$$Z_{oc} \times Z_{sc} = \frac{4Z_2^2Z_1}{4Z_2 + Z_1}$$

Multiplying both the numerator and the denominator by $Z_1/4$ gives

$$Z_{oc}Z_{sc} = \frac{Z_1^2Z_2^2}{\frac{Z_1^2}{4} + Z_1Z_2}$$

and so

$$Z_{0\pi} = \frac{Z_1Z_2}{\sqrt{\left(\frac{Z_1^2}{4} + Z_1Z_2\right)}} \tag{8.5}$$

or $$Z_{0\pi} = \frac{Z_1Z_2}{Z_{0T}} \tag{8.6}$$

Image Transfer Coefficient

The **image transfer coefficient** γ of a network is defined by

$$\gamma = \tfrac{1}{2}\log_e \frac{I_S V_S}{I_R V_R} \tag{8.7}$$

$$= \log_e \frac{I_S}{I_R}\sqrt{\frac{Z_A}{Z_B}} \tag{8.8}$$

where I_S and V_S are the current and voltage at the input terminals and I_R and V_R are the current and voltage at the output terminals. The term "image" implies that the network is connected between its image impedances.

The real part of γ is known as the **image attenuation coefficient** α, expressed in nepers. The imaginary part of γ is known as the image phase-change coefficient β, expressed in radians. Thus,

$$\gamma = \alpha + j\beta \tag{8.9}$$

In the case of a symmetrical network, the ratio of the input and output currents is equal to the ratio of the input and output voltages and then

$$\gamma = \tfrac{1}{2}\log_e (I_S^2/I_R^2) = \log_e (I_S/I_R) \tag{8.10a}$$

or $$\gamma = \log_e (V_S/V_R) \tag{8.10b}$$

γ is then known as the **propagation coefficient** of the network.

Expressions for the propagation coefficient of a symmetrical T or π network are easily obtained and are of considerable importance in transmission line theory (p. 120). General equations for asymmetrical networks are not so easily obtained and for such networks it is probably best to substitute numerical values as soon as possible and then work from basic principles.

Propagation Coefficient of a Symmetrical T Network

Consider Fig. 8.8 which shows a symmetrical T network fed by a source of e.m.f. E_S and internal impedance Z_{0T} and terminated by a load of impedance Z_{0T}.

Since $V_S = I_S Z_{0T}$ and $V_R = I_R Z_{0T}$,

$$V_S I_S / V_R I_R = I_S^2 Z_{0T} / I_R^2 Z_{0T} = I_S^2 / I_R^2$$

and $$\gamma = \log_e I_S/I_R \tag{8.11}$$

From the figure $$I_R = \frac{I_S Z_2}{Z_2 + (Z_1/2) + Z_{0T}}$$

$$e^{\gamma} = I_S/I_R = \frac{Z_2 + (Z_1/2) + Z_{0T}}{Z_2} = 1 + \frac{Z_1}{2Z_2} + \frac{Z_{0T}}{Z_2} \tag{8.11}$$

and $$\gamma = \log_e \left(1 + \frac{Z_1}{2Z_2} + \frac{Z_{0T}}{Z_2}\right) \tag{8.12}$$

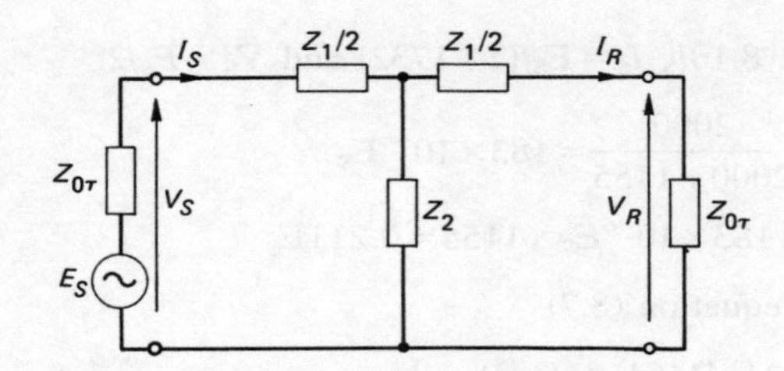

Fig. 8.8 Calculation of the propagation coefficient of a T network

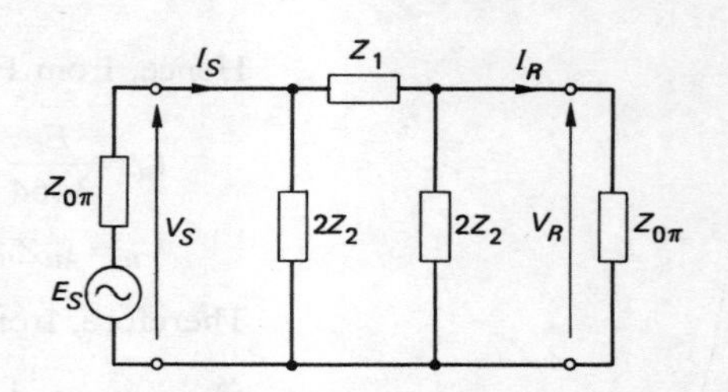

Fig. 8.9 Calculation of the propagation coefficient of a π network

Propagation Coefficient of a Symmetrical π Network

Fig. 8.9 shows a symmetrical π network connected between source and load impedances which are both equal to the characteristic impedance of the network. From the figure,

$$V_R = \frac{V_S 2Z_2 Z_{0\pi}/(2Z_2 + Z_{0\pi})}{Z_1 + \dfrac{2Z_2 Z_{0\pi}}{2Z_2 + Z_{0\pi}}} = \frac{V_S 2Z_2 Z_{0\pi}}{2Z_1 Z_2 + Z_1 Z_{0\pi} + 2Z_2 Z_{0\pi}}$$

$$e^{\gamma} = V_S/V_R = 1 + (Z_1/2Z_2) + (Z_1/Z_{0\pi}) \tag{8.13}$$

Since $Z_{0\pi} = Z_1 Z_2 / Z_{0T}$ the equation for γ can be written as

$$\gamma = \log_e \left(1 + \frac{Z_1}{2Z_2} + \frac{Z_{0T}}{Z_2}\right)$$

which is the same as equation (8.12). This means that symmetrical T and π networks having the same values of total series and shunt impedances have the same propagation coefficient.

Example 8.4

Calculate the image attenuation coefficient of the network shown in Fig. 8.10*a*.

Fig. 8.10

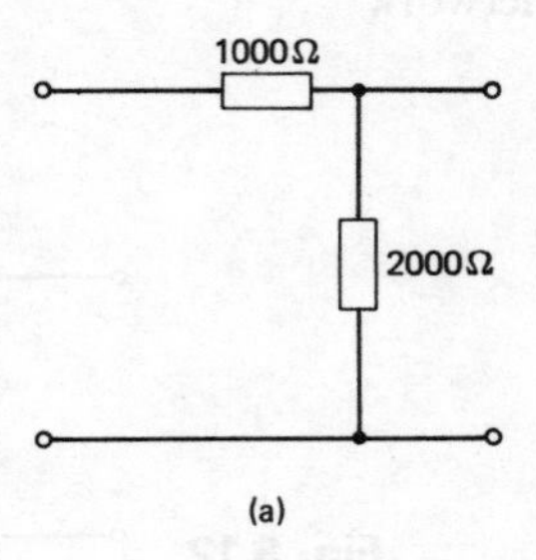

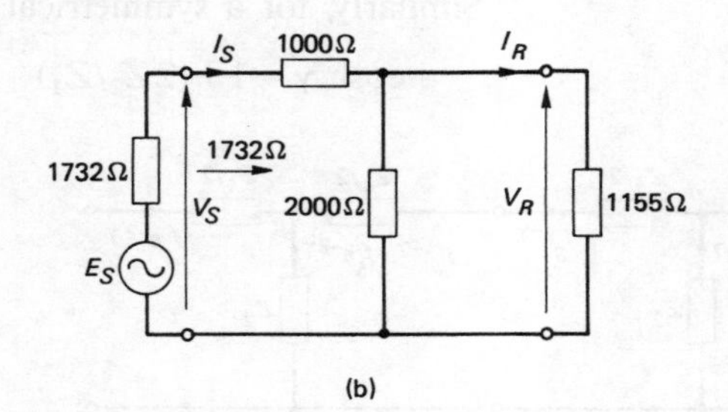

Solution

Image impedance $Z_A = \sqrt{(3000 \times 1000)} = 1732\ \Omega$

Image impedance $Z_B = \sqrt{(2000 \times 667)} = 1155\ \Omega$

Hence, from Fig. 8.10*b*, $I_S = E_S/(2\times 1732)$ and $V_S = E_S/2$.

$$I_R = \frac{E_S}{3464}\times\frac{2000}{2000+1155} = 183\times 10^{-6}E_S$$

$$V_R = I_R Z_B = 183\times 10^{-6}E_S\times 1155 = 0.211E_S$$

Therefore, from equation (8.7)

$$\gamma = \tfrac{1}{2}\log_e\left[\frac{(E_S/3464)\times(E_S/2)}{183\times 10^{-6}E_S\times 0.211E_S}\right]\text{ nepers}$$

$$= \tfrac{1}{2}\log_e 3.738 = 0.66\text{ N}\quad (Ans)$$

$$= 0.66\times 8.686 = 5.73\text{ dB}\quad (Ans)$$

To show that $\cosh\gamma = 1+(Z_1/2Z_2)$ From equation (8.11)

$$e^{\gamma} = 1+(Z_1/2Z_2)+(Z_{0T}/Z_2)$$

Hence

$$e^{-\gamma} = \frac{1}{1+(Z_1/2Z_2)+(Z_{0T}/Z_2)}$$

Now $\cosh\gamma = (e^{\gamma}+e^{-\gamma})/2$ and therefore,

$$\cosh\gamma = \frac{1}{2}\left[\frac{Z_2+(Z_1/2)+Z_{0T}}{Z_2}+\frac{Z_2}{Z_2+(Z_1/2)+Z_{0T}}\right]$$

or $$\cosh\gamma = 1+(Z_1/2Z_2) \quad (8.14)$$

Alternatively, consider Fig. 8.11 which shows two symmetrical T networks connected in cascade. Applying Kirchhoff's law to the middle loop,

$$0 = I_S e^{-\gamma}Z_1 + (I_S e^{-\gamma} - I_S e^{-2\gamma})Z_2 - (I_S - I_S e^{-\gamma})Z_2$$

$$I_S Z_2 = I_S e^{-\gamma}(Z_1+2Z_2) - e^{-2\gamma}Z_2 I_S$$

$$Z_2 e^{\gamma} + Z_2 e^{-\gamma} = Z_1 + 2Z_2 \quad\text{and so}$$

$$\frac{e^{\gamma}+e^{-\gamma}}{2} = \cosh\gamma = 1+(Z_1/2Z_2)\quad\text{as before}$$

Similarly, for a symmetrical π network

$$\cosh\gamma = 1+(2Z_2/Z_1) \quad (8.15)$$

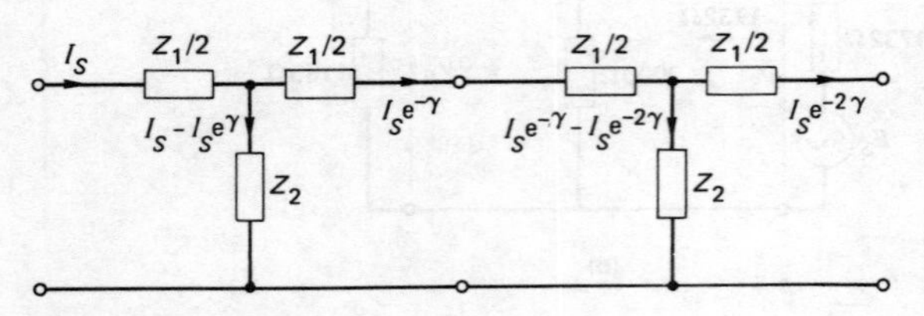

Fig. 8.11 Determination of $\cosh\gamma = 1+(Z_1/Z_2)$

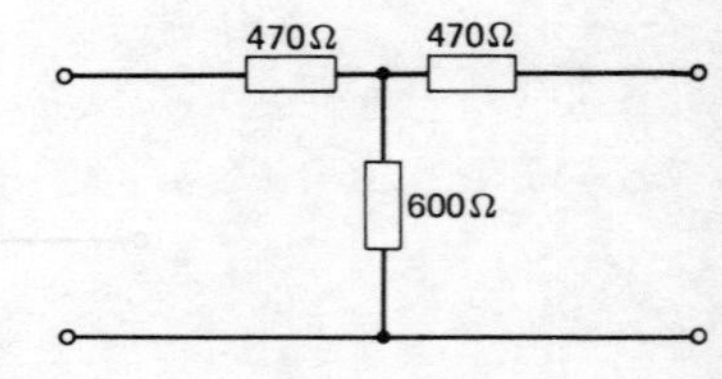

Fig. 8.12

Example 8.5

Calculate the attenuation of the circuit shown in Fig. 8.12.

Solution From equation (8.15)

$$\cosh \gamma = 1 + (470/600) = 1.783$$

$$\cosh \alpha \cos \beta + j \sinh \alpha \sin \beta = 1.783$$

There is no imaginary part and hence $\sin \beta = 0$ and $\cos \beta = 1$. Therefore

$$\cosh \alpha = (e^{\alpha} + e^{-\alpha})/2 = 1.783$$

Solving for e^{α} gives

$$e^{\alpha} = 3.26 \quad \text{or} \quad 0.31 \quad \text{therefore} \quad \alpha = 1.182\,\text{N} = 10.26\,\text{dB} \quad (Ans)$$

Alternatively, the characteristic impedance of the network is

$$Z_{OT} = \sqrt{\left(\frac{940^2}{4} + (940 \times 600)\right)} = 886\,\Omega$$

Hence, from Fig. 8.13,

$$I_R = I_S 600/1956 \quad \text{and} \quad I_S/I_R = e^{\alpha} = 3.26 \quad \text{as before}$$

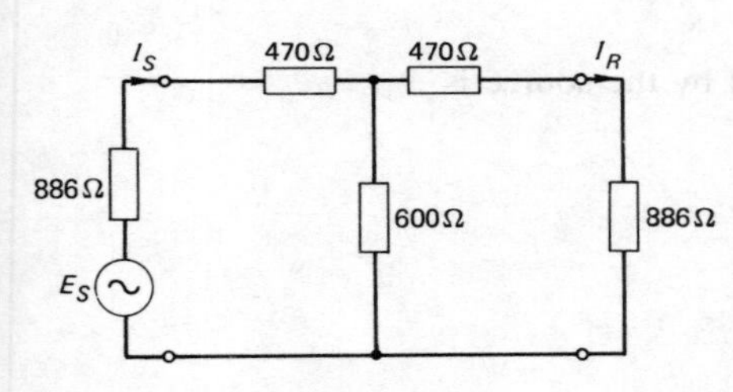

Fig. 8.13

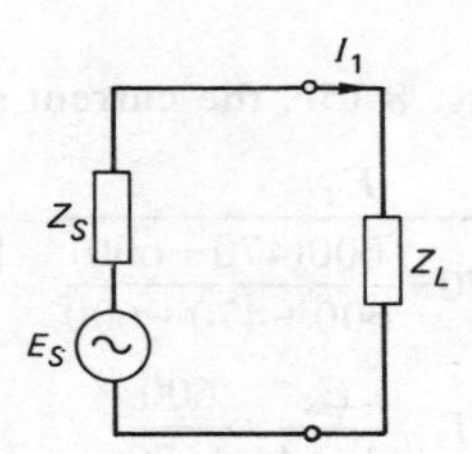

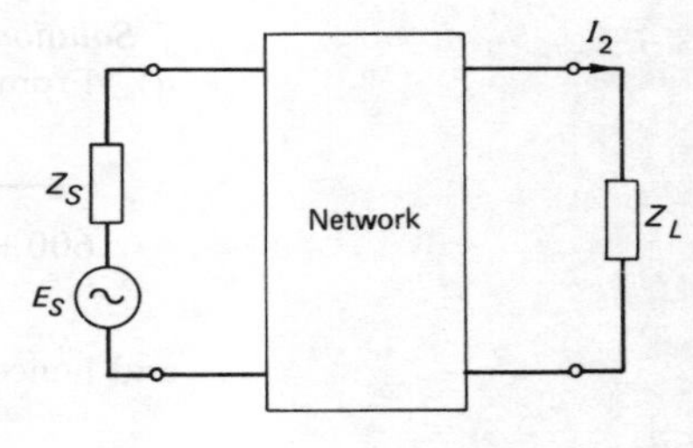

Fig. 8.14 Insertion loss (a) (b)

Insertion Loss

The **insertion loss** of a network is the ratio, expressed in dB, of the powers dissipated in a load before and after the insertion of the network in between the source and the load.

When (Fig. 8.14*a*) a source of e.m.f. E_S and internal impedance Z_S is directly connected to a load Z_L, the current that flows in the load is I_1. When (Fig. 8.14*b*) the network is inserted in between the source and the load, the load current is reduced to a new value I_2. The insertion loss of the network is then equal to

$$10 \log_{10}\left[\frac{|I_1|^2 R_L}{|I_2|^2 R_L}\right] \text{dB}$$

$$\text{Insertion loss} = 20 \log_{10} \frac{|I_1|}{|I_2|} \text{dB} \quad (8.16)$$

Exercise 8.6

Calculate the insertion loss of the network shown in Fig. 8.13 when it is connected *a*) between 600 Ω impedances and *b*) between its image impedances.

Fig. 8.15

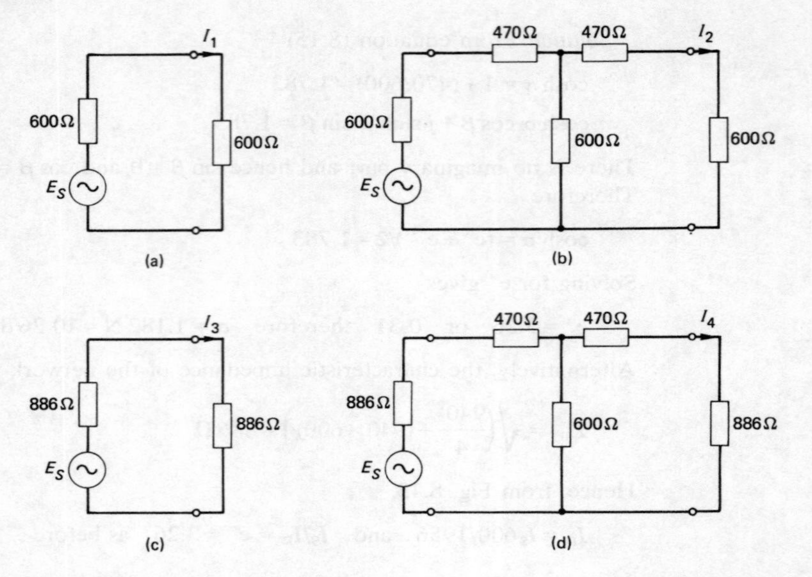

Solution

a) From Fig. 8.15*b*, the current supplied by the source is

$$\frac{E_S}{600+470+\dfrac{600(470+600)}{600+470+600}}=\frac{E_S}{1454.43}$$

and hence $I_2=\dfrac{E_S}{1454}\times\dfrac{600}{1670}$

From Fig. 8.15*a* $I_1=E_S/1200$
Therefore, the insertion loss is

$$20\log_{10}\left[\frac{E_S\times 1454\times 1670}{1200\times E_S\times 600}\right]=10.56\text{ dB}\quad(Ans)$$

b) From Fig. 8.15*c* $I_3=E_S/1772$
From Fig. 8.15*d*, the current supplied by the source to the network is $E_S/1772$. Hence

$$I_4=\frac{E_S}{1772}\times\frac{600}{1956}$$

Therefore, the insertion loss is

$$20\log_{10}\left[\frac{E_S\times 1772\times 1956}{1772\times E_S\times 600}\right]=10.26\text{ dB}\quad(Ans)$$

It should be noted that when a network is image-matched at both input and output terminals its insertion loss is at its minimum possible value and is equal to its image attenuation.

Example 8.7

Calculate the insertion loss of the 0.1 μF capacitor shown in Fig. 8.16.

Solution Without the capacitor in circuit the current flowing in the 1200 Ω resistor is $I_1 = 10/1800$ A. The reactance of the capacitor is $-j2000\ \Omega$. Thus, when the capacitor is connected in circuit, the current taken from the source is

$$I_S = \frac{10}{600 + \dfrac{1200 \times -j2000}{1200 - j2000}} = \frac{10}{1482 - j529}$$

The load current is

$$I_2 = \frac{10}{1482 - j529} \times \frac{-j2000}{1200 - j2000}$$

$$|I_2| = 5.44 \times 10^{-3}\ \text{A}$$

Therefore, the insertion loss of the capacitor is

$$20 \log_{10}\left[\frac{10}{1800 \times 5.45 \times 10^{-3}}\right] = 0.18\ \text{dB} \quad (Ans)$$

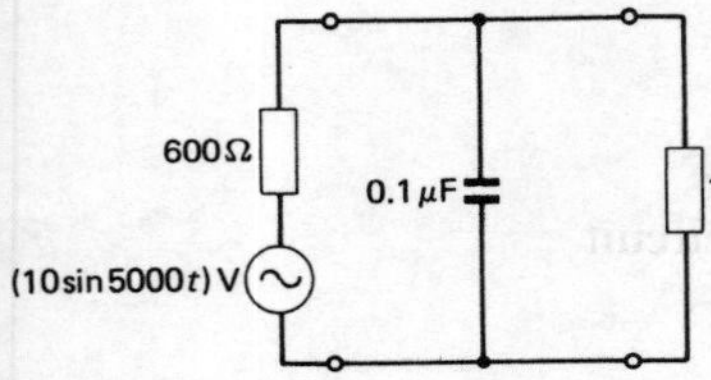

Fig. 8.16

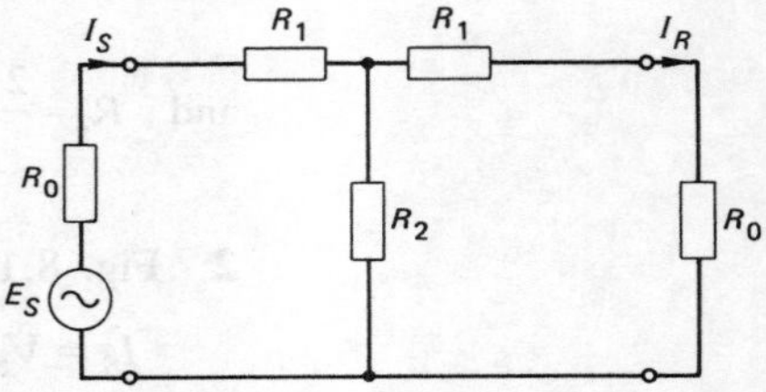

Fig. 8.17 T attenuator

Attenuators

An **attenuator** is a purely resistive network designed to introduce a specified loss into a circuit usually without a change in impedance. The majority of attenuators employ either the T or the π configuration although other arrangements may also be met.

1 Fig. 8.17 shows a **T attenuator** connected between a source and a load both of which have an impedance equal to the characteristic impedance of the network. From the figure,

$$I_R = \frac{I_S R_2}{R_1 + R_2 + R_0}$$

Therefore $$I_S/I_R = M = \frac{R_1 + R_2 + R_0}{R_2} \qquad (8.17)$$

Also,

$$R_0 = R_1 + \frac{R_2(R_1 + R_0)}{R_1 + R_2 + R_0} = R_1 + \frac{R_1 + R_0}{M}$$

$$R_0 M = R_1 M + R_1 + R_0 \qquad R_0(M-1) = R_1(M+1)$$

$$R_1 = \frac{R_0(M-1)}{M+1} \qquad (8.18)$$

From equation (8.17) $R_2M = R_1 + R_2 + R_0$

$$R_2(M-1) = R_1 + R_0 = \frac{R_0(M-1)}{M+1} + R_0$$

$$R_2(M-1)(M+1) = R_0(M-1) + R_0(M+1) = 2R_0M$$

and so $$R_2 = \frac{2R_0M}{M^2-1} \tag{8.19}$$

Example 8.8

Design a T attenuator to have a characteristic impedance of 600 Ω and an attenuation of 20 dB.

Solution 20 dB is a voltage or current ratio of 10:1, and therefore,

$$R_1 = \frac{600(10-1)}{10+1} = 491\ \Omega \quad (Ans)$$

and $$R_2 = \frac{2\times 600\times 10}{100-1} = 121\ \Omega \quad (Ans)$$

2 Fig. 8.18 shows a **π attenuator**. For this circuit

$$I_S = V_S/R_0 = (V_S/R_2) + (V_R/R_2) + (V_R/R_0)$$

Dividing throughout by V_R gives

$$M/R_0 = (M/R_2) + (1/R_2) + (1/R_0)$$

$$\frac{1}{R_0}(M-1) = \frac{1}{R_2}(M+1)$$

or $$R_2 = \frac{R_0(M+1)}{M-1} \tag{8.20}$$

Also $I_S = V_S/R_0 = (V_S/R_2) + [(V_S - V_R)/R_1]$

$$M/R_0 = (M/R_2) + (M/R_1) - (1/R_1)$$

$$\frac{1}{R_1}(M-1) = (M/R_0) - (M/R_2)$$

$$= \frac{M}{R_0} - \frac{M}{\dfrac{R_0(M+1)}{M-1}}$$

$$= \frac{1}{R_0}\left[M - \frac{M(M-1)}{M+1}\right] = \frac{1}{R_0}\left(\frac{2M}{M+1}\right)$$

$$\frac{1}{R_1} = \frac{1}{R_0}\left[\frac{2M}{(M+1)(M-1)}\right]$$

$$R_1 = \frac{R_0(M^2-1)}{2M} \tag{8.21}$$

Fig. 8.18 π attenuator

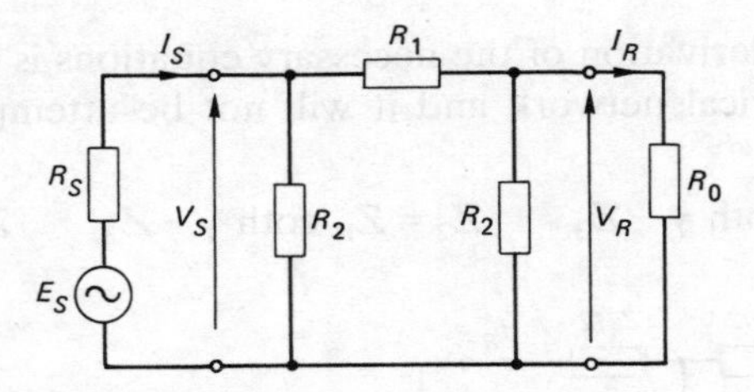

Example 8.9

Design a π attenuator to have a characteristic impedance of 600 Ω and an attenuation of 20 dB.

Solution From equation (8.20),

$$R_2 = \frac{600(10+1)}{10-1} = 733\ \Omega \quad (Ans)$$

From equation (8.21) $\quad R_1 = \dfrac{600(100-1)}{2\times 10} = 2970\ \Omega \quad (Ans)$

If a number of attenuators of identical characteristic impedances are connected in cascade, the overall attenuation is the *sum* of the attenuations of the individual attenuators.

Equivalent T Networks

Any network, of whatever configuration, can always be represented by an **equivalent T network**. The component values of the equivalent network can be determined by means of three measurements followed by some calculations.

The necessary measurements are of

A The input impedance with the output terminals open-circuited, Z_{oc1}
B The input impedance with the output terminals short-circuited, Z_{sc1}
C The output impedance with the input terminals open-circuited, Z_{oc2}

Referring to Fig. 8.1*a*,

$$Z_{oc1} = Z_1 + Z_2 \tag{8.22}$$

$$Z_{sc1} = Z_1 + \frac{Z_2 Z_3}{Z_2 + Z_3} \tag{8.23}$$

$$Z_{oc2} = Z_2 + Z_3 \tag{8.24}$$

From equations (8.22) and (8.24)

$$Z_1 = Z_{oc1} - Z_2 \quad \text{and} \quad Z_3 = Z_{oc2} - Z_2$$

Substituting into equation (8.23) gives

$$Z_{sc1} = Z_{oc1} - Z_2 + \frac{(Z_{oc2} - Z_2)Z_2}{Z_{oc2} - Z_2 + Z_2}$$

$$Z_{sc1}Z_{oc2} = Z_{oc1}Z_{oc2} - Z_2 Z_{oc2} + Z_{oc2}Z_2 - Z_2^2$$

$$Z_2^2 = \pm\sqrt{[Z_{oc2}(Z_{oc1} - Z_{sc1})]} \tag{8.25}$$

Z_1 and Z_3 can then be obtained using equations (8.22) and (8.24).

The components of the equivalent T network can also be expressed in terms of the image impedances and the image transfer coefficient of the network. The derivation of the necessary equations is complex in the case of a non-symmetrical network and it will not be attempted here. The results are

$$Z_1 = Z_A \coth \gamma - Z_3 \qquad Z_2 = Z_B \coth \gamma - Z_3 \qquad Z_3 = \sqrt{(Z_A Z_B)}/\sinh \gamma \tag{8.26}$$

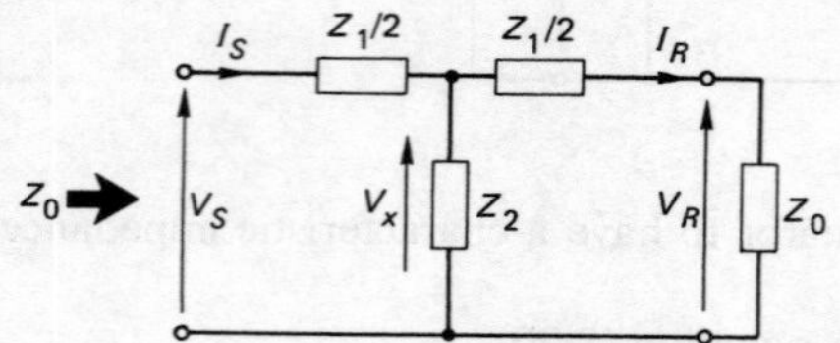

Fig. 8.19

For a symmetrical network the equivalent T network is shown in Fig. 8.19. From the figure,

$$I_R = \frac{V_x}{(Z_1/2)+Z_o} = \frac{V_S-(I_S Z_1/2)}{(Z_1/2)+Z_o} = \frac{I_S Z_o-(I_S Z_1/2)}{(Z_1/2)+Z_o}$$

$$I_R/I_S = e^{-\gamma} = \frac{Z_o-(Z_1/2)}{Z_o+(Z_1/2)}$$

$$\frac{Z_1}{2}(1+e^{-\gamma}) = Z_o(1-e^{-\gamma})$$

$$Z_1/2 = \frac{Z_o(1-e^{-\gamma})}{1+e^{-\gamma}} = \frac{Z_o(e^{\gamma/2}-e^{-\gamma/2})}{e^{\gamma/2}+e^{-\gamma/2}}$$

$$Z_1/2 = Z_o \tanh \gamma/2 \tag{8.27}$$

Also $I_R = \dfrac{I_S Z_2}{(Z_1/2)+Z_2+Z_o}$

$$I_R/I_S = e^{-\gamma} = \frac{Z_2}{(Z_1/2)+Z_2+Z_o}$$

$$e^{-\gamma}[(Z_1/2)+Z_o] = Z_2(1-e^{-\gamma})$$

$$e^{-\gamma}Z_o\left(\frac{2}{1+e^{-\gamma}}\right) = Z_2(1-e^{-\gamma})$$

$$Z_2 = \frac{Z_o 2e^{-\gamma}}{(1+e^{-\gamma})(1-e^{-\gamma})} = \frac{2Z_o e^{-\gamma}}{1-e^{-2\gamma}} = \frac{Z_o}{(e^{\gamma}-e^{-\gamma})/2}$$

or $Z_2 = Z_o/\sinh \gamma$ (8.28)

Example 8.10

A symmetrical network has a characteristic impedance of 600 Ω and an image transfer coefficient of $2+j0$. Calculate the component values of the equivalent T network.

Solution From equation (8.27)

$Z_1/2 = Z_o \tanh \gamma/2$ Here $\gamma = \alpha + j\beta = 2 + j0$.

Hence $Z_1/2 = 600 \tanh 1$.

$$\tanh 1 = \frac{e^1 - e^{-1}}{e^1 + e^{-1}} = \frac{2.7183 - 0.3679}{2.7183 + 0.3679} = 0.76$$

Therefore $Z_1/2 = 600 \times 0.76 = 456\ \Omega$ (*Ans*)
From equation (8.28) $Z_2 = 600/\sinh 2$

$$\sinh 2 = (e^2 - e^{-2})/2 = (7.389 - 0.1353)/2 = 3.63$$

Therefore $Z_2 = 600/3.63 = 165\ \Omega$ (*Ans*)

Use of Transmission Parameters

Equations can be derived for the iterative and image impedances, and the image transfer coefficient, of a network using the transmission parameters of that network.

A Image impedances $Z_A = \sqrt{\dfrac{AB}{CD}}$ (8.29)

$$Z_B = \sqrt{\frac{BD}{AC}} \quad (8.30)$$

B Iterative impedances complex and not given.
C Image transfer coefficient $\gamma = \log_e[\sqrt{(AD)} + \sqrt{(BC)}]$ (8.31)

Example 8.11

A network has the following transmission parameters. Calculate its image impedances.

$A = 1.6\angle -150°$, $B = 200\angle -143°\ \Omega$, $C = 6.32 \times 10^{-3}\angle -71.6°$ S, $D = 0.45\angle -117°$

Solution

$$Z_A = \sqrt{\frac{1.6\angle -150° \times 200\angle -143°}{6.32 \times 10^{-3}\angle -71.6° \times 0.45\angle -117°}} = 335.4\angle -52.2°\ \Omega \quad (Ans)$$

$$Z_B = \sqrt{\frac{200\angle -143° \times 0.45\angle -117°}{1.6\angle -150° \times 6.32 \times 10^{-3}\angle -71.6°}} = 94.3\angle -19.2°\ \Omega \quad (Ans)$$

Exercises 8

8.1 For the network shown in Fig. 8.20 calculate the image impedances and the image transfer coefficient. Calculate the insertion loss of the network when it is connected in between its image impedances.

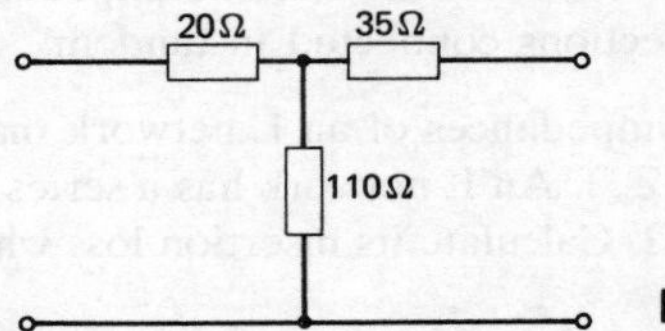

Fig. 8.20

8.2 Measurements on a network produced the following data:
input resistance with output open-circuited = 1100 Ω
input resistance with output short-circuited 550 Ω
output resistance with input open-circuited = 1650 Ω
Calculate the component values of the equivalent T network.

8.3 For the network shown in Fig. 8.21 calculate *a*) its image impedances, *b*) its iterative impedances, and *c*) its insertion loss when connected between its image impedances.

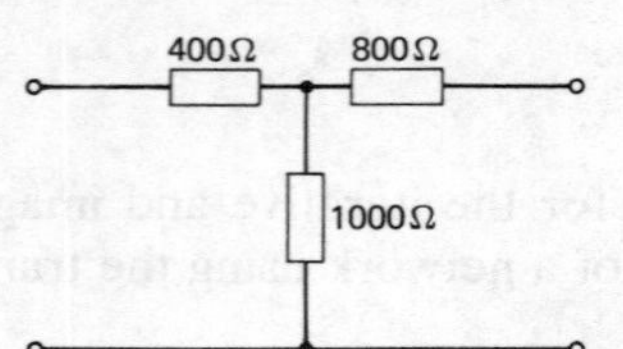

Fig. 8.21

8.4 A T network has a total series impedance Z_1 of 2000 Ω and a shunt impedance Z_2 of 5600 Ω. Calculate *a*) the characteristic impedance and *b*) the propagation coefficient of the network. Find also the current in a matched load when five such networks are connected in cascade if the input current is 10 mA.

8.5 Explain with the aid of a sketch the difference between an image impedance and an iterative impedance. The T network shown in Fig. 8.22 is connected between its image impedances. Calculate its insertion loss.

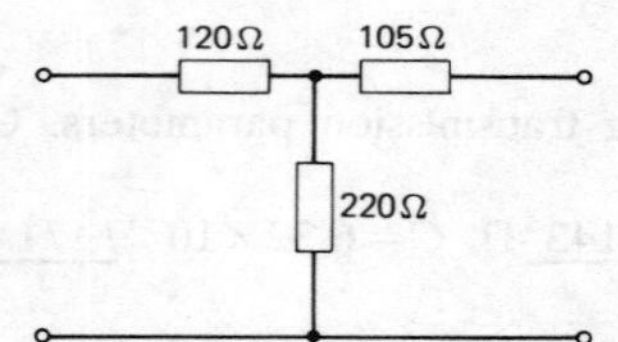

Fig. 8.22

8.6 Calculate the insertion loss of the network given in Fig. 8.23 when it is connected between *a*) its image impedances and *b*) its iterative impedances.

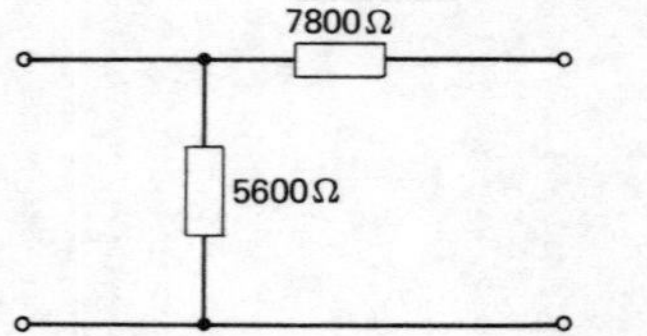

Fig. 8.23

8.7 Derive an expression for the characteristic impedance of a symmetrical T network. A symmetrical T attenuator has a total series impedance of 400 Ω and a shunt impedance of 380 Ω. Calculate the characteristic impedance of the network and the overall loss of four such sections connected in tandem.

8.8 Prove that the two image impedances of an L network may both be calculated using the expression $Z_{im} = \sqrt{(Z_{oc}Z_{sc})}$. An L network has a series resistance of 1200 Ω and a shunt impedance of 3300 Ω. Calculate its insertion loss when it is connected between its image impedances.

8.9 A 4-terminal network has an input resistance of 1200 Ω when the output terminals are open-circuited. When the output terminals are short-circuited the input impedance is 480 Ω. Lastly, when the input terminals are short-circuited the output impedance is 560 Ω. Calculate the component values of the equivalent T network.

8.10 Design a T attenuator to have a characteristic impedance of 600 Ω and an attentuation of 12 dB. Derive any formulae used.

8.11 Design a T attenuator to have a characteristic impedance of 140 Ω and an attenuation of 15 dB. Derive any formulae used.

Short Exercises

8.12 With the aid of appropriate sketches explain the difference between image and iterative impedances.

8.13 Design a T attenuator to have an attenuation of 30 dB and a characteristic impedance of 600 Ω.

8.14 A network has input and output currents of 250 mA and 122 mA respectively and input and output voltages of 14 V and 6.2 V respectively. Calculate the image transfer coefficient of the network.

8.15 When are the image impedances of a network equal to its iterative impedances? When are the iterative impedances of a network equal to its characteristic impedance? What is the difference between the image transfer coefficient and the propagation coefficient of a network?

8.16 A network is connected between its image impedances. The source has an e.m.f. of 6 V and an internal impedance of 350 Ω and the load is 400 Ω. *a*) What is the input impedance of the network? *b*) Is the network symmetrical? *c*) What is the input current to the network? *d*) If the load voltage is 0.5 V what is the image attenuation of the network? *e*) What is the insertion loss of the network?

8.17 The input and output currents of a non-symmetrical network are 80 mA and 10 mA respectively. The input and output voltages are 100 V and 25 V respectively. Calculate *a*) the input and load impendances, *b*) the image transfer coefficient of the network. State whether the input and load impedances are equal to the image impedances of the network.

9 Matched Transmission Lines

Introduction

Transmission lines are widely used in both telecommunication engineering and electric power engineering for the transmission of electrical energy from one point to another. Telephone cables operate at either audio-frequencies or in frequency bands of some 12 kHz upwards to a maximum of about 12 MHz for cables used as an integral part of the trunk telephone system, and many MHz for cables used in radio systems. Very often transmission lines are also used to simulate the action of a component, such as an inductor or a capacitor, or a tuned circuit. In such cases the frequencies involved may be as high as a few GHz. In a power system the frequency is usually fixed at 50 Hz.

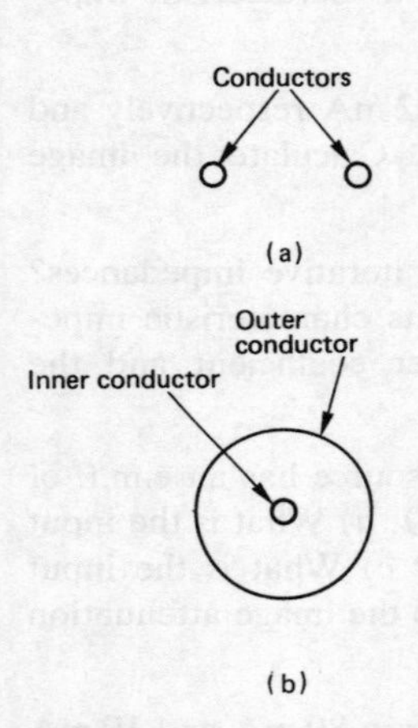

Fig. 9.1 *a*) Twin and *b*) coaxial line

The transmission performance of a line depends upon its *secondary coefficients* and these, in turn, are determined by the values of its *primary coefficients.*

Two main types of line are in common usage; the two-wire or twin, and the coaxial, shown respectively in Figs. 9.1*a* and *b*. The coaxial line is only used at the higher frequencies, say some 60 kHz and upwards.

The length of a line may be many kilometres in the case of power lines and audio-frequency telecommunication lines; and only a few tens of metres for some radio applications of lines such as a feeder to connect an aerial to a receiver; or even just a fraction of a metre when a line is employed to simulate a component.

Primary Coefficients of a Line

Resistance

The **resistance** R of a line is the sum of the resistances of the two conductors comprising a pair. The unit of resistance is ohms per kilometre loop.

At zero frequency the resistance of a pair is merely the d.c. resistance R_{dc} but at frequencies greater than a few kilohertz a phenomenon known as *skin effect* comes into play. This effect ensures that current flows only in a thin layer or "skin" at the surface of the conductors. The thickness of this layer reduces as the frequency is increased and this means that the effective cross-sectional area of the conductor is reduced. Since resistance is equal to $\rho l/a$ the *a.c. resistance* of a conductor will increase with increase in frequency. While the skin effect is developing, the relationship between the a.c. resistance and the frequency is rather complicated, but once skin effect is fully developed (usually at about 12 kHz) the a.c. resistance becomes directly proportional to the square root of the frequency, i.e.

$$R_{ac} = k_1\sqrt{f} \tag{9.1}$$

where k_1 is a constant.

Fig. 9.2*a* shows how the resistance of a line varies with frequency. Initially, little variation from the d.c. value is observed but at higher frequencies the shape of the graph is determined by equation (9.1).

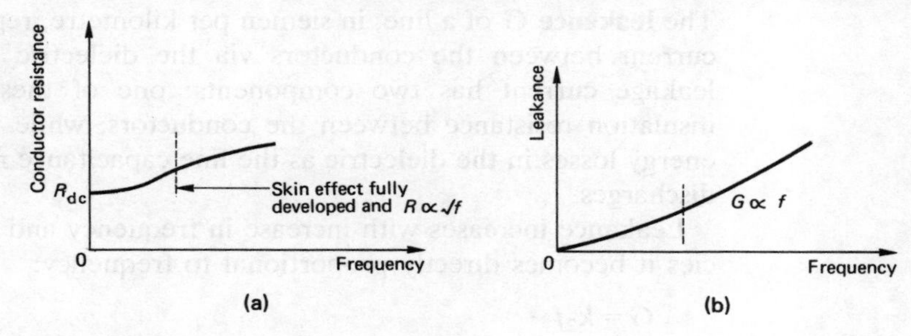

Fig. 9.2 *a*) Variation of line resistance with frequency, *b*) variation of line leakance with frequency

Inductance

The **inductance** of a line, in henrys/kilometre, depends upon the dimensions and the spacing of its two conductors. Equations for the inductance of both two-wire and coaxial pairs were derived in Chapter 7. These are repeated here for convenience.

A Two-wire or twin

$$L = \frac{\mu_0}{\pi} \log_e d/r \text{ H/km} \tag{9.2}$$

B Coaxial

$$L = \frac{\mu_0}{2\pi} \log_e R/r \text{ H/km} \tag{9.3}$$

Neither the dimensions of a line nor the absolute permeability are functions of frequency and consequently the inductance of a line is not a frequency-dependent parameter.

Capacitance

Equations were also derived in Chapter 7 for the **capacitance** of both kinds of line. These equations are repeated here for convenience.

C Twin

$$C = \frac{\pi\varepsilon}{\log_e d/r} \text{ F/km} \tag{9.4}$$

D Coaxial

$$C = \frac{2\pi\varepsilon}{\log_e R/r} \text{ F/km} \tag{9.5}$$

None of the terms in equations (9.4) or (9.5) are frequency dependent and so the capacitance of a line can also be considered to be reasonably constant with change in frequency.

Leakance

The **leakance** G of a line, in siemen per kilometre, represents the leakage of current between the conductors via the dielectric separating them. The leakage current has two components: one of these passes through the insulation resistance between the conductors, while the other supplies the energy losses in the dielectric as the line capacitance repeatedly charges and discharges.

Leakance increases with increase in frequency and at the higher frequencies it becomes directly proportional to frequency:

$$G = k_2 f \tag{9.6}$$

where k_2 is another constant.

Fig. 9.2*b* shows how the leakance of a line varies with frequency. Some typical figures for the primary coefficients of a line are given in Table 9.1. Clearly, the values of the four primary coefficients may differ considerably between pairs in different types of cable.

Table 9.1 Line primary coefficients

Line type	*R* (Ω/km)	*L* (mH/km)	*C* (μF/km)	*G* (μS/km)
Twin (800 Hz)	55	0.6	0.033	0.6
Coaxial (1 MHz)	34	0.28	0.05	1.4

Secondary Coefficients of a Line

The secondary coefficients of a transmission line are its

characteristic impedance Z_0
propagation coefficient γ
velocity of propagation v_p

The propagation coefficient has both a real part and an imaginary part; the former is known as the *attenuation coefficient* α and the latter is known as the *phase-change coefficient* β.

Characteristic Impedance

When a symmetrical T network is terminated in its characteristic impedance Z_0, the input impedance of the network is also equal to Z_0 (p. 118). Similarly, if five or more identical T networks are connected in cascade, the input impedance of the combination will also be equal to Z_0. Since a transmission line can be considered to consist of the tandem connection of a very large number of very short lengths δl of line, as shown by Fig. 9.3, the concept of characteristic impedance is also applicable to a line.

Each elemental length δl of line has a total series impedance of $(R + j\omega L)\,\delta l$ and a total shunt admittance of $(G + j\omega C)\,\delta l$. The **characteristic impedance** Z_0 of a line can therefore be defined as being the input

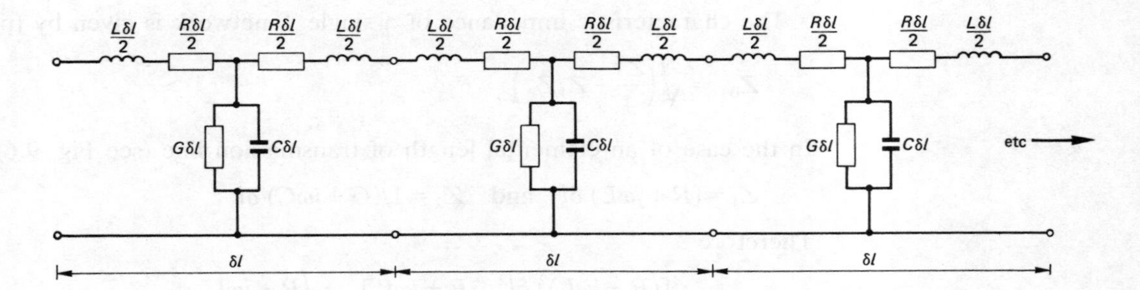

Fig. 9.3 Representation of a line by the cascade connection of δl sections

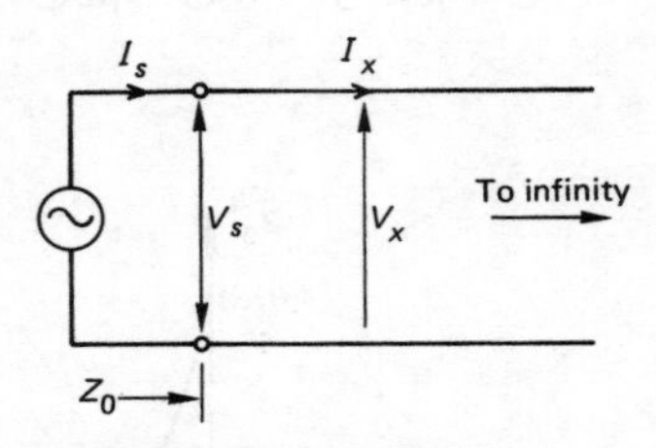

Fig. 9.4 Definition of the characteristic impedance of a line

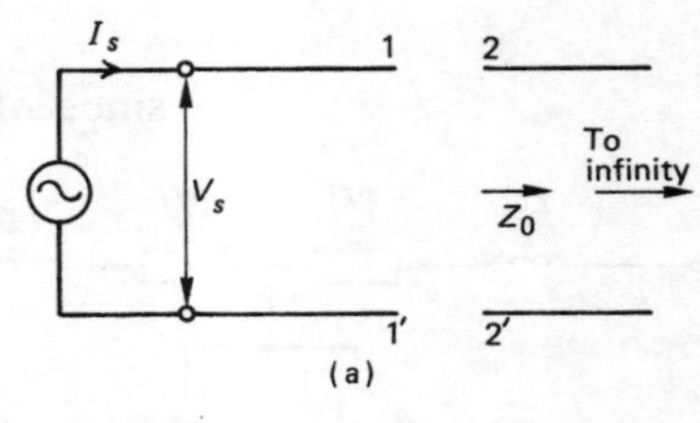

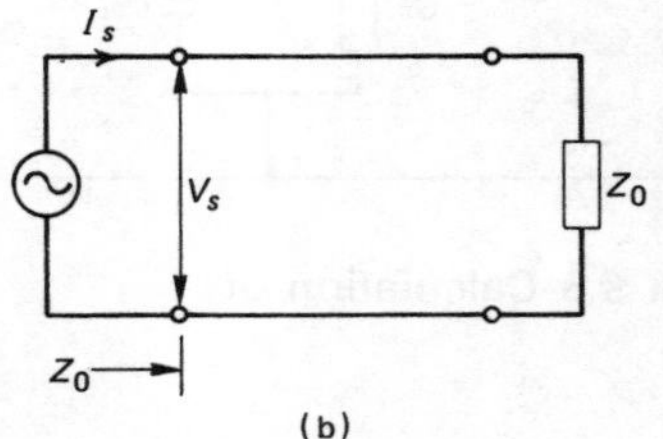

Fig. 9.5 Alternative definition of the characteristic impedance of a line

impedance of a long length of that line. Fig. 9.4 shows a very long length of line; its input impedance is the ratio of the voltage V_S impressed across the sending-end terminals of the line to the current I_S that flows into those terminals, i.e.

$$Z_0 = V_S/I_S \text{ ohms} \tag{9.7}$$

Similarly, at any point x along the line the ratio V_x/I_x is always equal to Z_0. Suppose that the line is cut a finite distance from its sending-end terminals as shown by Fig. 9.5*a*. The remainder of the line is still very long and so the impedance measured at terminals 2–2 is equal to the characteristic impedance. Thus, before the line was cut, terminals 1–1 were effectively terminated in impedance Z_0. The conditions at the input terminals will not be changed if the terminals 1–1 are closed in a physical impedance equal to Z_0, as in Fig. 9.5*b*. This leads to a more practical definition: the characteristic impedance of a transmission line is the input impedance of a line that is itself terminated in the characteristic impedance. This result can be used to derive an expression for the characteristic impedance. A line that is terminated in its characteristic impedance is generally said to be **correctly terminated**.

The characteristic impedance of a single T network is given by (p. 119)

$$Z_{0T}=\sqrt{\left(\frac{Z_1^2}{4}+Z_1Z_2\right)} \tag{8.4}$$

In the case of an elemental length of transmission line (see Fig. 9.6),

$$Z_1=(R+j\omega L)\,\delta l \quad \text{and} \quad Z_2=1/(G+j\omega C)\,\delta l$$

Therefore

$$Z_0=\sqrt{\left[\frac{(R+j\omega L)^2\,\delta l^2}{4}+\frac{R+j\omega L}{G+j\omega C}\right]}\simeq\sqrt{\frac{R+j\omega L}{G+j\omega C}} \tag{9.8}$$

since δl^2 is *very* small.

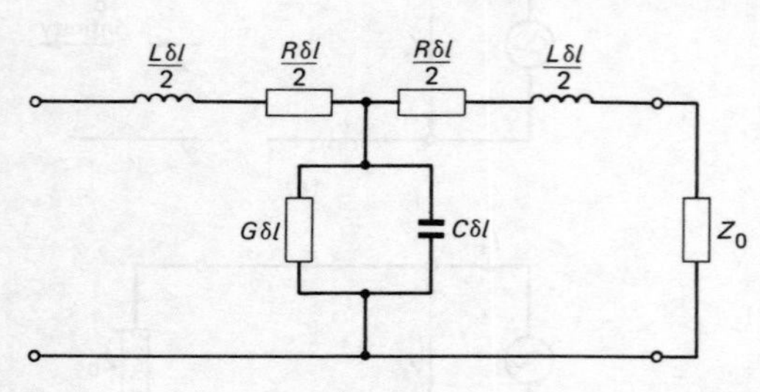

Fig. 9.6 Calculation of Z_0

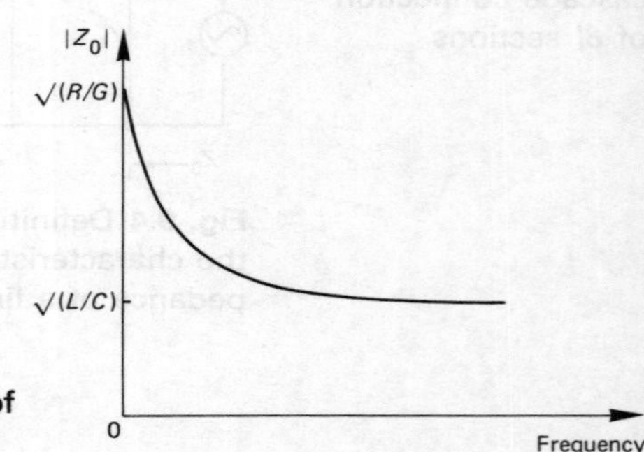

Fig. 9.7 Variation of Z_0 with frequency

At zero, and very low, frequencies, $R \gg \omega L$ and $G \gg \omega C$ and then $Z_0 \simeq \sqrt{(R/G)}$ ohms. With increase in frequency both the magnitude and the phase angle of Z_0 decrease in value and, when the frequency is reached at which the statements $\omega L \gg R$ and $\omega C \gg G$ are valid, then

$$Z_0=\sqrt{(L/C)} \text{ ohms} \tag{9.9}$$

Since both L and C are both very nearly constant with frequency, Z_0 remains at this value with any further increase in frequency. Furthermore, Z_0 is now a purely resistive quantity. The approximate expression (9.9) for Z_0 is always used for radio-frequency lines since these are only used at fairly high frequencies. Fig. 9.7 shows how the magnitude of the characteristic impedance of a line varies with frequency.

Example 9.1

A transmission line has the following primary coefficients: $R=40\ \Omega$/km, $L=48$ mH/km, $C=0.06\ \mu$F/km and $G=4\times10^{-6}$ S/km. Calculate its characteristic impedance at a frequency of $5000/2\pi$ Hz.

Solution From equation (9.8)

$$Z_0=\sqrt{\frac{40+j5000\times48\times10^{-3}}{(4+j5000\times0.06)\times10^{-6}}}=\sqrt{\frac{40+j240}{(4+j300)\times10^{-6}}}$$

Clearly, $G \ll \omega C$ and so it may be neglected with very little error. Therefore,

$$Z_0=\sqrt{\frac{243.3\,\underline{/80.5^\circ}}{300\,\underline{/90^\circ}\times 10^{-6}}}\simeq 900\,\underline{/-4.8^\circ}\;\Omega\quad (Ans)$$

The characteristic impedance of a transmission line is a function of the physical dimensions of the line and the permittivity of the dielectric. At higher frequencies when the approximate expression (9.9) for Z_0 is valid, the characteristic impedance can be written as

A Twin line

$$Z_0=\sqrt{(L/C)}=\sqrt{\left[\frac{\mu_0\log_e d/r}{\pi}\times\frac{\log_e d/r}{\pi\varepsilon}\right]}$$

$$=\frac{\log_e d/r}{\pi}\sqrt{\frac{\mu_0}{\varepsilon_0\varepsilon_r}}=120\log_e\frac{d}{r}\sqrt{\frac{1}{\varepsilon_r}}$$

$$=\frac{276\log_{10} d/r}{\sqrt{\varepsilon_r}} \tag{9.10}$$

B Coaxial line

$$Z_0=\sqrt{(L/C)}=\sqrt{\left[\frac{\mu_0\log_e R/r}{2\pi}\times\frac{\log_e R/r}{2\pi\varepsilon}\right]}$$

$$=\frac{\log_e R/r}{2\pi}\sqrt{\frac{\mu_0}{\varepsilon_0\varepsilon}}=\frac{60\log_e R/r}{\sqrt{\varepsilon_r}}$$

$$Z_0=\frac{138\log_{10} R/r}{\sqrt{\varepsilon_r}} \tag{9.11}$$

Propagation Coefficient

1 The **propagation coefficient** γ of a 1 km length of correctly terminated transmission line is defined as

$$\gamma=\log_e V_S/V_R=\log_e I_S/I_R \tag{9.12a}$$

or $$V_R=V_S e^{-\gamma} \tag{9.12b}$$

For a line which is 2 km in length the received voltage will be

$$V_R=V_S e^{-\gamma}e^{-\gamma}=V_S e^{-2\gamma}$$

Similarly, for a line of length l kilometres

$$V_R=V_S e^{-\gamma l} \tag{9.13}$$

In similar fashion the current received at the end of a line of length l can be written as

$$I_R=I_S e^{-\gamma l} \tag{9.14}$$

An expression for the propagation coefficient of a symmetrical T network has been derived on page 120 and is

$$\gamma=\log_e[1+(Z_1/2Z_2)+(Z_0/Z_2)] \tag{9.15}$$

For an elemental length of line

$$Z_1 = (R + j\omega L)\,\delta l \qquad Z_2 = 1/(G + j\omega C)\,\delta l$$

and, of course,

$$Z_0 = \sqrt{\frac{R + j\omega L}{G + j\omega C}}$$

Hence

$$\begin{aligned} e^{\gamma} &= 1 + \frac{(R + j\omega L)\,\delta l(G + j\omega C)\,\delta l}{2} + \sqrt{\left[\frac{R + j\omega L}{G + j\omega C}\right]}(G + j\omega C)\,\delta l \\ &= 1 + \sqrt{[(R + j\omega L)(G + j\omega C)]}\,\delta l + \frac{(R + j\omega L)(G + j\omega C)\,\delta l^2}{2} \end{aligned} \qquad (9.16)$$

The series form of e^x is

$$e^x = 1 + x + \frac{x^2}{2!} + \frac{x^3}{3!} + \cdots \qquad (9.17)$$

Comparing equations (9.16) and (9.17) term by term it should be evident that for an elemental length of line

$$\gamma = \sqrt{[(R + j\omega L)(G + j\omega C)]}\,\delta l$$

In a 1 kilometre length of line there are $1/\delta l$ such elemental sections and so

$$\gamma = \sqrt{[(R + j\omega L)(G + j\omega C)]} \text{ per km} \qquad (9.18)$$

It should be noted that γ has no units.

2 In general, γ is a complex quantity and it therefore has both a real part and an imaginary part. The real part of γ is known as the **attenuation coefficient** α, measured in units of nepers/kilometre. The imaginary part of γ is called the **phase-change coefficient** β, measured in terms of units of radians/kilometre.

Thus $\gamma = \alpha + j\beta$.

The voltage V_x at any point along a line, distance x from the sending end, can be written as

$$V_x = V_S e^{-\gamma x} = V_S e^{-(\alpha + j\beta)x} = V_S e^{-\alpha x} \cdot e^{-j\beta x} = V_S e^{-\alpha x} \underline{/-\beta x} \qquad (9.19)$$

The voltage V_R received at the far end of the line can now be written as

$$V_R = V_S e^{-\alpha l} \underline{/-\beta l} \qquad (9.20)$$

Equation (9.19) demonstrates that the magnitude of the line voltage (or current) decreases exponentially with the distance x from the sending end of the line. The phase angle of the line voltage is *always* a *lagging* angle and this angle increases in direct proportion to the length of the line.

Expressions for the attenuation coefficient and for the phase-change coefficient *can* be derived from equation (9.18) but the results are unwieldy and not worthwhile. When the values of α and β are to be calculated it is better to use equation (9.18) and take the real and imaginary parts of the result.

Example 9.2

Calculate, for a frequency of $5000/2\pi$ Hz, the propagation coefficient of a line whose primary coefficients are: $R=55\ \Omega$/km, $L=28$ mH/km, $C=0.07\ \mu$F/km and $G=2.0\ \mu$S/km. Determine also the values of α in dB and β in degrees.

Solution From equation (9.18),

$$\gamma=\sqrt{[(55+j5000\times 28\times 10^{-3})(2\times 10^{-6}+j5000\times 0.07\times 10^{-6})]}$$

$$=\sqrt{[(55+j140)(2\times 10^{-6}+j3.5\times 10^{-4})]}$$

$$=\sqrt{[150.4\underline{/68.6^\circ}\times 3.5\times 10^{-4}\underline{/89.7^\circ}]}$$

$$=0.229\underline{/79.2^\circ}/\text{km} \quad (Ans)$$

$$\alpha=0.229\cos 79.1^\circ=0.043 \text{ N/km}$$

$$=0.043\times 8.686=0.37 \text{ dB/km} \quad (Ans)$$

$$\beta=0.229\sin 79.1^\circ=0.225 \text{ R/km}=13^\circ/\text{km} \quad (Ans)$$

3 Approximate expressions for the attenuation coefficient and the phase-change coefficient can be obtained at *a*) very low frequencies where $R\gg\omega L$ and $G\gg\omega C$, *b*) low frequencies where $R\gg\omega L$ and $\omega C\gg G$, and *c*) high frequencies where $\omega L\gg R$ and $\omega C\gg G$.

A Very low frequencies If $R\gg\omega L$ and $G\gg\omega C$ then

$$\gamma=\sqrt{(RG)} \quad \text{and is wholly real} \tag{9.21}$$

B Low frequencies If $R\gg\omega L$ and $\omega C\gg G$ then

$$\gamma\simeq\sqrt{(j\omega CR)}=\sqrt{(\omega CR\underline{/90^\circ})}=\sqrt{(\omega CR)}\underline{/45^\circ}$$

Taking the real and the imaginary parts,

$$\gamma=\sqrt{\frac{\omega CR}{2}}+j\sqrt{\frac{\omega CR}{2}} \tag{9.22}$$

C High frequencies If $\omega L\gg R$ and $\omega C\gg G$ then

$$\gamma=\sqrt{\left[\left(1+\frac{R}{j\omega L}\right)j\omega L\left(1+\frac{G}{j\omega C}\right)j\omega C\right]}$$

$$=j\omega\sqrt{(LC)}\left[1+\frac{R}{j\omega L}\right]^{1/2}\left[1+\frac{G}{j\omega C}\right]^{1/2}$$

$$=j\omega\sqrt{(LC)}\left[1+\frac{R}{2j\omega L}\right]\left[1+\frac{G}{2j\omega C}\right]$$

$$\simeq j\omega\sqrt{(LC)}\left[1+\frac{R}{2j\omega L}+\frac{G}{2j\omega C}\right]$$

$$=\frac{R}{2}\sqrt{\frac{C}{L}}+\frac{G}{2}\sqrt{\frac{L}{C}}+j\omega\sqrt{(LC)}$$

Hence $\alpha = \dfrac{R}{2Z_0} + \dfrac{GZ_0}{2}$ N/km (9.23)

and $\beta = \omega\sqrt{(LC)}$ R/km (9.24)

Example 9.3

A radio-frequency line has an attenuation of 2.5 dB/km at 2 MHz, 10% of which is dielectric loss. Determine the attenuation of the line at 4 MHz.

Solution At 2 MHz the conductor loss is 2.25 dB/km and the dielectric loss is 0.25 dB/km. From equation (9.23) the conductor loss is proportional to the resistance of the line and so it is proportional to the square root of the frequency. Therefore, the conductor loss at 4 MHz is

$$2.25\sqrt{(4/2)} = 3.18 \text{ dB/km}$$

The dielectric loss is proportional to the leakance of the line and hence to the frequency. Therefore, the dielectric loss at 4 MHz is $0.25 \times 4/2 = 0.5$ dB/km.

Therefore, the attenuation of the line at 4 MHz is

$$3.18 + 0.5 = 3.68 \text{ dB/km} \quad (\textit{Ans})$$

Velocity of Propagation

The **phase velocity** v_p of a line is the velocity with which a sinusoidal wave travels along that line. Any sinusoidal wave travels with a velocity of one wavelength per cycle. There are f cycles per second and so a wave travels with a velocity of λf kilometres per second, i.e.

$$v_p = \lambda f \text{ kilometres/second} \qquad (9.25)$$

In one wavelength, a phase change of 2π radians occurs. Hence the phase change per kilometre is $2\pi/\lambda$ radians and this is also equal to the phase-change coefficient β of the line. Thus

$$\beta = 2\pi/\lambda \quad \text{or} \quad \lambda = 2\pi/\beta \quad \text{and}$$

$$v_p = 2\pi f/\beta = \omega/\beta \text{ km/sec} \qquad (9.26)$$

The *phase delay* of a line is the product of the line length and the reciprocal of its phase velocity.

Any repetitive non-sinusoidal waveform contains components at a number of different frequencies. Each of these components will be propagated along a line with a phase velocity given by equation (9.26). For all of these components to travel with the same velocity and arrive at the far end of the line together it is necessary for β to be a *linear* function of frequency. Unfortunately, it is only at the higher frequencies where $\omega L \gg R$ and $\omega C \gg G$ that this requirement is satisfied. At these frequencies $\beta = \omega\sqrt{(LC)}$ and hence

$$v_p = \omega/\beta = 1/\sqrt{(LC)} \text{ km/sec} \qquad (9.27)$$

When the component frequencies of a complex wave travel with different velocities, their relative phase relationships will be altered and the resultant waveform will be changed, or *distorted*.

It is customary to consider the **group velocity** of a complex wave rather than the phase velocities of its individual components. Group velocity is the velocity with which the envelope of a complex wave is propagated along a line. The group velocity v_g of a line is given by

$$v_g = d\omega/d\beta \tag{9.28}$$

The *group delay* of a line is the product of the length of the line and the reciprocal of its group velocity. Group delay measures the time taken for the envelope of a complex wave to propagate over a transmission line [DT&S].

The Current and Voltage at the End of a Line

If a transmission line is *correctly terminated*, i.e. the load impedance is equal to the characteristic impedance Z_0, then the input impedance of the line will also be equal to Z_0. This point is emphasized by Fig. 9.8 and it is true regardless of the length of the line.

Fig. 9.9 shows a source of e.m.f. E_S and internal impedance Z_S connected across the input terminals of a line. The current I_S flowing into the line is

$$I_S = E_S/(Z_S + Z_0)$$

and the voltage appearing at the input terminals is

$$V_S = E_S Z_0/(Z_S + Z_0)$$

Note that $V_S/I_S = Z_0$.

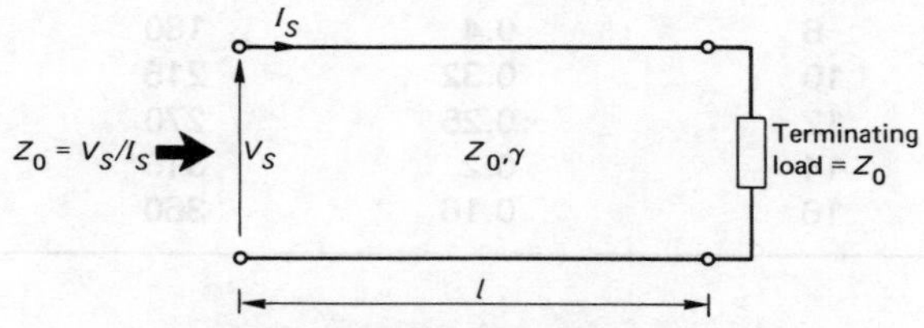

Fig. 9.8 Matched line

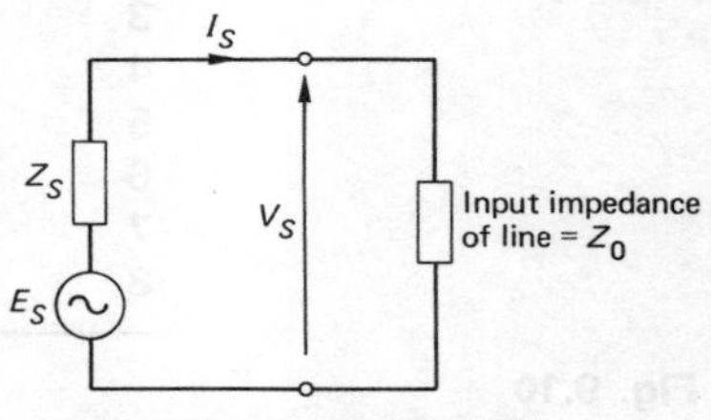

Fig. 9.9 Input circuit of a matched line

The input current and voltage propagate along the line to its far end with a phase velocity v_p. The waves are subject to both attenuation and phase shift (lagging) as they travel. At any point along the line, distance x from the sending end, the current and voltage are equal to

$$I_x = I_S e^{-\gamma x} \quad \text{and} \quad V_x = V_S e^{-\gamma x}$$

respectively. Note that once again the ratio V_x/I_x is equal to Z_0.

At the far end of the line the received current I_R is equal to $I_S e^{-\gamma l}$ and the received voltage is $V_R = V_S e^{-\gamma l}$. Once again the ratio V_R/I_R is equal to Z_0. The power dissipated in the load can be worked out in two different ways, e.g.

$$P_R = |I_R|^2 \times (\text{real part of } Z_0)$$

or $$P_R = |V_R|\,|I_R| \cos\theta$$

where θ is the phase angle between the received current and voltage.

A *polar diagram* can be drawn to show the magnitude and the phase, relative to the sending end, of the line voltage (or current) at any point along a line. Such a diagram consists of a series of phasors drawn to represent the line voltage at various points along the line. Each phasor is drawn with an appropriate length and angle to indicate the magnitude and the phase of the voltage at that point.

Consider an 8 km length of line that has an attenuation of 2 dB/km and a phase-change coefficient of 45°/km at a particular frequency. If the sending-end voltage is 1 volt then the magnitude and phase of the line voltage at 1 km steps from the sending end of the line are given in Table 9.2.

The polar diagram is shown plotted in Fig. 9.10.

At the instant when the sending-end voltage is zero and just about to go positive, the line voltage at various points along the line can be plotted by projecting from the tips of the phasors to the corresponding points on axis A.

Table 9.2

Distance from sending end	*Attenuation* (dB)	*Voltage V* (V)	*Phase lag* (deg)
0	0	1	0
1	2	0.79	45
2	4	0.63	90
3	6	0.5	135
4	8	0.4	180
5	10	0.32	215
6	12	0.25	270
7	14	0.2	315
8	16	0.16	360

Fig. 9.10

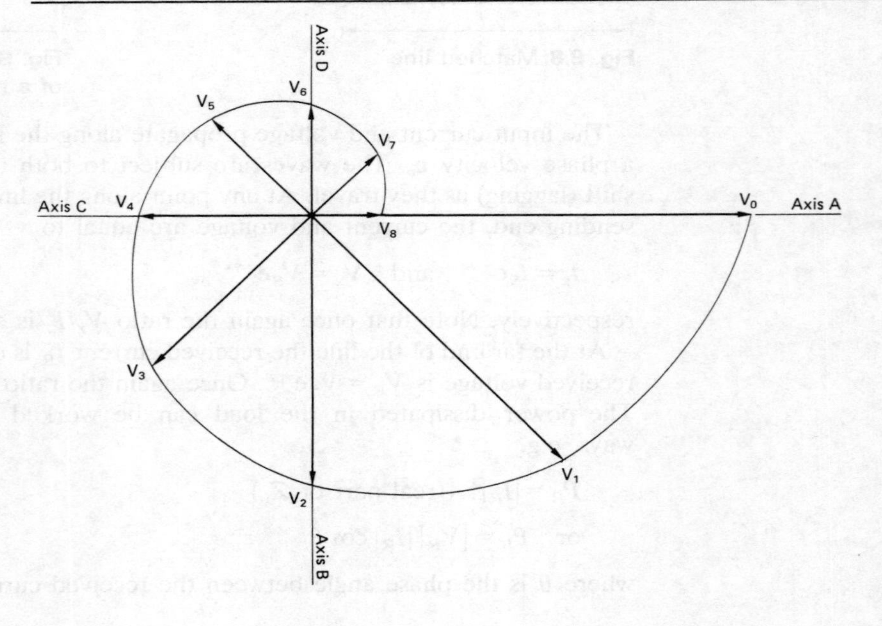

A quarter of the period time later, the sending-end voltage is instantaneously at its peak positive value of 1 V. The line voltages at this instant in time can be plotted by projecting from the tips of the phasors onto axis B.

The waveforms of the line voltages at the instants in time when the input voltage is, first, zero and about to go negative and, secondly, at its peak negative value are plotted on axes C and D respectively.

Fig. 9.11

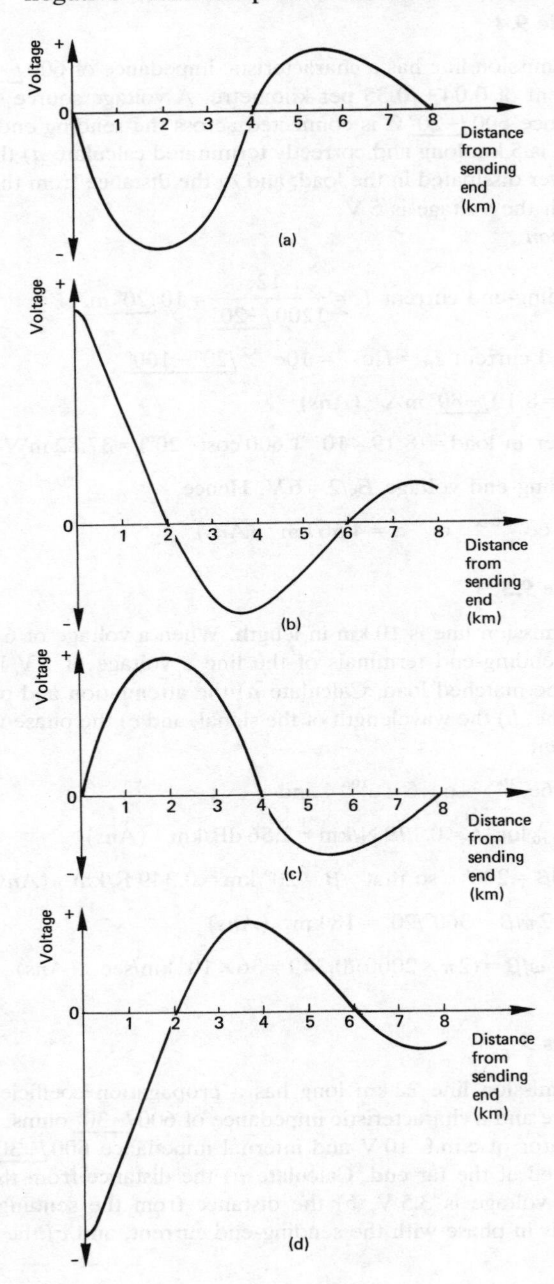

Fig. 9.11 shows the four line voltage waveforms drawn beneath one another and it illustrates clearly that a particular part of the applied voltage waveform moves along the line. For example, note that the negative peak voltage moves progressively through distances of approximately 1.5 km, 3.5 km, 5.5 km and 7.5 km from the sending-end of the line.

Example 9.4

A transmission line has a characteristic impedance of $600\angle{-20^\circ}\ \Omega$ and a propagation coefficient of $0.04+j0.35$ per kilometre. A voltage source of e.m.f. 12 V r.m.s. and impedance $600\angle{-20^\circ}\ \Omega$ is connected across the sending end terminals of the line. If the line is 5 km long and correctly terminated calculate *a*) the current in the load, *b*) the power dissipated in the load, and *c*) the distance from the sending end of the line at which the voltage is 5 V.

Solution

a) Sending-end current $I_S = \dfrac{12}{1200\angle{-20^\circ}} = 10\angle{20^\circ}$ mA

Received current $I_R = I_S e^{-\gamma l} = 10e^{-0.2}\angle{20^\circ - 100^\circ}$

$I_R = 8.19\angle{-80^\circ}$ mA (*Ans*)

b) Power in load $= (8.19\times10^{-3})^2 600\cos(-20^\circ) = 37.82$ mW (*Ans*)

c) Sending-end voltage $E_S/2 = 6$ V. Hence,

$5 = 6e^{-0.04x}$ or $x = 4.56$ km (*Ans*)

Example 9.5

A transmission line is 10 km in length. When a voltage of 6 V at 2000 Hz is applied to the sending-end terminals of the line a voltage of 1 V lagging by 200° appears across the matched load. Calculate *a*) the attenuation and phase-change coefficients of the line, *b*) the wavelength of the signal, and *c*) the phase velocity of propagation.

Solution

a) $1 = 6e^{-10\alpha}$ so $6 = e^{10\alpha}$ and

$\alpha = \frac{1}{10}\log_e 6 = 0.179$ N/km $= 1.56$ dB/km (*Ans*)

and $10\beta = 200^\circ$ so that $\beta = 20^\circ/\text{km} = 0.349$ R/km (*Ans*)

b) $\lambda = 2\pi/\beta = 360^\circ/20^\circ = 18$ km (*Ans*)

c) $v_p = \omega/\beta = (2\pi\times2000)/0.349 = 36\times10^3$ km/sec (*Ans*)

Exercises 9

9.1 A transmission line 22 km long has a propagation coefficient of $(0.05+j\pi/10)$ per kilometre and a characteristic impedance of $600\angle{-30^\circ}$ ohms. The line is connected to a generator of e.m.f. 10 V and internal impedance $600\angle{-30^\circ}$ ohms and is correctly terminated at the far end. Calculate *a*) the distance from the sending end at which the line voltage is 3.5 V, *b*) the distance from the sending end at which the line current is in phase with the sending-end current, and *c*) the power dissipated in the load.

9.2 Explain what is meant by the propagation coefficient of a line. A transmission line has a propagation coefficient of $(0.015+j\pi/60)$ per kilometre and a characteristic impedance of $550\angle-25°$ ohms. The line is terminated in its characteristic impedance 20 km from the sending end. The voltage across the sending end terminals is 12 V. Calculate the power dissipated in the load and the phase difference between the sent and received voltages.

9.3 A voltage source of e.m.f. 20 V and internal impedance $500\angle 0°$ ohms is connected across the input terminals of a loss-free line of $Z_0 = 500\angle 0°$ ohms. Calculate the power dissipated in the correctly terminated load. If the phase-change coefficient of the line is $\pi/3$ radians/m calculate the phase velocity on the line when the frequency of the signal is 30 MHz.

9.4 A transmission line has a characteristic impedance of $600\angle-30°$ ohms and a propagation coefficient of $(0.1+j\pi/2)$ per km. If the line is terminated in its characteristic impedance calculate the current in the load when the sending-end voltage is 10 V and the line is 12 km long.

9.5 A correctly terminated transmission line is 10 km long and has the following primary coefficients: $R=20\ \Omega$/km, $L=16$ mH/km, $C=0.045\ \mu$F/km and $G\simeq 0$ at $\omega =$ 5000 R/s. A voltage generator with an e.m.f. of 12 V and an internal impedance of 500 Ω is connected across the sending-end terminals of the line. Calculate the magnitude and phase, relative to the sending-end voltage, of the load current.

9.6 Explain what is meant by the term "characteristic impedance" when applied to a transmission line. A 2 m length of line is terminated in its characteristic impedance of 75 ohms and has a voltage of 10 V at 100 MHz applied to the sending-end terminals. If the line has negligible loss calculate *a*) the wavelength of the signal on the line, *b*) the power dissipated in the load, and *c*) the current 1 m from the sending-end terminals.

9.7 A 3 km loss-free line has a characteristic impedance of 50 Ω. It is fed by a source of e.m.f. 6 V and impedance 50 Ω and is terminated by a 50 Ω resistor. If the wavelength of the signal on the line is 200 m calculate *a*) the sent voltage, *b*) the received voltage, *c*) the power in the 50 Ω resistor, and *d*) the phase-change coefficient of the line.

9.8 A transmission line has the following primary coefficients: $R=25\ \Omega$/km, $L=$ 1 mH/km, $C=0.05\ \mu$F/km and $G\simeq 0$. If the frequency is $10\,000/2\pi$ Hz calculate the characteristic impedance, the attenuation coefficient, and the phase-change coefficient of the line.

9.9 A line has an attenuation coefficient of 6 dB/km, a phase-change coefficient of 0.62 R/km and a characteristic impedance of $1800\angle-28°\ \Omega$ at a particular frequency. A voltage of 3 V is maintained across the input terminals of a 2 km length of this line. Calculate *a*) the current in the load and *b*) the length of the line in wavelengths.

9.10 A transmission line has the following primary coefficients: $R=28\ \Omega$/km, $L=$ 0.6 mH/km, $C=0.055\ \mu$F/km and $G\simeq 0$ at $\omega = 18\,000$ R/s. Calculate its velocity of propagation.

9.11 A line has a characteristic impedance of $650\angle-20°$ ohms and a propagation coefficient of $(0.025+j\pi/3)$/km. If the line is 10 km long calculate the power dissipated in its correctly terminated load when the sending-end voltage is 60 V. State the phase difference between *a*) the voltages at each end of the line and *b*) the current at the load and the sending-end voltage.

9.12 A line has the following primary coefficients: $R = 22\ \Omega/\text{km}$, $L = 0.6\ \text{mH/km}$, $C = 0.09\ \mu\text{F/km}$ and $G \simeq 0$. Calculate its propagation coefficient at $\omega = 20\,000\ \text{R/s}$.

9.13 A line has the following primary coefficients: $R = 40\ \Omega/\text{km}$, $L = 1.2\ \text{mH/km}$, $C = 0.1\ \mu\text{F/km}$, and $G = 1.3 \times 10^{-3}\ \text{S/km}$ at $\omega = 10\,000\ \text{R/s}$. Calculate its characteristic impedance and its propagation coefficient.

Short Exercises

9.14 A line has negligible losses and inductance and capacitance values of $1.5\ \mu\text{H}$ and $12.2\ \text{pF}$ respectively per metre. If this line is terminated by its characteristic impedance calculate *a*) the power in the load when the sending-end voltage is 30 V and *b*) the velocity of propagation.

9.15 A line has $Z_0 = 500\angle{-15°}$ ohms at 1500 Hz and is correctly terminated. If a voltage of 10 V is maintained across the sending-end terminals calculate the input power to the line. If the load power is 10 mW calculate the attenuation of the line.

9.16 An air-spaced line has $Z_0 = 75\ \Omega$. What will Z_0 become if the space between the conductors is filled with a dielectric whose relative permittivity is 4?

9.17 The propagation coefficient of a line is $(0.25 + j0.32)$ per km. Determine the loss in dB and the phase-shift in degrees of a 4 km length of this line.

9.18 A line has an input current of 100 mA and the current in the distant load is 20 mA. If the line is 10 km long calculate its attenuation coefficient.

9.19 A line has a phase-change coefficient of $\pi/6$ R/km and is 6 km in length. Calculate *a*) the wavelength on the line and *b*) the phase lag introduced by the line.

9.20 A line of negligible loss has $L = 1.8\ \text{mH/km}$ and $C = 0.08\ \mu\text{F/km}$. The line is connected to a voltage generator of 10 V e.m.f. and is matched at both ends. Calculate the power dissipated in the load.

9.21 A $50\ \Omega$ loss-free coaxial line has a phase velocity of 2.2×10^8 m/s. Calculate its inductance and capacitance per metre.

10 Mismatched Transmission Lines

Introduction

Very often a transmission line is operated with a load impedance that is not equal to its characteristic impedance. In some instances this state of affairs is intentional but in many others it is because the correct load impedance is not available for one reason or another. When a line is **mismatched**, the load impedance is unable to absorb all of the power incident upon it and so some of this power is *reflected* back towards the sending end of the line. If the sending-end terminals are matched to the source impedance, all of the reflected energy will be absorbed by the source impedance and there will be no further reflections. On the other hand, if the sending end of the line is also mismatched, some of the energy reflected by the load will be further reflected at the sending-end terminals and will be returned towards the load again. In such situations *multiple reflections* will take place.

Whether reflections on a line are desirable or not depends upon the intended use of the line. If the line is to be employed for the transmission of energy from one point to another, reflections are undesirable for a number of reasons that will be discussed later in this chapter. If, however, the intended application of the line is to simulate a component of some kind, reflections will be essential.

1 Consider Fig. 10.1 which shows a loss-free line whose output terminals are **OPEN-CIRCUITED.** The line has an electrical length of one wavelength and its input terminals are connected to a source of e.m.f. E_S volts and impedance Z_0 ohms.

When the generator is first connected to the line, the input impedance of the line is equal to its characteristic impedance Z_0. An *incident* current of $E_S/2Z_0$ then flows into the line and an *incident* voltage of $E_S/2$ appears across the input terminals. These are, of course, the same values of sending-end current and voltage that flow into a correctly-terminated line. The incident current and voltage waves propagate along the line, being phase-shifted as they travel. Since the electrical length of the line is one wavelength, the overall phase shift experienced is 360°.

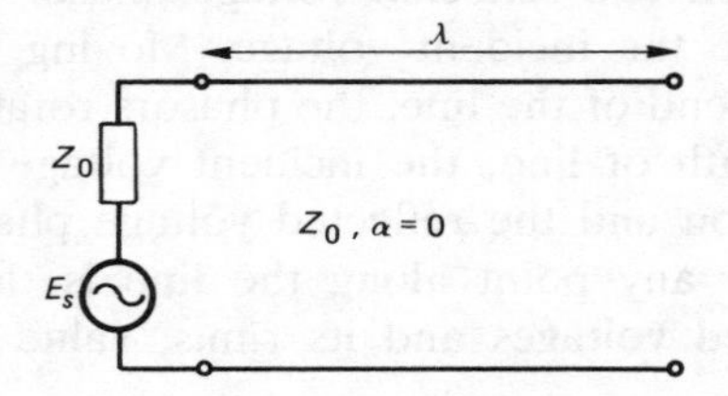

Fig. 10.1 Open-circuited loss-free line

Since the output terminals of the line are open-circuited, no current can flow between them. This means that *all* of the incident current must be *reflected* at the open circuit. The total current at the open-circuit is the phasor sum of the incident and reflected currents, and since this must be zero the current must be reflected with 180° phase shift. The incident voltage is also totally reflected at the open circuit but with zero phase shift. The total voltage across the open-circuited terminals is twice the voltage that would exist if the line were correctly terminated. The reflected current and voltage waves propagate along the line towards its sending end, being phase-shifted as they go. When the reflected waves reach the sending end, they are completely absorbed by the impedance of the matched source.

At any point along the line, *the total current and voltage is the phasor sum of the incident and reflected currents and voltages.* Consider Fig. 10.2. At the open circuit the phasors representing the incident and reflected currents are of equal length (since *all* the incident current is reflected) and point in opposite directions. The current flowing in the open circuit is the sum of these two phasors and is therefore zero as expected. At a distance of $\lambda/8$ from the open circuit, the incident current phasor is 45° leading, and the reflected current phasor is 45° lagging on the open-circuit phasors. The lengths of the two phasors are equal since the line loss is zero but they are 90° out of phase with one another. The total current at this point is $\sqrt{2}$ times the incident current. Moving a further $\lambda/8$ along the line, the incident and reflected current phasors have rotated, in opposite directions, through another 45° and are now in phase with one another. The total current $\lambda/4$ from the open circuit is equal to twice the incident current. A further $\lambda/8$ along the line finds the two phasors once again at right angles to one another so that the total line current is again $\sqrt{2}$ incident current. At a point $\lambda/2$ from the end of the line, the incident and reflected current phasors are in antiphase with one another and the total line current is zero. Over the next half-wavelength of line the phasors continue to rotate in opposite directions, by 45° in each $\lambda/8$ distance, and the total line current is again determined by their phasor sum.

It is usual to consider the r.m.s. values of the total line current and then its phase need not be considered. The way in which the r.m.s. line current varies with distance from the open circuit is shown by Fig. 10.2*c*. The points at which maxima (*antinodes*) and minima (*nodes*) of current occur are always the same and do not vary with time. Because of this the waveform of Fig. 10.2 is said to be a **standing wave**.

If, now, the voltages existing on the line of Fig. 10.1 are considered, the phasors shown in Fig. 10.3 are obtained. At the open-circuited output terminals, the incident and reflected voltage phasors are in phase and the total voltage is twice the incident voltage. Moving from the open circuit towards the sending end of the line, the phasors rotate through an angle of 45° in each $\lambda/8$ length of line; the incident voltage phasors rotate in the anticlockwise direction and the reflected voltage phasors rotate clockwise. The total voltage at any point along the line is the phasor sum of the incident and reflected voltages and its r.m.s. value varies in the manner shown in Fig. 10.3.

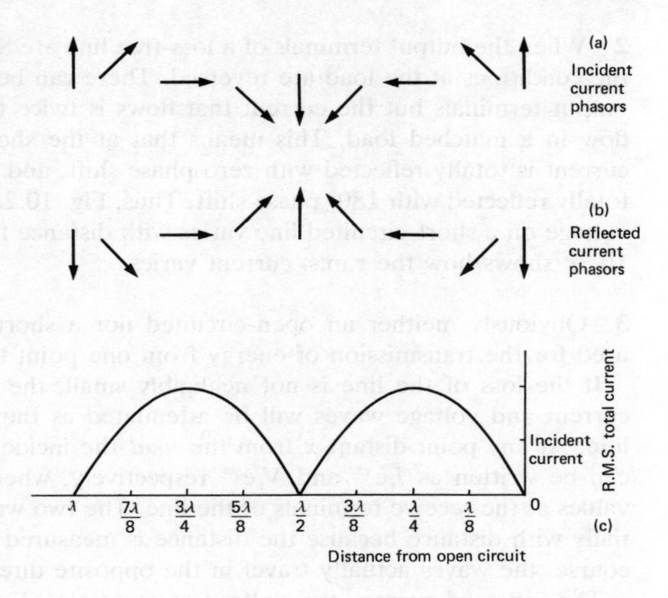

Fig. 10.2 R.M.S. current on an open-circuited loss-free line

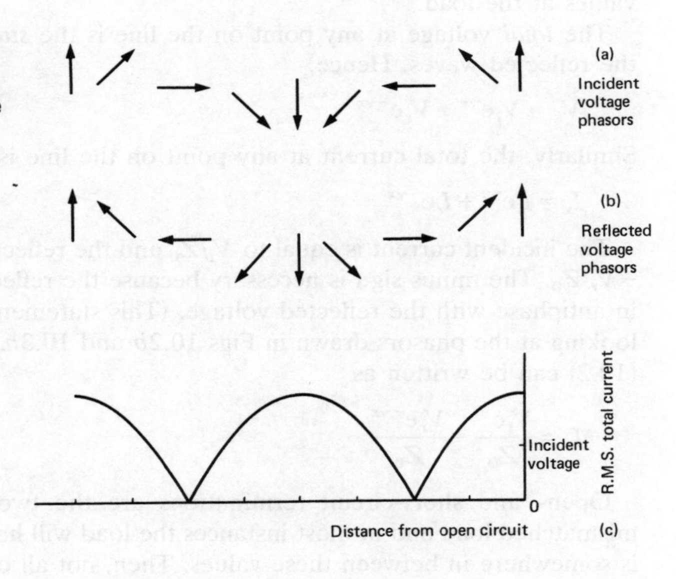

Fig. 10.3 R.M.S. voltage on an open-circuited loss-free line

Two things should be noted from Figs. 10.2 and Fig. 10.3. Firstly, the *voltage standing-wave* pattern is displaced by λ/4 from the current standing wave pattern, i.e. a current antinode occurs at the same place as a voltage node and vice versa. Secondly, the current and voltage values at the open-circuit are repeated at λ/2 intervals along the length of the line; this remains true for any longer length of loss-free line.

2 When the output terminals of a loss-free line are SHORT-CIRCUITED, the conditions at the load are reversed. There can be no voltage across the output terminals but the current that flows is twice the current that would flow in a matched load. This means that at the short circuit the incident current is totally reflected with zero phase shift, and the incident voltage is totally reflected with 180° phase shift. Thus, Fig. 10.2*c* shows how the r.m.s. voltage on a short-circuited line varies with distance from the load, and Fig. 10.3*c* shows how the r.m.s. current varies.

3 Obviously, neither an open-circuited nor a short-circuited line can be used for the transmission of energy from one point to another.

If the loss of the line is not negligibly small, the incident and reflected current and voltage waves will be attenuated as they propagate along the line. At any point distant x from the *load* the incident current and voltage can be written as $I_i e^{\gamma x}$ and $V_i e^{\gamma x}$ respectively, where I_i and V_i are their values at the receive terminals of the line. The two waves *increase* exponentially with distance because the distance is measured from the load and, of course, the waves actually travel in the opposite direction.

The reflected current and voltage at any point distance x from the load can be written as $I_r e^{-\gamma x}$ and $V_r e^{-\gamma x}$ respectively, where I_r and V_r are their values at the load.

The *total* voltage at any point on the line is the *sum* of the incident and the reflected waves. Hence,

$$V_x = V_i e^{\gamma x} + V_r e^{-\gamma x} \tag{10.1}$$

Similarly, the total current at any point on the line is given by

$$I_x = I_i e^{\gamma x} + I_r e^{-\gamma x} \tag{10.2}$$

The incident current is equal to V_i/Z_0 and the reflected current is equal to $-V_r/Z_0$. The minus sign is necessary because the reflected current is *always* in antiphase with the reflected voltage. (This statement can be confirmed by looking at the phasors drawn in Figs 10.2*b* and 10.3*b*.) Therefore, equation (10.2) can be written as

$$I_x = \frac{V_i e^{\gamma x}}{Z_0} - \frac{V_r e^{-\gamma x}}{Z_0} \tag{10.3}$$

Open- and short-circuit terminations are the two extreme cases of a mismatched load and in most instances the load will have an impedance that is somewhere in between these values. Then, not all of the incident current and voltage will be reflected but only some fractional value. The fraction of the incident current and voltage reflected by the mismatched load is determined by the *reflection coefficient* of the load.

Voltage and Current Reflection Coefficients

The **voltage reflection coefficient** ρ_v of a mismatched load is the ratio

$$\rho_v = \frac{\text{reflected voltage}}{\text{incident voltage}} \tag{10.4}$$

and the **current reflection coefficient** ρ_i is the ratio

$$\rho_i = \frac{\text{reflected current}}{\text{incident current}} \tag{10.4a}$$

Always $\rho_i = -\rho_v$

At the load $x = 0$ so that

$$V_R = V_i + V_r \quad \text{and} \quad I_R = \frac{V_i}{Z_0} - \frac{V_r}{Z_0}$$

The load impedance Z_L is equal to the ratio $\frac{\text{load voltage}}{\text{load current}}$, i.e.

$$Z_L = V_R/I_R = \frac{V_i + V_r}{(V_i/Z_0) - (V_r/Z_0)} = \frac{[1 + (V_r/V_i)]Z_0}{1 - (V_r/V_i)}$$

i.e. $$\frac{Z_L}{Z_0} = \frac{1 + \rho_v}{1 - \rho_v}$$

On re-arranging

$$\rho_v = \frac{Z_L - Z_0}{Z_L + Z_0} \tag{10.5}$$

Hence $$\rho_i = \frac{Z_0 - Z_L}{Z_0 + Z_L} \tag{10.6}$$

It should be noted that these equations give $\rho_v = +1$ and $\rho_i = -1$ for an open-circuited line and $\rho_v = -1$, $\rho_i = +1$ for a short-circuited line.

The expression for the current reflection coefficient can also be derived by applying Thevenin's theorem to the end of the line.

Assume that the line is matched at its sending end terminals so that the impedance measured at the open-circuited output terminals is the characteristic impedance of the line. Then, from Fig. 10.4*a*, the current I_L flowing in the mismatched load Z_L is $V_{oc}/(Z_0 + Z_L)$. If the load impedance were equal to the characteristic impedance of the line, the incident current would flow in it. Hence, from Fig. 10.4*b*,

$$I_i = V_{oc}/2Z_0$$

The load current is the phasor sum of the incident and the reflected currents at the load. i.e. $I_L = I_i + I_r$. Hence,

$$I_r = I_L - I_i = \frac{V_{oc}}{Z_0 + Z_L} - \frac{V_{oc}}{2Z_0} = \frac{V_{oc}(Z_0 - Z_L)}{2Z_0(Z_0 + Z_L)}$$

Therefore,

$$\rho_i = I_r/I_i = \frac{Z_0 - Z_L}{Z_0 + Z_L} \tag{10.6 (again)}$$

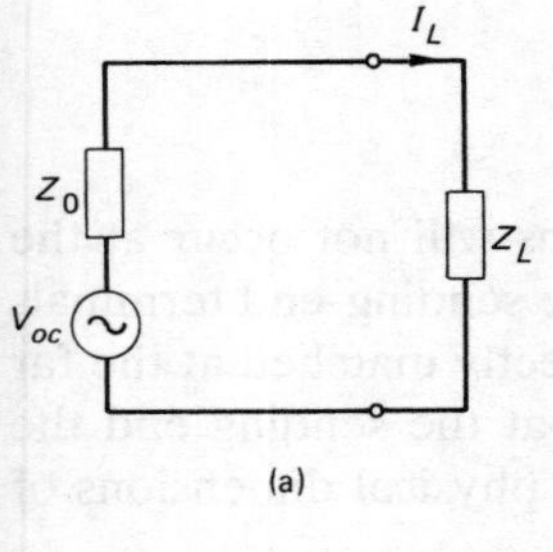

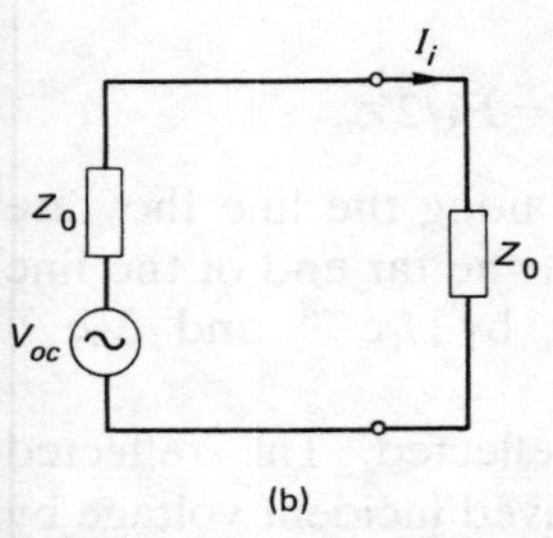

Fig. 10.4 Calculation of current reflection coefficient

Example 10.1

A line has a characteristic impedance of $600\angle{-20°}\ \Omega$ and is terminated in a load of $300\angle{0°}\ \Omega$. Calculate its current and voltage reflection coefficients.

Solution

$600\angle{-20°} = 563.8 - j205.2$ Therefore,

$$\rho_v = \frac{300 - 563.8 + j205.2}{300 + 563.8 - j205.2} = \frac{-263.8 + j205.2}{863.8 - j205.2}$$

$$= 0.38\angle{155.5°} \quad (Ans)$$

Voltage and Current on a Mismatched Line

With dealing with mismatched line calculations it is generally helpful to draw a sketch of the line and mark it with the currents and voltages determined at relevant points.

Consider Fig. 10.5 which shows a line of length l having secondary coefficients Z_0 and γ, terminated in a load impedance that is not equal to Z_0.

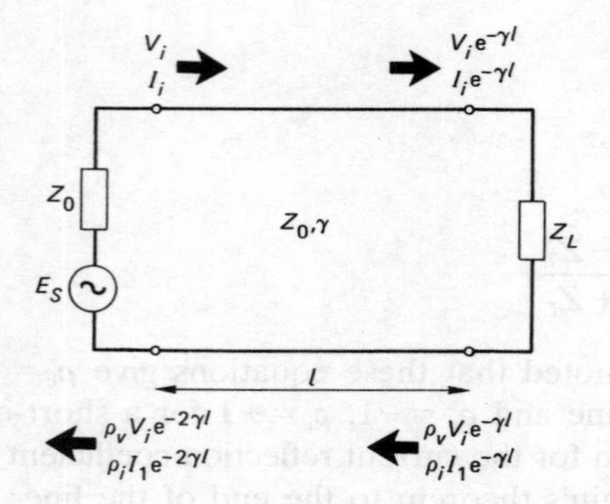

Fig. 10.5 Currents and voltages on a mismatched line

The line is matched to its source so that reflections will not occur at the sending end of the line. The incident voltage V_i at the sending-end terminals is the voltage that would appear if the line were correctly matched at the far end. (Since there are initially no reflections present at the sending end the input impedance is initially determined solely by the physical dimensions of the line, i.e. it is Z_0.) Therefore,

$$V_i = E_S Z_0 / 2Z_0 = E_S/2 \quad \text{and} \quad \text{incident current } I_i = E_S/2Z_0$$

As the incident current and voltage waves travel along the line they are each subjected to both attenuation and phase lag. At the far end of the line the incident current and voltage waves are given by $I_i e^{-\gamma l}$ and $V_i e^{-\gamma l}$ respectively.

At the load $Z_L \neq Z_0$ and so both waves are reflected. The reflected voltage wave is found by merely multiplying the received incident voltage by the voltage reflection coefficient (see equation 10.4). Hence the reflected voltage at the load is equal to $\rho_v V_i e^{-\gamma l}$. Similarly, the reflected current at the load is given by $\rho_i I_i e^{-\gamma l}$.

The reflected current and voltage waves now travel along the line towards the sending end and are subject to exactly the same attenuation and phase shift as the incident waves. At the sending end of the line the reflected current is $\rho_i I_i e^{-2\gamma l}$ and the reflected voltage is $\rho_v V_i e^{-2\gamma l}$.

The total voltage at any point on the line is the *phasor sum* of the incident and reflected voltages at that point. Hence

Sending-end voltage

$$V_S = V_i + \rho_v V_i e^{-2\gamma l} = V_i(1+\rho_v e^{-2\gamma l}) \tag{10.7}$$

Load voltage

$$V_R = V_i e^{-\gamma l} + \rho_v V_i e^{-\gamma l} = V_i e^{-\gamma l}(1+\rho_v) \tag{10.8}$$

Voltage at distance x from the sending end:

$$V_x = V_i e^{-\gamma x} + \rho_v V_i e^{-\gamma(2l-x)} \tag{10.9}$$

Similar expressions apply for the total current at any point; it is merely necessary to replace V_i and ρ_v with I_i and ρ_i respectively.

Example 10.2

At a particular frequency a transmission line is $\lambda/2$ long and has an attenuation of 3 dB. The characteristic impedance of the line is $600\angle 0°$ ohms. A voltage source of e.m.f. 60 V and $600\angle 0°$ ohms impedance is connected across the input terminals of the line. Calculate *a*) the voltage across the 300 ohm load, *b*) the voltage across the sending-end terminals.

Solution

a) *Method A* 3 dB = 0.345 nepers $= \alpha l$

$$\rho_v = \frac{300-600}{300+600} = -1/3$$

$V_i = 60/2 = 30$ V. Substituting into equation (10.8),

$$V_R = 30e^{-0.345}(1-\tfrac{1}{3}) = 14.16 \text{ V} \quad (Ans)$$

Method B 3 dB is a voltage ratio of $1/\sqrt{2}$ and hence, referring to Fig. 10.6,

$$V_R = \frac{30}{\sqrt{2}} - \frac{30}{3\sqrt{2}} = 14.14 \text{ V} \quad (Ans)$$

The slight difference in the results arises because 3 dB is not exactly equal to 0.345 N and neither is it exactly equal to $1/\sqrt{2}$.

Fig. 10.6

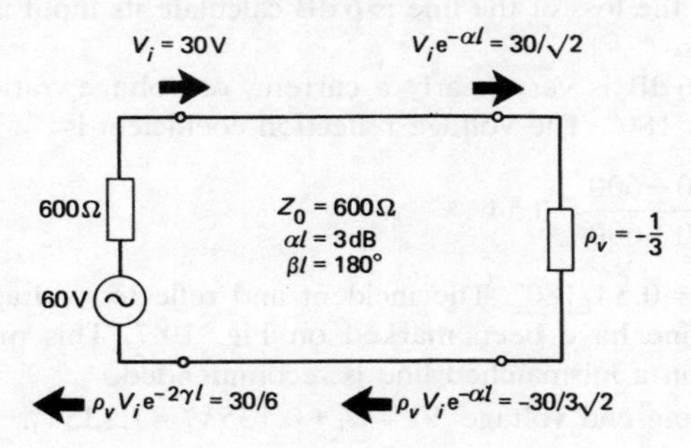

b) *Method A* Substituting into equation (10.7),

$$V_S = 30(1-\tfrac{1}{3}e^{-0.69}) = 25\text{ V} \quad (Ans)$$

Method B From Fig. 10.6,

$$V_S = 30 - \frac{30}{6} = 25\text{ V} \quad (Ans)$$

Input Impedance of a Mismatched Line

The impedance at any point on a line is the ratio of total voltage to total current at that point. Thus the **input impedance** of a line is the ratio V_S/I_S. Therefore,

$$Z_S = V_S/I_S = \frac{V_i(1+\rho_v e^{-2\gamma l})}{I_i(1+\rho_i e^{-2\gamma l})}$$

$$Z_S = \frac{Z_0(1+\rho_v e^{-2\gamma l})}{1+\rho_i e^{-2\gamma l}} = \frac{Z_0(1+\rho_v e^{-2\gamma l})}{1-\rho_v e^{-2\gamma l}} \qquad (10.10)$$

since $\rho_i = -\rho_v$.

Equation (10.10) is often written in another form. Re-writing by substituting for ρ_v

$$Z_S = Z_0\left[\frac{Z_L+Z_0+(Z_L-Z_0)e^{-2\gamma l}}{Z_L+Z_0-(Z_L-Z_0)e^{-2\gamma l}}\right]$$

$$= Z_0\left[\frac{(Z_L+Z_0)e^{\gamma l}+(Z_L-Z_0)e^{-\gamma l}}{(Z_L+Z_0)e^{\gamma l}-(Z_L-Z_0)e^{-\gamma l}}\right]$$

$$= Z_0\left[\frac{Z_L(e^{\gamma l}+e^{-\gamma l})+Z_0(e^{\gamma l}-e^{-\gamma l})}{Z_0(e^{\gamma l}+e^{-\gamma l})+Z_L(e^{\gamma l}-e^{-\gamma l})}\right]$$

$$Z_S = Z_0\frac{Z_L\cosh\gamma l+Z_0\sinh\gamma l}{Z_0\cosh\gamma l+Z_L\sinh\gamma l} \qquad (10.11)$$

Generally, equation (10.11) is not as convenient to use as equation (10.10) but, in any case, the input impedance can always be found from basic principles.

Example 10.3

A line has a characteristic impedance of 600 Ω and is λ/2 long at a particular frequency. If the loss of the line is 6 dB calculate its input impedance when the load is 2000 ohms.

Solution 6 dB is very nearly a current, or voltage, ratio of 2:1 and the overall phase shift is 180°. The voltage reflection coefficient is

$$\rho_v = \frac{2000-600}{2000+600} = 0.54$$

Therefore $\rho_i = 0.54\underline{/180^\circ}$. The incident and reflected voltages and currents at each end of the line have been marked on Fig. 10.7. This method of laying out the calculations on a mismatched line is recommended.

Total sending-end voltage $V_S = V_i + 0.135V_i = 1.135V_i$.

Fig. 10.7

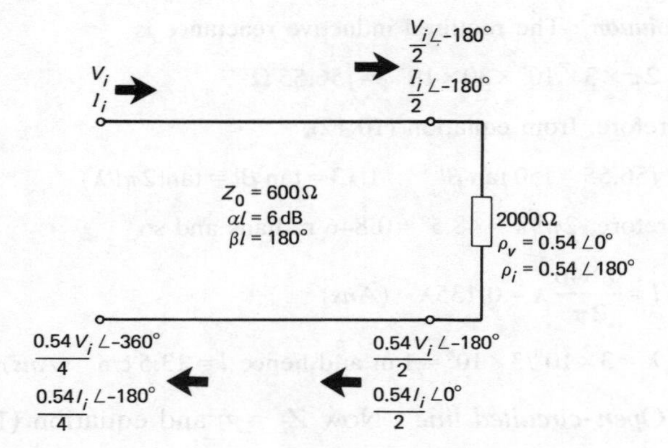

Total sending-end current $I_S = I_i - 0.135I_i = 0.865I_i$. Therefore, the input impedance of the line is

$$Z_S = V_S/I_S = \frac{1.135\,V_i}{0.865I_i} = 1.312 \times 600 = 787\ \Omega \quad (Ans)$$

Low-loss Lines

A line whose attenuation is very small is often said to be a low-loss or even a loss-free line. When the loss of a line is small enough to be neglected

$$\alpha l = 0 \quad \text{and} \quad \gamma l = j\beta l$$

Then $\cosh \gamma l = \cosh j\beta l = \cos \beta l$

$$\sinh \gamma l = \sinh j\beta l = j \sin \beta l$$

The use of these approximations considerably simplifies the application of equation (10.11).

There are several special cases of importance.

1 *Short-circuited line* If the far-end terminals of a line are short-circuited, $Z_L = 0$. Equation (10.11) then becomes

$$Z_S = Z_0 \frac{Z_0 \sinh \gamma l}{Z_0 \cosh \gamma l} = Z_0 \frac{jZ_0 \sin \beta l}{Z_0 \cos \beta l}$$

$$Z_S = jZ_0 \tan \beta l \tag{10.12}$$

This means that the input impedance of a loss-free line short-circuited at its output terminals is a pure reactance whose magnitude and sign depend upon the electrical length of the line. Such a line can therefore be used to simulate either an inductor or a capacitor or, if a resonant length ($\lambda/2$ or $\lambda/4$) is employed, a series- or parallel-tuned circuit.

Example 10.4

A length of 50 Ω low-loss short-circuited line is to be used to simulate an inductance of 30 nH at a frequency of 300 MHz. Calculate the necessary length of the line.

Solution The required inductive reactance is

$$2\pi \times 3 \times 10^8 \times 30 \times 10^{-9} = j56.55\ \Omega$$

Therefore, from equation (10.12),

$$j56.55 = j50 \tan \beta l \qquad 1.13 = \tan \beta l = \tan(2\pi l/\lambda)$$

Therefore, $2\pi l/\lambda = 48.5° = 0.846$ radians and so

$$l = \frac{0.846}{2\pi}\lambda = 0.135\lambda \quad (\textit{Ans})$$

But $\lambda = 3 \times 10^8/3 \times 10^8 = 1$ m and hence $l = 13.5$ cm (*Ans*)

2 *Open-circuited line* Now $Z_L = \infty$ and equation (10.11) becomes

$$Z_S = -jZ_0 \cot \beta l \tag{10.13}$$

The open-circuited line is not often used to simulate a component because at very high frequencies an open-circuited line will tend to radiate energy.

3 *λ/4 length of line* When the length of a loss-free line is exactly λ/4, $\gamma l = j\beta = j\pi/2$. Hence

$$\cosh \gamma l = \cos \beta l = \cos \pi/2 = 0$$

$$\sinh \gamma l = j \sin \beta l = j \sin \pi/2 = j$$

Substituting these values into equation (10.11) gives

$$Z_S = Z_0 \frac{jZ_0}{jZ_L} = Z_0^2/Z_L \tag{10.14}$$

This result can just as easily be obtained from equation (10.10). Thus

$$Z_S = Z_0 \frac{1 + \rho_v e^{-2j\beta l}}{1 - \rho_v e^{-2j\beta l}} = Z_0 \left[\frac{1 + \rho_v \angle 180°}{1 - \rho_v \angle 180°}\right]$$

$$= Z_0 \frac{1 - \dfrac{Z_L - Z_0}{Z_L + Z_0}}{1 + \dfrac{Z_L - Z_0}{Z_L + Z_0}} = Z_0 \left(\frac{Z_0}{Z_L}\right)$$

or $Z_S = Z_0^2/Z_L$ (10.15) (10.14 again)

This result means that a λ/4 section of loss-free line can be employed to match two impedances together. The necessary characteristic impedance for the λ/4 matching section can be obtained from equation (10.15). Rearranging

$$Z_0 = \sqrt{(Z_S Z_L)} \tag{10.16}$$

The device is often known as a λ/4 *transformer.*

Example 10.5

A low-loss transmission line of characteristic impedance 600 ohms is to be used to connect a 600 Ω source to a 300 Ω load. Calculate the required characteristic impedance for the λ/4 line matching section.

Solution From equation (10.16)

$$Z_0 = \sqrt{(600 \times 300)} = 424\ \Omega \quad (Ans)$$

The arrangement is shown by Fig. 10.8.

Fig. 10.8

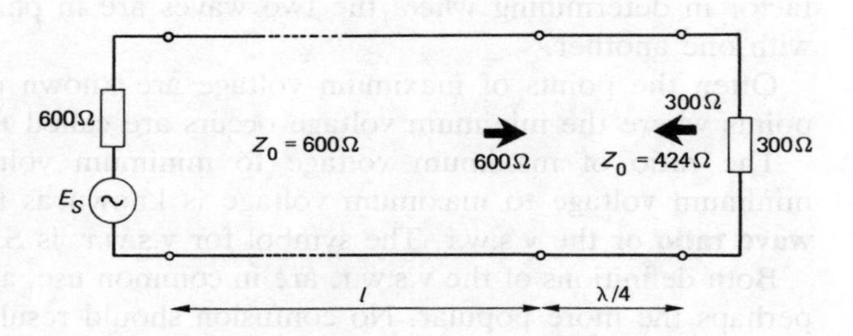

4 *λ/2 length of line* For this length of loss-free line

$$\cosh \gamma l = \cos \beta l = \cos\pi - 1 \qquad \sinh \gamma l = j \sin \beta l = \ j \sin \pi = \ 0$$

$$\text{so} \quad Z_S = Z_0(-Z_L/-Z_0) = Z_L \tag{10.17}$$

Thus, the input impedance of a λ/2 length of loss-free line is equal to the load impedance. This means that the line will act like a transformer with a 1:1 turns ratio.

5 *λ/8 length of line* A λ/8 length of loss-free line introduces a phase lag of 45° and hence $\cos \beta l = \sin \beta l = 1/\sqrt{2}$ and, from equation (10.11)

$$Z_S = Z_0\left[\frac{(Z_L/\sqrt{2}) + (jZ_0/\sqrt{2})}{(Z_0/\sqrt{2}) + (jZ_L/\sqrt{2})}\right]$$

$$\underline{Z_S = Z_0/\tan^{-1}[(Z_0/Z_L) - (Z_L/Z_0)]} \tag{10.18}$$

Thus the input impedance of a λ/8 length of loss-free line is equal to the characteristic impedance of the λ/8 section.

Standing Waves

When reflections occur at the mismatched load of a transmission line, both incident and reflected current and voltage waves propagate along the line. At any point on the line the total voltage or current is the phasor sum of the incident and reflected waves at that point.

If the r.m.s. values of the total voltage and current are plotted to a base of distance from the load, **standing waves** are obtained. Examples of these have already been plotted for low-loss open- and short-circuited lines (see Figs. 10.2 and 10.3).

When the voltage and current reflection coefficients have a magnitude other than unity, not all of the incident energy is reflected by the load. The maximum voltage V_{max} on the line occurs each time the incident and reflected voltages are *in phase* with one another. The minimum voltage V_{min} occurs when the incident and reflected voltages are in anti-phase with one another. Therefore, for a low-loss line

$$V_{max} = V_i(1+|\rho_v|) \qquad V_{min} = V_i(1-|\rho_v|) \tag{10.19}$$

Only the magnitude of ρ_v appears in equation (10.19) because its angle is a factor in determining *where* the two waves are in phase, or in anti-phase, with one another.

Often the points of maximum voltage are known as *antinodes* and the points where the minimum voltage occurs are called *nodes.*

The ratio of maximum voltage to minimum voltage *or* the ratio of minimum voltage to maximum voltage is known as the **voltage standing-wave ratio** or the v.s.w.r. The symbol for v.s.w.r. is *S.*

Both definitions of the v.s.w.r. are in common use, although the former is perhaps the more popular. No confusion should result since the v.s.w.r. is always either greater than, or less than, unity depending upon which of the two definitions is employed. Consider as an example a loss-free line having a characteristic impedance of 50 Ω that is terminated by a resistance of 150 Ω. Suppose a voltage source at the sending end of the line supplies an incident voltage of 10 V.

The voltage reflection coefficient is

$$\rho_v = \frac{150-50}{150+50} = 0.5\angle 0^\circ$$

The incident voltage is reflected by the load with zero phase change and so the maximum voltage occurs at the load, and at $\lambda/2$ intervals from the load. The maximum voltage is $10(1+0.5) = 15$ V. The minimum voltage occurs at *odd* multiples of $\lambda/4$ from the load and is equal to $10(1-0.5) = 5$ V.

The voltage standing-wave pattern for this line is shown in Fig. 10.9. The incident and reflected currents are $10/50 = 0.2$ A and $0.2 \times 0.5 = 0.1$ A respectively, Hence, $I_{max} = 300$ mA and $I_{min} = 100$ mA. The maximum current occurs at the same points as the minimum voltage and vice versa; this is because $\rho_i = -\rho_v$.

The impedance at any point on the line is the ratio of total voltage to total current at that point. Clearly, this means that the impedance of the line will vary both above and below the characteristic impedance at different points on the line.

At the load

$$Z_{max} = V_{max}/I_{min} = 15/0.1 = 150\ \Omega$$

$\lambda/4$ from the load

$$Z_{min} = V_{min}/I_{max} = 5/0.3 = 16.67\ \Omega$$

Note that from equation (10.14) $Z_{min} = 50^2/150 = 16.67\ \Omega$

It should be evident from Fig. 10.9 that the maximum and minimum values of the line impedance repeat at $\lambda/2$ intervals along the line.

$$\text{v.s.w.r.} = S = V_{max}/V_{min} = 15/5 = 3$$

Fig. 10.9

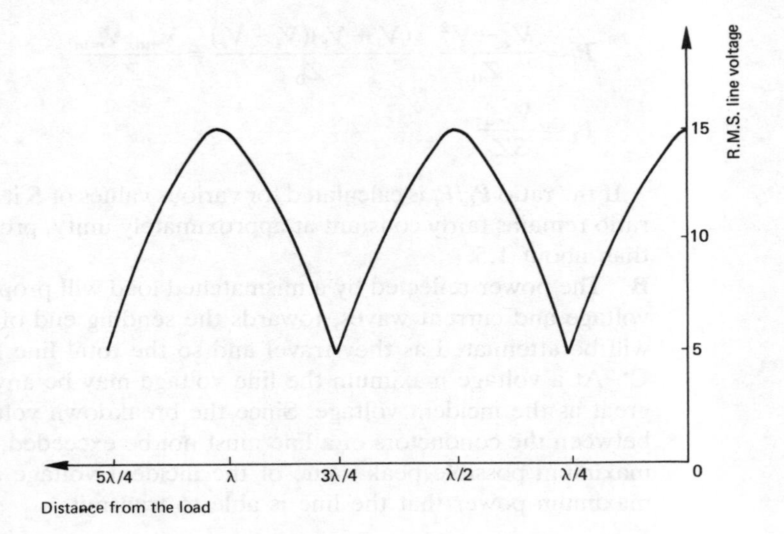

Relationship between V.S.W.R. and Voltage Reflection Coefficient

$$\text{v.s.w.r.} = S = \frac{V_{max}}{V_{min}} = \frac{V_i(1+|\rho_v|)}{V_i(1-|\rho_v|)} = \frac{1+|\rho_v|}{1-|\rho_v|} \quad (10.20)$$

$$|\rho_v| = \frac{S-1}{S+1} \quad (10.21)$$

Example 10.6

A low-loss line of characteristic impedance 1000 Ω is terminated in a load of $800 - j100\ \Omega$. Calculate the v.s.w.r. on the line.

Solution

$$\rho_v = \frac{(800-j100)-1000}{(800-j100)+1000} = \frac{-200-j100}{1800-j100}$$

or $|\rho_v| = 0.124$ Therefore,

$$S = \frac{1+0.124}{1-0.124} = 1.283 \quad (Ans)$$

The presence of a standing wave on a line which is used to transmit energy from one point to another is undesirable for a number of reasons. These reasons are as follows:

A Maximum power is transferred from a source to a load when the load impedance is equal to the characteristic impedance. When a load mismatch exists some of the incident power is reflected at the load and so the transfer efficiency is reduced.

The power in the incident wave is V_i^2/Z_0 and the power reflected by the load is V_r^2/Z_0. Therefore, the power P_L dissipated in the load is

$$P_L = \frac{V_i^2 - V_r^2}{Z_0} = \frac{(V_i + V_r)(V_i - V_r)}{Z_0} = \frac{V_{max}V_{min}}{Z_0}$$

$$P_L = \frac{V_{max}^2}{SZ_0} \tag{10.22}$$

If the ratio P_L/P_i is calculated for various values of S it will be found that the ratio remains fairly constant at approximately unity, provided S is not higher than about 1.5.

B The power reflected by a mismatched load will propagate, in the form of voltage and current waves, towards the sending end of the line. The waves will be attenuated as they travel and so the total line loss is increased.

C At a voltage maximum the line voltage may be anything up to twice as great as the incident voltage. Since the breakdown voltage of the dielectric between the conductors of a line must not be exceeded, this factor limits the maximum possible peak value of the incident voltage and hence limits the maximum power that the line is able to transmit.

Measurement of V.S.W.R.

The v.s.w.r. on a mismatched line can be determined by measuring the maximum and minimum voltages on the line. Often the measurement is carried out using an instrument known as a *standing-wave indicator.* The measurement of v.s.w.r. not only shows up the presence of any reflections on a line but it also offers a way of determining the value of an unknown load impedance.

The measurement procedure is briefly as follows. The v.s.w.r. is measured and the distance in wavelengths from the load to the nearest voltage minimum is found. Referring to Fig. 10.10, suppose that the v.s.w.r. on the line is S.

Then $|\rho_v| = (S-1)/(S+1)$.

Suppose that the first voltage minimum is at a distance of x from the load. For this point to be a voltage minimum the incident and reflected phasors

Fig. 10.10

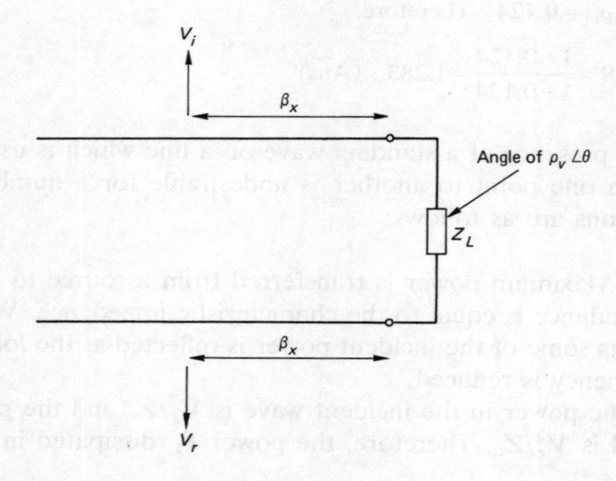

must be in *anti-phase* with one another. This means that the total phase lag between V_i and V_r, i.e. $(\beta x + \theta + \beta x)$, must be equal to π radians. Hence,

$$\theta = \pi - 2\beta x \qquad (10.23)$$

Once the value of $\rho_v\underline{/\theta}$ has been found it must be substituted into equation (10.5) in order to calculate the value of Z_L.

Example 10.7

In a measurement of the v.s.w.r. on a loss-free line it was found that the ratio V_{max}/V_{min} was 5 and the first voltage minimum from the load occurred 20 cm from the load. If the characteristic impedance of the line is 50 Ω and the frequency of the test oscillator is 300 MHz calculate the magnitude and angle of the load impedance.

Solution From equation (10.21)

$|\rho_v| = 4/6 = 0.67 \qquad \lambda = 3\times 10^8/3\times 10^8 = 1\text{m}$

Hence 20 cm $= 0.2\lambda$ and $\beta x = 0.4\pi$

From equation (10.23) $\underline{/\theta} = \pi - 0.8\pi = 0.2\pi = 36°$ *lag*

Therefore, from equation (10.5)

$$0.67\underline{/-36°} = (Z_L - 50)/(Z_L + 50)$$

$$Z_L = 75 - j108\,\Omega \quad (\textit{Ans})$$

The calculation of Z_L using equation (10.5) is rather lengthy and tedious and it is usual to employ a graphical aid, known as a **Smith Chart,** to greatly simplify the problem. If, however, the load impedance is known to be purely resistive an easier method becomes available. When $R_L > R_0$ and $S > 1$

$$\rho_v = \frac{R_L - R_0}{R_L + R_0}$$

so that $$S = \frac{1 + \dfrac{R_L - R_0}{R_L + R_0}}{1 - \dfrac{R_L - R_0}{R_L + R_0}} = R_L/R_0$$

Hence $$R_L = SR_0 \qquad (10.24)$$

When $R_L < R_0$,

$$S = \frac{1 - \dfrac{R_L - R_0}{R_0 + R_L}}{1 + \dfrac{R_L - R_0}{R_L + R_0}} = \frac{R_0}{R_L}$$

Hence $$R_L = R_0/S \qquad (10.25)$$

Impedance Matching on Lines

In nearly all cases when a transmission line is used to convey energy from one point to another, the requirement is for the load to be matched, or nearly matched, to the line. Whenever possible the load impedance is selected to satisfy this requirement. Very often, however, this is not possible and in such instances some kind of matching device is used.

The use of a loss-free line of $\lambda/4$ length as an impedance matching device has already been discussed. Here it will suffice to state that more elaborate arrangements than a single $\lambda/4$ section are practical and are commonly employed.

An alternative method of eliminating mismatch from a line is known as **stub matching**. The impedance and hence the admittance of a mismatched line varies as the distance from the load is increased. At some particular distance x from the load the admittance of the line will be equal to the *characteristic admittance* ($Y_0 = 1/Z_0$) in parallel with some value of susceptance. Thus. at this particular point in the line $Y_{in} = Y_0 \pm jB$. If the susceptance can be cancelled out by another susceptance of equal magnitude but opposite sign, the resulting total admittance would be equal to Y_0. The line would then be matched at this point.

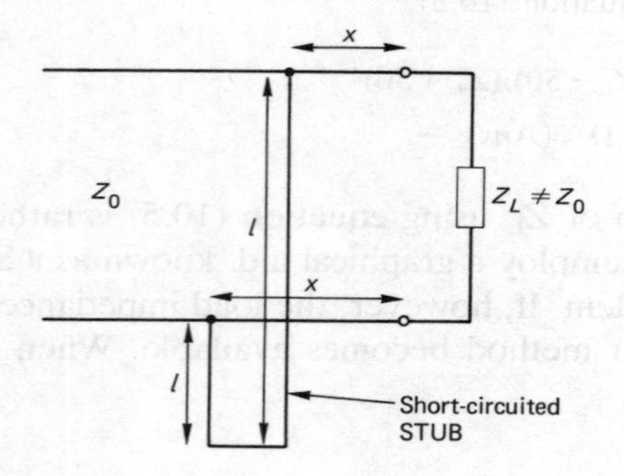

Fig. 10.11 Stub matching

The necessary shunt admittance is provided by a short-circuited line of length l—known as a *stub*—connected in parallel with the line at a distance x from the load (see Fig. 10.11). The required length l for the stub is calculated using equation (10.12) but the determination of the distance x is beyond the scope of this book. The design of a stub matching system is another calculation that is best performed using a Smith Chart.

Exercises 10

10.1 A loss-free line is short-circuited at its output terminals and is to be used to provide a reactance of $-j160\ \Omega$ at a frequency of 120 MHz. If the characteristic impedance of the line is 300 Ω calculate the length of line needed.

10.2 A loss-free transmission line has a characteristic impedance of 300 Ω and a load impedance of $(200 - j300)\ \Omega$. Calculate the v.s.w.r. on the line and the impedance of the line at a distance of $\lambda/2$ from the load.

10.3 A loss-free line has a characteristic impedance of 450 Ω and a load impedance of $(150+j120)$ Ω. Calculate the voltage reflection coefficient and the v.s.w.r.

10.4 A transmission line has a characteristic impedance of 600 Ω, 6 dB loss, and is 1.3 wavelengths long. If the line is terminated in an impedance of $(400-j350)$ Ω calculate its input impedance.

10.5 A loss-free line has a characteristic impedance of 600 Ω. When the load impedance is 200 Ω the voltage across the load is 12 V. Calculate *a*) the voltage reflection coefficient, *b*) the v.s.w.r., *c*) the voltage at a distance of $\lambda/4$ from the load, and *d*) the voltage at a distance of $\lambda/2$ from the load,

10.6 A transmission line has a characteristic impedance of 1000 Ω, 6 dB loss and is $\lambda/2$ long at a particular frequency. Calculate the input impedance of the line when the far-end terminals are short-circuited.

10.7 A transmission line has a characteristic impedance of 300 Ω, negligible loss, is $\lambda/4$ long and is terminated by a load impedance of $(300-j100)$Ω. Calculate the input impedance of the line.

10.8 A 2λ length of transmission line has a characteristic impedance of 50 Ω and is connected to a source of e.m.f. 12 V and internal impedance 50 Ω. If the load impedance is 100 Ω calculate *a*) the voltage across the load, *b*) the voltage across the input terminals if the line loss is 3 dB.

10.9 A transmission line has a characteristic impedance of 600 Ω and is correctly terminated at its far end. A resistor of 300 Ω is connected across the line $\lambda/2$ from the load. Calculate the voltage reflection coefficient at this point. If the line is 3λ long and its losses are negligible calculate the input impedance of the line.

10.10 Derive expressions for the input impedance of *a*) an open-circuited line that is $\lambda/4$ long, *b*) a short-circuited line that is $\lambda/2$ long. A line has a loss of 3 dB per $\lambda/4$ length and a characteristic impedance of 75 Ω. Calculate its input impedance when the line is (i) $\lambda/4$, (ii) $\lambda/2$ long and is short-circuited at its output terminals.

10.11 A transmission line is 10 km long and has a propagation coefficient of $(0.11+j0.157)$ per km and a characteristic impedance of 900 Ω. The source has an e.m.f. of 50 V and negligible internal impedance and the load has an impedance of 600 Ω. Calculate the values of the load current and voltage.

10.12 A transmission line is $\lambda/4$ long and has an attenuation of 6 dB and a characteristic impedance of 100 Ω. Calculate the total voltage across the sending-end terminals when the load impedance is $75\angle 90°$ Ω and the voltage across the load is $250\angle 0°$ V.

Short Exercises

10.13 A loss-free line has a characteristic impedance of 600 Ω and is terminated in a load of 1800 Ω. Sketch the current and voltage standing-wave patterns on the line.

10.14 A loss-free line has a characteristic impedance of 500 Ω and is terminated in a pure reactance of $+j250$ Ω. Sketch the current and voltage standing patterns on the line.

10.15 A loss-free line has a characteristic impedance of 300 Ω and is terminated in a load of 450 Ω. Draw phasor diagrams to show the incident, reflected and total currents and voltages at the load.

10.16 Explain how a length of coaxial cable could be used to act as a series tuned circuit. Why is this technique only available at very high frequencies?

10.17 An impedance of 1200 Ω is to be used as the load for a line whose characteristic impedance is 400 Ω. To avoid reflections a λ/4 matching section is to be employed. Calculate the required characteristic impedance of the section and sketch the arrangement.

10.18 List the reasons why it is desirable that a line used for transmitting signals from one point to another should have a matched load.

10.19 Derive equations (10.17) and (10.18) from equation (10.10).

10.20 A loss-free line has a characteristic impedance of 75 Ω and is terminated in a load of 150 Ω. What is the impedance of the line at distances of *a*) λ/8, *b*) λ/4 and *c*) λ/2 from the load?

10.21 A line having a characteristic impedance of 400 Ω is mismatched and as a result the v.s.w.r. on the line is 1.6. Calculate the maximum and minimum values of the line impedance.

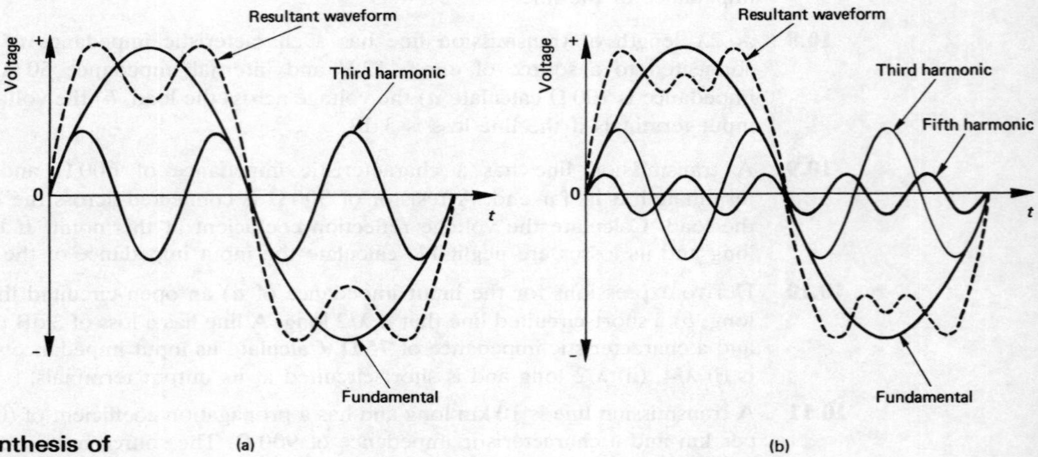

Fig. 11.1 Synthesis of a square waveform

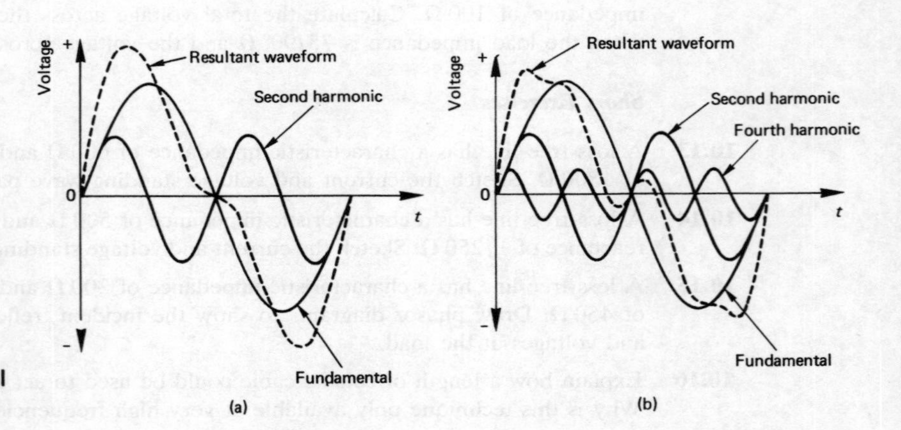

Fig. 11.2 Fundamental plus even harmonics

11 Complex Waveforms

All of the a.c. calculations carried out so far in this book have been based upon the assumption that the signal waveform was sinusoidal. Very often, however, the signal waveform is *not* sinusoidal; examples springing immediately to mind include rectangular waveforms produced by astable multivibrators [EIV] and ramp waveforms produced by sawtooth generators [EIV]. Many other voltages are initially of sinusoidal waveform but, at some stage, have been applied to a *non-linear device* and have had their waveform altered. Examples of *non-linearity* in electrical and electronic circuits are many; some of them are desirable and are used for such purposes as modulation and mixing [RSIII]. Others are undesirable and may produce, for example, distortion in an audio-frequency amplifier [EIV].

Synthesis and Analysis of Repetitive Waveforms

Fig. 11.1*a* shows a sinusoidal voltage $v = V_m \sin \omega t$, of peak value V_m and frequency $\omega/2\pi$, and another, smaller voltage at three times the frequency, i.e. at $3\omega/2\pi$ Hz. The first voltage is known as the *fundamental* and the other voltage is the *third harmonic*. At time $t = 0$, the two voltages are in phase with one another and the waveform produced by summing their instantaneous values is shown by the dotted line. The resultant waveform is clearly non-sinusoidal.

If the fifth harmonic, also with zero phase angle at time $t = 0$, is also added to the fundamental, the resultant waveform becomes an even better approximation to a square wave (Fig. 11.1*b*). If, first, the seventh, then the ninth, and then further higher-order odd harmonics are also added, the resultant waveform or *envelope* will become progressively nearer and nearer to the truly square waveform.

If the second harmonic, and then the fourth harmonic also, are added to the fundamental, a different resultant waveshape is obtained, as is shown by Figs. 11.2*a* and *b*. The effect of a phase shift, at time $t = 0$, of the second and/or third harmonic is to alter the resultant waveform.

The equation for the resultant waveform of Fig. 11.1*a* is

$$v = V_1 \sin \omega t + V_3 \sin 3\omega t \qquad (11.1)$$

When the fifth harmonic is added (Fig. 11.1*b*),

$$v = V_1 \sin \omega t + V_3 \sin 3\omega t + V_5 \sin 5\omega t \qquad (11.2)$$

where V_1, V_3, V_5 are the amplitudes of the fundamental and the third and fifth harmonics respectively, and ω is 2π times the fundamental frequency.

Similarly, when higher-order odd harmonics are also added to the fundamental

$$v = V_1 \sin \omega t + V_3 \sin 3\omega t + \cdots + V_n \sin n\omega t \tag{11.3}$$

$$= \sum_{n=1}^{n=\infty} V_n \sin n\omega t \tag{11.4}$$

Equations (11.1) through to (11.4) provide an indication that the reverse process to the synthesis of a waveform is also possible. The reverse process is, of course, the analysis of a non-sinusoidal waveform to determine its component frequencies.

Any repetitive waveform consists of the sum of a fundamental and a number of harmonics, in general of both even- and odd-order. Most repetitive non-sinusoidal waveforms can be analysed by the use of a mathematical process known as the **Fourier Series**. Essentially this states that a repetitive waveform can be represented by an equation of the form

$$v = A_0 + A_1 \cos \omega t + A_2 \cos 2\omega t + A_3 \cos 3\omega t \cdots$$
$$+ B_1 \sin \omega t + B_2 \sin 2\omega t + B_3 \sin 3\omega t \cdots \tag{11.5}$$

or

$$v = A_0 + \sum_{n=0}^{n=\infty} (A_n \cos n\omega t + B_n \sin n\omega t) \tag{11.6}$$

where $$A_0 = \frac{1}{T} \int_{-T/2}^{T/2} f(t)\, dt \tag{11.7}$$

$$A_n = \frac{2}{T} \int_{-T/2}^{T/2} f(t) \cos n\omega t\, dt \tag{11.8}$$

$$B_n = \frac{2}{T} \int_{-T/2}^{T/2} f(t) \sin n\omega t\, dt \tag{11.9}$$

Examples of the use of the Fourier series to analyse non-sinusoidal waveforms will not be given in this chapter since such work is considered to be the province of a mathematical text.

For many waveforms it is possible to recognize that only *either* cosine *or* sine terms are present and then a considerable reduction in the work involved in an analysis can be achieved.

An **odd function** is one for which $f(t) = -f(-t)$. Such a function passes through the zero voltage axis at the chosen origin and will *only* contain sine terms. Some examples of odd functions are given in Figs. 11.3*a*, *b*, and *c*. Fig. 11.3*a* shows a sawtooth waveform that is clearly symmetrical either side of the origin ($t = 0$). Figs. 11.3*b* and *c* show two different rectangular waveforms for which $f(t) = -f(-t)$. Both of these waveforms can be made into *even* functions instead by a suitable change of the origin. Lastly, Fig. 11.3*d* shows a waveform which, although it passes through the origin, is *not* symmetrical about it. This means that this waveform is not an odd function.

An **even function** is one for which $f(t) = f(-t)$ and which contains only cosine terms. Some examples are given in Figs. 11.4*a*, *b*, and *c*. Note that

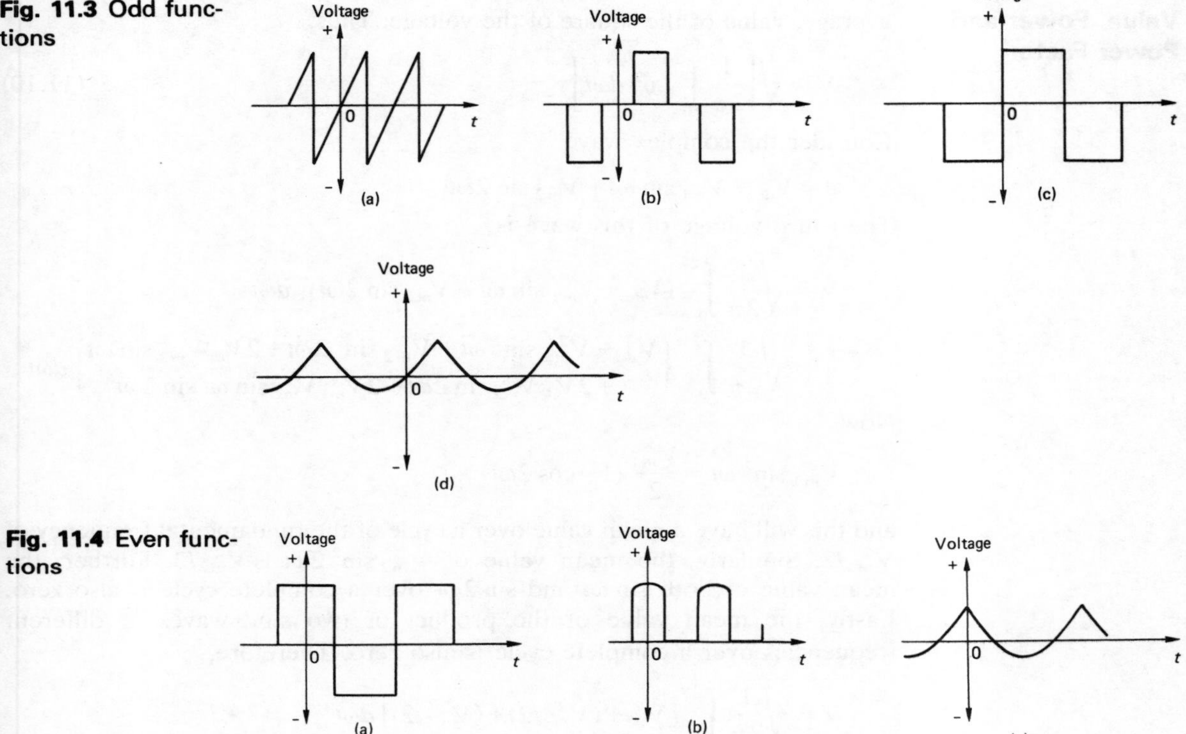

Fig. 11.3 Odd functions

Fig. 11.4 Even functions

Fig. 11.5 *a*) and *b*) Waves containing only odd harmonics, *c*) wave containing only even harmonics

the waveforms *a* and *c* appear in Fig. 11.3 also and have both been made into odd functions by a suitable choice of the origin, i.e. making the origin at the middle of a pulse.

It is also often possible to recognize whether a waveform contains any even or any odd harmonics. Any complex wave whose positive and negative half-cycles are identical, i.e. $f(t) = -f(t + T/2)$, will not contain any even harmonics. Some examples of waveforms that consist of a fundamental frequency plus a number of *odd* harmonics are shown in Figs. 11.5*a* and *b*.

Other waveforms satisfy the equation $f(t) = f(t + T/2)$ and these are waveforms that repeat themselves after each half-cycle of the fundamental frequency. Such waveforms consist of a fundamental plus a number of *even* harmonics. No odd harmonics are present in such a waveform. An example of such a waveform is given in Fig. 11.5*c*.

Root Mean Square Value, Power and Power Factor

The **r.m.s. value** of a complex wave is the square root of the mean, or average, value of the square of the voltage. Thus,

$$V=\sqrt{\left[\frac{1}{2\pi}\int_0^{2\pi} v^2\, d\omega t\right]} \tag{11.10}$$

Consider the complex wave

$$v=V_{dc}+V_{m1}\sin\omega t+V_{m2}\sin 2\omega t$$

The r.m.s. voltage of this wave is

$$V=\sqrt{\frac{1}{2\pi}\int_0^{2\pi}(V_{dc}+V_{m1}\sin\omega t+V_{m2}\sin 2\omega t)^2\, d\omega t}$$

$$=\sqrt{\frac{1}{2\pi}\int_0^{2\pi}\left\{\begin{matrix}V_{dc}^2+V_{m1}^2\sin^2\omega t+V_{m2}^2\sin^2 2\omega t+2V_{dc}V_{m1}\sin\omega t\\ +2V_{dc}V_{m2}\sin 2\omega t+2V_{m1}V_{m2}\sin\omega t\sin 2\omega t\end{matrix}\right\}d\omega t}$$

Now

$$V_{m1}^2\sin^2\omega t=\frac{V_{m1}^2}{2}(1-\cos 2\omega t)$$

and this will have a mean value over a cycle of the fundamental frequency of $V_{m1}^2/2$. Similarly, the mean value of $V_{m2}^2\sin^2 2\omega t$ is $V_{m2}^2/2$. Further, the mean value of both $\sin\omega t$ and $\sin 2\omega t$ over a complete cycle is also zero. Lastly, the mean value of the product of two sine waves at different frequencies over a complete cycle is also zero. Therefore,

$$V=\sqrt{\frac{1}{2\pi}\int_0^{2\pi}[V_{dc}^2+(V_{m1}^2/2)+(V_{m2}^2/2)]\, d\omega t}$$

$$=\sqrt{\frac{1}{2\pi}\left[V_{dc}^2\omega t+\frac{V_{m1}^2}{2}\omega t+\frac{V_{m2}^2}{2}\omega t\right]_0^{2\pi}}$$

$$V=\sqrt{\left[V_{dc}^2+\frac{V_{m1}^2}{2}+\frac{V_{m2}^2}{2}\right]} \tag{11.11}$$

It should be noted that $V_{m1}^2/2$ is the square of the r.m.s. voltage of the fundamental frequency component and $V_{m2}^2/2$ is the square of the r.m.s. value of the second harmonic component. Thus, equation (11.11) can be re-written as

$$V=\sqrt{(V_{dc}^2+V_1^2+V_2^2)} \tag{11.12}$$

where V_1 and V_2 are respectively the r.m.s. values of the fundamental and the second harmonic frequency components.

Should any of the harmonic components of a complex wave have some phase angle other than zero, relative to the fundamental at time $t=0$, it will have no effect upon the r.m.s. value of the wave and equation (11.12) will still apply.

Example 11.1

Calculate the r.m.s. value of the complex wave that has an instantaneous voltage given by

$$v = 4 \sin 5000t + 1 \sin 10\,000t + 0.5 \sin 15\,000t$$

Solution From equation (11.12),

$$V = \sqrt{\left[\left(\frac{4}{\sqrt{2}}\right)^2 + \left(\frac{1}{\sqrt{2}}\right)^2 + \left(\frac{0.5}{\sqrt{2}}\right)^2\right]} = 2.937 \text{ V} \quad (Ans)$$

When a complex wave is applied to a circuit the current that flows will consist of a number of components at the same frequencies as the components in the voltage wave. The magnitude and angle of each component of the current will be equal to the corresponding voltage divided by the impedance of the circuit at that frequency. Thus, if the voltage

$$v = V_1 \sin \omega t + V_2 \sin 2\omega t$$

is applied to a circuit, the current that flows will be equal to

$$i = \frac{V_1}{Z_1} \sin \omega t + \frac{V_2}{Z_2} \sin 2\omega t$$

where Z_1 is the impedance of the circuit at frequency $\omega/2\pi$ and Z_2 is the impedance of the circuit at frequency $2\omega/2\pi$.

Example 11.2

Determine the expression for the current flowing in the circuit given in Fig. 11.6.

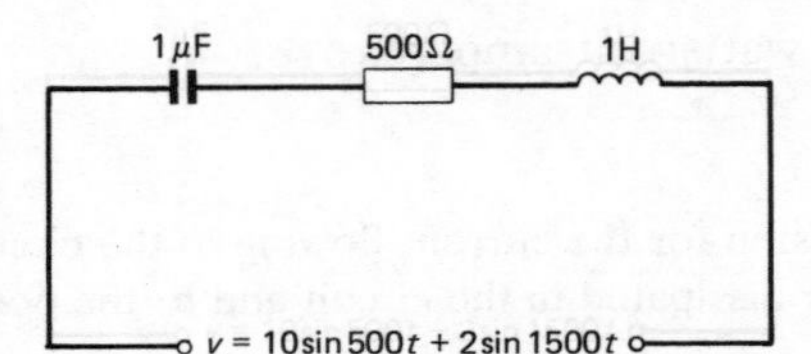

Fig. 11.6

Solution At the fundamental frequency of $500/2\pi$ Hz the reactance of the capacitor is

$$X_{C1} = 1/500 \times 10^{-6} = -j2000 \ \Omega$$

The inductive reactance is $X_{L1} = j500 \ \Omega$. Therefore, at this frequency the impedance of the circuit is

$$Z_1 = 500 - j1500 = 1581\underline{/-71.6^\circ} \ \Omega$$

At the third harmonic frequency

$$X_{C3} = -j2000/3 = -j666.7 \ \Omega$$

and

$$X_{L3} = j500 \times 3 = j1500 \ \Omega$$

Therefore the impedance of the circuit at the third harmonic is

$$Z_3 = 500 + j833.3 = 971.8\underline{/59^\circ} \ \Omega$$

The current flowing in the circuit is

$$i = \frac{10 \sin 500t}{1581\angle{-71.6^\circ}} + \frac{2 \sin 1500t}{971.8\angle{59^\circ}}$$

$$i = 6.33 \sin(500t + 71.6^\circ) + 2.06 \sin(1500t - 59^\circ) \text{ mA} \quad (Ans)$$

Power and Power Factor

When a complex wave v is applied to a circuit and causes a current i to flow, the instantaneous power dissipated is $p = iv$ watts. The average or **mean power** is

$$P = \frac{1}{2\pi} \int_0^{2\pi} iv \, d\omega t \qquad (11.13)$$

$$P = \frac{1}{2\pi} \int_0^{2\pi} [I_1(\sin \omega t + \theta_1) + I_2 \sin(2\omega t + \theta_2) + \cdots] \times [V_1 \sin \omega t + V_2 \sin 2\omega t + \cdots] \, d\omega t$$

or $P = V_1 I_1 \cos \theta_1 + V_2 I_2 \cos \theta_2 + \cdots \qquad (11.14)$

since most of the terms are equal to zero.

This result means that the total power dissipated in a circuit is the algebraic sum of the powers dissipated at each component frequency—including any d.c. term that may be present.

The **power factor** is *not* $\cos \theta$ (*which* θ would be used?) but is given by

$$\text{Power factor} = \text{watts/volt-amperes} \qquad (11.15)$$

Example 11.3

Determine the expression for the current flowing in the circuit given in Fig. 11.7 and calculate *a*) the power dissipated in the circuit and *b*) the power factor of the circuit.

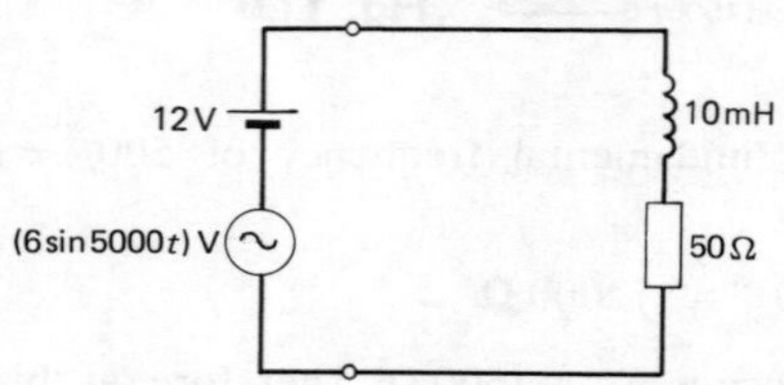

Fig. 11.7

Solution At d.c. the impedance of the circuit is merely the resistance of 50 Ω. Therefore $I_{dc} = 12/50 = 0.24$ A.

At $\omega = 5000$ R/s, $Z = 50 + j50 = 70.7\angle{45^\circ}$ Ω and so $I_{ac} = 6/70.7 = 0.085$ A.

The d.c. power dissipated $= 0.24^2 \times 50 = 2.88$ W.

The a.c. power dissipated $= \left(\frac{0.085}{\sqrt{2}}\right)^2 \times 50 = 0.181$ W.

The total power dissipated $= 2.88 + 0.181 = 3.061$ W (*Ans*)

The r.m.s. voltage $V = \sqrt{[12^2 + (6/\sqrt{2})^2]} = 12.727$ V.

The r.m.s. current $I = \sqrt{[0.24^2 + (0.085/\sqrt{2})^2]} = 0.247$ A.

Therefore,

$$\text{Power factor} = 3.061/12.727 \times 0.247 = 0.974 \quad (Ans)$$

Example 11.4

Determine an expression for the current taken from the voltage source in the circuit of Fig. 11.8. Also calculate the power dissipated in the circuit and the power factor.

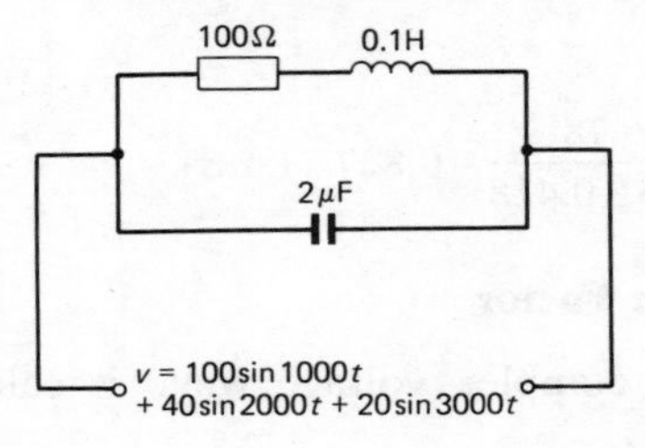

Fig. 11.8

Solution

At $\omega = 1000$ $X_{L1} = 100\ \Omega$ and $X_{C1} = 500\ \Omega$

$\omega = 2000$ $X_{L2} = 200\ \Omega$ and $X_{C2} = 250\ \Omega$

$\omega = 3000$ $X_{L3} = 300\ \Omega$ and $X_{C3} = 166.67\ \Omega$

The impedance of the circuit at $\omega = 1000$ is

$$Z_1 = \frac{(100 + j100)(-j500)}{100 - j400} = 171.5\angle 31^\circ\ \Omega$$

At $\omega = 2000$

$$Z_2 = \frac{(100 + j200)(-j250)}{100 - j50} = 500\angle 0^\circ\ \Omega$$

At $\omega = 3000$

$$Z_3 = \frac{(100 + j300)(-j166.67)}{100 + j133.33} = 316.2\angle -71.5^\circ\ \Omega$$

Therefore,

$$|I_1| = |V_1|/Z_1 = 100/171.5 = 0.583\ \text{A}$$
$$|I_2| = |V_2|/Z_2 = 40/500 = 0.08\ \text{A}$$
$$|I_3| = |V_3|/Z_3 = 20/316.2 = 0.063\ \text{A}$$

The required expression for the current flowing in the circuit is

$$i = 0.583\sin(1000t - 31^\circ) + 0.08\sin(2000t) + 0.063\sin(3000t + 71.5^\circ)\ \text{A} \quad (Ans)$$

The power dissipated in the circuit is

$$P = \left(\frac{100}{\sqrt{2}} \times \frac{0.583}{\sqrt{2}}\cos 31^\circ\right) + \left(\frac{40}{\sqrt{2}} \times \frac{0.08}{\sqrt{2}}\cos 0^\circ\right) + \left(\frac{20}{\sqrt{2}} \times \frac{0.063}{\sqrt{2}}\cos -71.5^\circ\right)$$
$$= 24.98 + 1.6 + 0.2 = 26.78\ \text{W} \quad (Ans)$$

The r.m.s. applied voltage is

$$V = \sqrt{\left[\left(\frac{100}{\sqrt{2}}\right)^2 + \left(\frac{40}{\sqrt{2}}\right)^2 + \left(\frac{20}{\sqrt{2}}\right)^2\right]} = 77.46\ \text{V}$$

The r.m.s. current is

$$I=\sqrt{\left[\left(\frac{0.583}{\sqrt{2}}\right)^2+\left(\frac{0.08}{\sqrt{2}}\right)^2+\left(\frac{0.063}{\sqrt{2}}\right)^2\right]}=0.418\text{ A}$$

Therefore,

$$\text{Power factor}=\frac{26.78}{77.45\times0.418}=0.827\quad(Ans)$$

Mean Value and Form Factor

The **mean value** of a complex voltage wave is calculated using equation (11.16), i.e.

$$V_{mean}=\frac{1}{\pi}\int_0^{\pi} v\,d\omega t \tag{11.16}$$

The value obtained for the mean value of a wave will be a function of the relative phase angles of the various harmonic components. This means that its computation may, in some cases, be a somewhat lengthy process.

The **form factor** of a complex wave is the ratio of the r.m.s. value to the mean value.

Example 11.5

Determine the form factor of the complex wave

$$v=50\sin 200\pi t+10\sin 400\pi t+5\sin 600\pi t\text{ V}$$

Solution

$$V=\sqrt{\left[\left(\frac{50}{\sqrt{2}}\right)^2+\left(\frac{10}{\sqrt{2}}\right)^2+\left(\frac{5}{\sqrt{2}}\right)^2\right]}=36.23\text{ V}$$

$$V_{mean}=(1/\pi)\int_0^{\pi}(50\sin\omega t+10\sin 2\omega t+5\sin 3\omega t)\,d\omega t$$

$$=(1/\pi)[-50\cos\omega t-5\cos 2\omega t-1.67\cos 3\omega t]_0^{\pi}$$

$$=(1/\pi)(50+50-5+5+1.67+1.67)=103.34/\pi$$

Therefore,

$$\text{Form factor}=(36.23\times\pi)/103.34=1.1\quad(Ans)$$

Effect of Harmonics on Component Measurement

When harmonics are present in a current or voltage waveform the measurement of an impedance will be subject to some error. The error arises because the percentage harmonic in the voltage wave will always differ from the percentage harmonic in the current wave. If the circuit is capacitive, the current wave will possess the larger harmonic content, whilst for an inductive circuit the voltage will have a higher percentage harmonic.

Example 11.6

The current $i=(10\sin 314t+3\sin 942t)$ A flows in an inductance. At the fundamental frequency of 50 Hz the ratio reactance/resistance for the inductance is 5. Calculate the percentage error involved in measuring the impedance of the inductance as the ratio V/I.

Solution At the fundamental frequency the impedance of the inductance is

$$Z_1 = R + j\omega L = R + j5R \quad \text{and} \quad |Z_1| = (\sqrt{26})R\ \Omega$$

At the third harmonic X_L will be $j15R$ and so

$$Z_3 = R + j15R \qquad |Z_3| = (\sqrt{226})R\ \Omega$$

Hence, the voltage v across the inductance is

$$v = 10(\sqrt{26})R \sin 314t + 3(\sqrt{226})R \sin 942t \text{ V}$$

Therefore,

$$V = \sqrt{\left[\left(\frac{10(\sqrt{26})R}{\sqrt{2}}\right)^2 + \left(\frac{3(\sqrt{226})R}{\sqrt{2}}\right)^2\right]} = 48.14R \text{ volts}$$

and $$I = \sqrt{\left[\left(\frac{10}{\sqrt{2}}\right)^2 + \left(\frac{3}{\sqrt{2}}\right)^2\right]} = 7.38 \text{ A}$$

The apparent impedance of the inductance is

$$V/I = 48.14/7.38 = 6.52R$$

The true impedance of the inductance is

$$V_1/I_1 = (\sqrt{26})R = 5.1R$$

Therefore, the percentage error in the measurement of the impedance of the inductor is

$$\%\ \text{error} = \frac{6.52 - 5.1}{5.1} \times 100 = 27.84\% \quad (\textit{Ans})$$

Selective Resonance

Another effect that the presence of harmonics in a waveform may have is known as **selective resonance**. This term means that a circuit containing both inductance and capacitance may resonate at any one of the harmonic frequencies. In general, this is an undesirable effect since it may result in a disturbingly high voltage appearing across the capacitance at some frequency. It is particularly undesirable in power circuits where the unwanted resonant current could be dangerously high. Selective resonance will occur at a harmonic frequency when

$$n\omega L = 1/n\omega C$$

where n is the order of the harmonic and ω is 2π times the fundamental frequency.

Example 11.7

The circuit shown in Fig. 11.9 is resonant at the third harmonic frequency of the applied voltage. Calculate *a*) the fundamental frequency of the applied voltage, *b*) the current flowing at the fundamental frequency, and *c*) the current flowing at the third harmonic frequency.

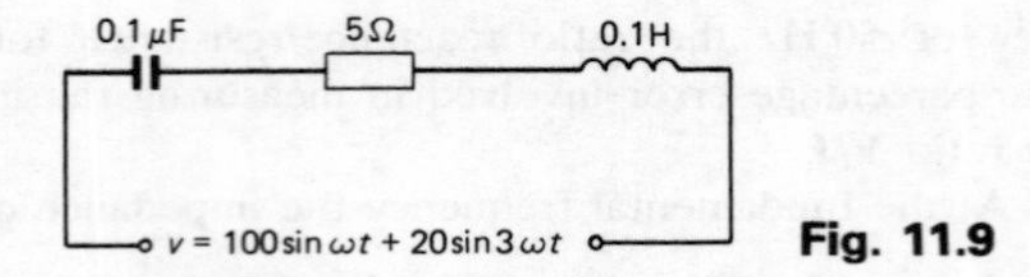

Fig. 11.9

Solution

a) $3\omega \times 0.1 = 1/3\omega \times 0.1 \times 10^{-6}$

Fundamental frequency = 530.5 Hz (*Ans*)

b) $X_{C1} = 1/(2\pi \times 530.5 \times 10^{-7}) = -j3000\ \Omega$

$X_{L1} = 2\pi \times 530.5 \times 0.1 = j333.3\ \Omega$

$Z_1 = 5 - j2666.7\ \Omega$

$I_1 = V_1/Z_1 = 100/(5 - j2666.7) = 37.5\underline{/89.9^\circ}$ mA (*Ans*)

c) $I_3 = V_3/R = 20/5 = 4$ A (*Ans*)

Non-linear Devices

The waveforms produced by a.c. generators are usually very nearly sinusoidal but may often contain some harmonics. The harmonic content of most sinusoidal oscillators is very small indeed and usually need only be taken into account when precise measurements are carried out. Most of the harmonic content of waveforms is produced when an originally sinusoidal waveform is applied, intentionally or not, to a device having a non-linear current/voltage characteristic.

The **non-linear device** may be a non-linear resistor, a semiconductor diode, a suitably biased transistor, or perhaps the core of a transformer, to mention just a few of the many possible examples.

1 Consider the iron core of a *transformer*. The relationship between the magnetizing current and the flux set up in the core is not linear. This means that a sinusoidal magnetizing current will *not* produce a sinusoidal variation of the core flux.

The point is illustrated by Fig. 11.10 which shows a sinusoidal magnetizing current applied to the hysteresis loop of a core. The hysteresis loop has been drawn to exaggerate the non-linearity produced. Because the core flux is not sinusoidal the e.m.f. induced into the secondary winding will also have a non-sinusoidal waveform. Since $e = N\,d\Phi/dt$ the waveform of the secondary e.m.f. can be deduced and it is shown in Fig. 11.11. Clearly the waveform is far from sinusoidal and this shows that it is essential for a core not to be taken into saturation if waveform distortion is to be minimized. If the magnetizing force is kept small enough, fairly linear operation is possible.

It should be evident from Fig. 11.10 that if a sinusoidal core flux is needed the waveform of the magnetizing current will have to be suitably distorted. The necessary waveform can be deduced by projecting from the hysteresis loop, as has been done in Fig. 11.12.

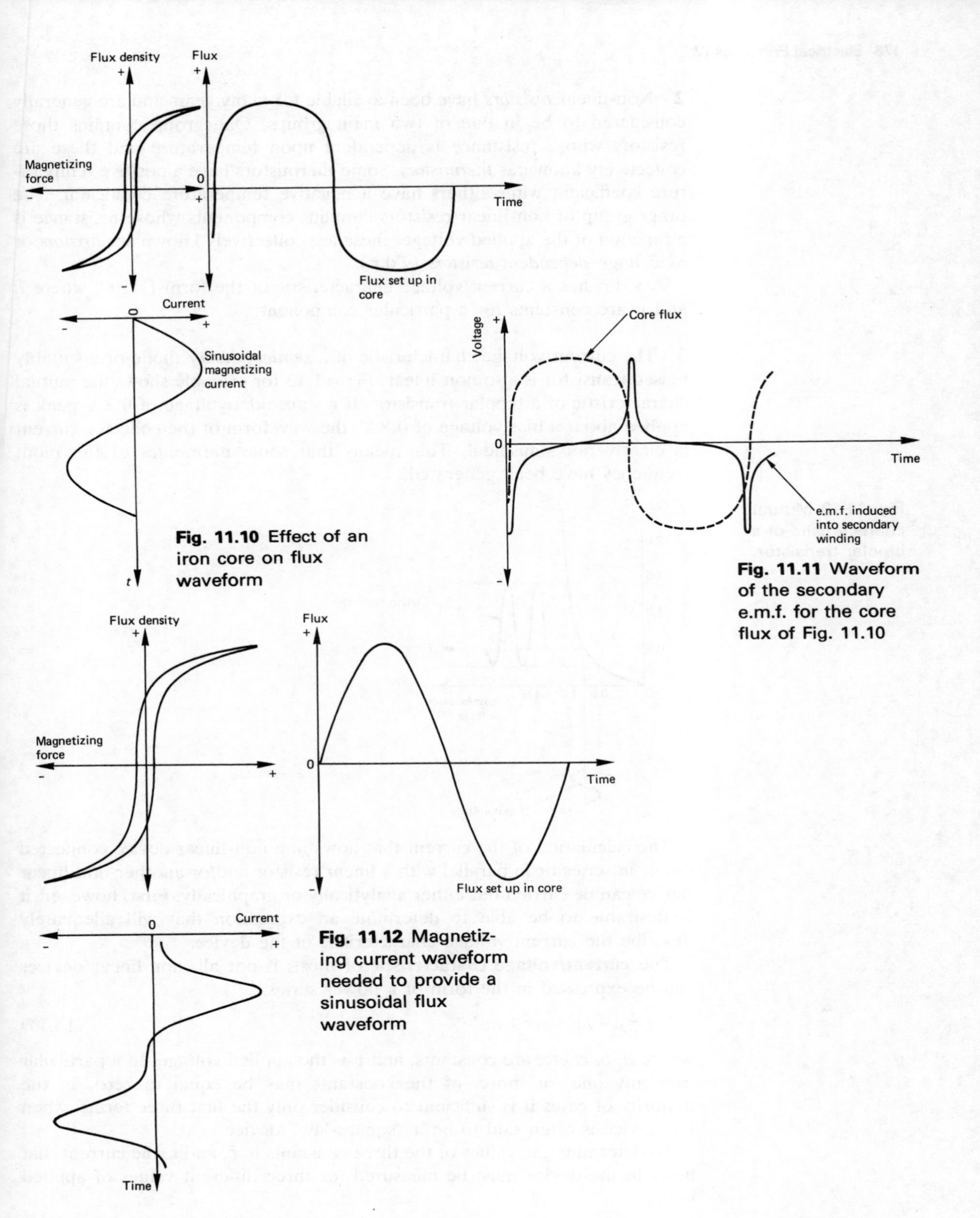

Fig. 11.10 Effect of an iron core on flux waveform

Fig. 11.11 Waveform of the secondary e.m.f. for the core flux of Fig. 11.10

Fig. 11.12 Magnetizing current waveform needed to provide a sinusoidal flux waveform

2 Non-linear *resistors* have been available for many years and are generally considered to be in one of two main groups. One group contains those resistors whose resistance is dependent upon temperature and these are collectively known as *thermistors.* Some thermistors have a positive temperature coefficient while others have a negative temperature coefficient. The other group of non-linear resistors contains components whose resistance is a function of the applied voltage; these are collectively known as *varistors* or as *voltage-dependent resistors* (v.d.r.).

A v.d.r. has a current/voltage characteristic of the form $i = av^n$ where a and n are constants for a particular component.

3 The current/voltage characteristic of a *semiconductor* diode or a suitably biased transistor is also non-linear. Fig. 11.13 for example shows the mutual characteristic of a bipolar transistor. If a sinusoidal voltage of 0.2 V peak is applied about a bias voltage of 0.8 V, the waveform of the collector current is clearly not sinusoidal. This means that some harmonics of the input frequency have been generated.

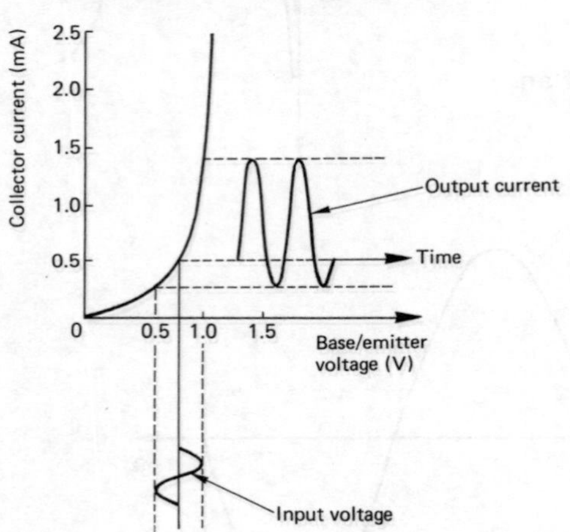

Fig. 11.13 Mutual characteristic of a bipolar transistor

The calculation of the current that flows in a non-linear device connected either in series or in parallel with a linear resistor and/or another non-linear device can be carried out either analytically or graphically. First, however, it is desirable to be able to determine an expression that will adequately describe the current/voltage characteristic of the device.

The **current/voltage characteristic** of most, if not all, non-linear devices can be expressed in the form of a power series:

$$i = a + bv + cv^2 + dv^3 \cdots \tag{11.17}$$

where a, b, c, etc. are constants, and v is the applied voltage. In a particular case any one, or more, of the constants may be equal to zero. In the majority of cases it is sufficient to consider only the first three terms, when the device is often said to be a "square-law" device.

To determine the values of the three constants a, b, and c, the current that flows in the device must be measured for three different values of applied voltage.

Example 11.8

Measurements of a non-linear device gave the data in Table 11.1. The characteristic may be assumed to be a square law. Determine the constants *a*, *b*, and *c*. Calculate *a*) the mean value of the current when the applied voltage is 1 sin ωt volts and *b*) the percentage second harmonic content of the current.

Table 11.1

Applied voltage (V)	1	2	3
Current (mA)	4.2	8	13.4

Solution From equation (11.17)

$$4.2 = a + b + c \tag{11.18}$$

$$8 = a + 2b + 4c \tag{11.19}$$

$$13.4 = a + 3b + 9c \tag{11.20}$$

Solving these equations gives $a = 2$, $b = 1.4$ and $c = 0.8$. Hence the current/voltage characteristic of the device is

$$i = 2 + 1.4v + 0.8v^2 \text{ mA} \tag{11.21}$$

a) When $v = 1 \sin \omega t$ volts,

$$i = 2 + 1.4 \sin \omega t + 0.8 \sin^2 \omega t \text{ mA}$$
$$= 2 + 1.4 \sin \omega t + 0.4 - 0.4 \cos 2\omega t$$

(A trig. identity is $\sin^2 A = (1 - \cos 2A)/2$.)

$$i = 2.4 + 1.4 \sin \omega t - 0.4 \cos 2\omega t \text{ mA}$$

Mean current = 2.4 mA (*Ans*)

b) $\%$ second harmonic content $= \dfrac{\text{amplitude of second harmonic}}{\text{amplitude of fundamental}} \times 100\%$

$= (0.4/1.4) \times 100 = 28.57\%$ (*Ans*)

This example shows how a second harmonic component is produced by an element with a non-linear current/voltage characteristic. If a cubic term dv^3 had been included in the problem a third harmonic term would also have made its appearance.

If two sinusoidal signals at frequencies f_1 and f_2 are applied to a square-law device, components at the sum $f_1 + f_2$ and the difference $f_1 - f_2$ would be generated. These are known as *intermodulation components*, and are of considerable importance in telecommunications [RSIII].

Calculation of Currents in Circuits Containing a Non-linear Device

1 The calculation of the current flowing in a circuit in which a non-linear device is connected in series with a linear resistor *R*, as in Fig. 11.14, can, in principle, be carried out either analytically or graphically. Depending upon the actual circuit and/or the non-linear characteristic one method or the other may turn out to be considerably easier in a particular case.

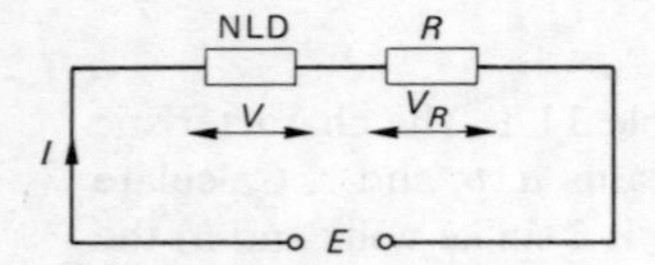

Fig. 11.14 A non-linear device in series with a linear resistor

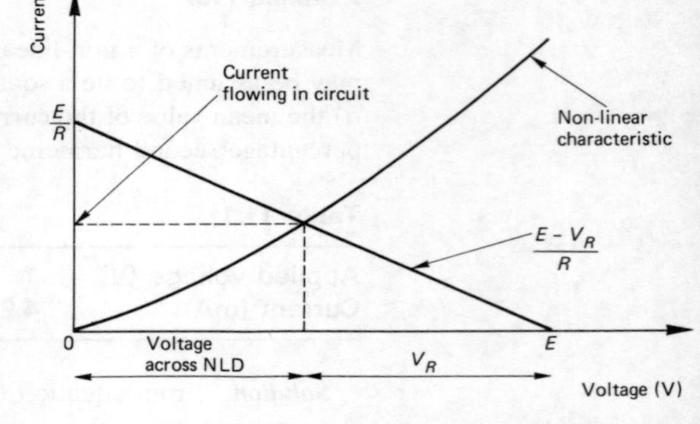

Fig. 11.15 Graphical determination of the current and voltages in Fig. 11.14

When a voltage E is applied to the circuit a current I will flow. The voltage V_R across the linear resistor R will then be $V_R = IR$ and the voltage V across the non-linear device will be $V = E - V_R$. Substitution of this value of V into the I/V characteristic of the non-linear device will then produce the value of the current I.

Alternatively, the current/voltage characteristic of the non-linear device can be plotted. The voltage across the linear resistor should then be taken as $V_R = E - V$ and the current/voltage characteristic of this component plotted on the same axes. Since $I = (E - V)/R$ this characteristic is linear and therefore only two points are needed (see Fig. 11.15). The point of intersection of the two graphs gives the current in the circuit and the voltages across each of the components as shown by the figure.

When a sinusoidal voltage is applied to a circuit similar to Fig. 11.14, a similar procedure should be followed if the peak current and/or voltages in the circuit are to be calculated.

Example 11.9

A non-linear device whose current/voltage characteristic is given by $i = 0.1V + 0.02V^2$ A is connected in series with a linear resistor of 10 Ω. A sinusoidal voltage of peak value 10 V is applied across the circuit. Calculate *a*) the peak current flowing in the circuit and *b*) the voltages across each of the two components when the peak current is flowing. Use first a graphical method, and then an analytical method, of solution.

Solution

a) *Graphical Method*

Using the figures given in Table 11.2 the current/voltage characteristic of the non-linear device has been plotted in Fig. 11.16. The linear characteristic $(E - V)/10$ is also plotted, for the peak supply voltage of 10 V.

Table 11.2

V (volts)	1	2	3	4	5	6	7	8	9	10
$0.1V$	0.1	0.2	0.3	0.4	0.5	0.6	0.7	0.8	0.9	1.0
V^2	1	4	9	16	25	36	49	64	81	100
$0.02V^2$	0.02	0.08	0.18	0.32	0.5	0.72	0.98	1.28	1.62	2.0
I (A)	0.12	0.28	0.48	0.72	1.0	1.32	1.68	2.08	2.52	3.0

Fig. 11.16

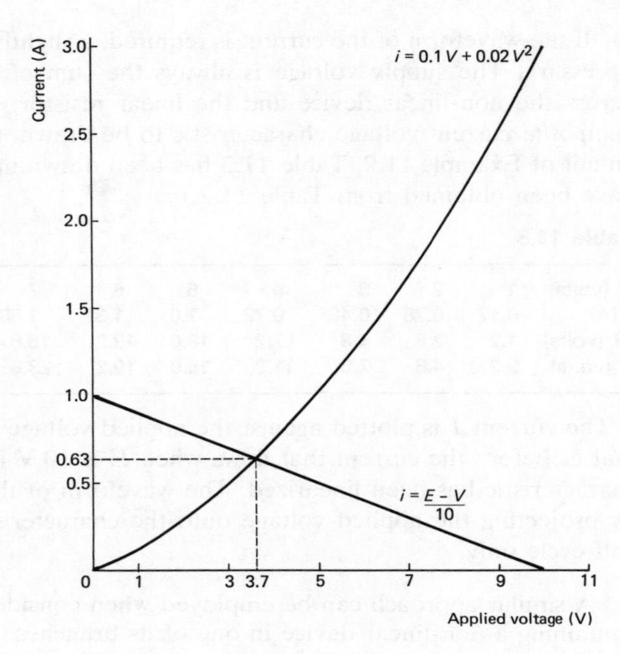

From the point of intersection of the two graphs:

a) Peak current = 0.63 A (*Ans*)

b) The voltage across the non-linear device = 3.7 V (*Ans*)

The voltage across the 10 Ω resistor = 6.3 V (*Ans*)

b) *Analytical Method*

The voltage across the non-linear device is

$$E - V = E - IR = 10 - 10I$$

Hence

$$I = 0.1(10 - 10I) + 0.02(10 - 10I)^2$$

$$= 1 - I + 2 - 4I + 2I^2$$

$$2I^2 - 6I + 3 = 0$$

$$I = \frac{6 \pm \sqrt{(36 - 24)}}{4} = 2.37 \text{ A or } 0.63 \text{ A}$$

Clearly, the first result cannot be correct since it gives $IR = 23.7$ V which is greater than the applied voltage. Therefore,

a) $I = 0.63$ A (*Ans*)

b) $V = 6.3$ V (*Ans*)

c) $V = 10 - 6.3 = 3.7$ V (*Ans*)

2 If the waveform of the current is required, a slightly different approach is necessary. The supply voltage is always the sum of the voltages dropped across the non-linear device and the linear resistor. This fact will allow a *composite* current/voltage characteristic to be drawn. Considering again the circuit of Example 11.9, Table 11.3 has been drawn up. (The current values have been obtained from Table 11.2.)

Table 11.3

V (volts)	1	2	3	4	5	6	7	8	9	10
I (A)	0.12	0.28	0.48	0.72	1.0	1.32	1.68	2.08	2.52	3.0
IR (volts)	1.2	2.8	4.8	7.2	10.0	13.2	16.8	20.8	25.2	30.0
E (volts)	2.2	4.8	7.8	11.2	15.0	19.2	23.8	28.8	34.2	40.0

The current I is plotted against the applied voltage E in Fig. 11.17; note that as before the current that flows when E is 10 V is 0.63 A and that the characteristic has been linearized. The waveform of the current is obtained by projecting the applied voltage onto the characteristic as shown for one half-cycle only.

3 A similar approach can be employed when considering a parallel circuit containing a non-linear device in one of its branches. Fig. 11.18 shows one possible arrangement. The total current I_T is the sum of the currents I_R and I_N flowing in the two branches of the circuit. If V is the voltage developed across the combination, then $I_R = V/R$ and I_N is given by the particular non-linear characteristic in use. Thus, in general,

$$I_T = (V/R) + a + bV + cV^2 + \cdots$$

If the total current I_T is plotted against voltage the voltage required to produce a given total current can be determined.

Example 11.10

A 100 Ω resistor is connected in parallel with a non-linear device whose I/V characteristic is given by $I = 5V^2$ mA. The peak current supplied to the circuit is to be limited to 100 mA when a voltage is applied across the circuit. Calculate *a*) the current flowing in each component, *b*) the maximum permissible value of the applied voltage.

Solution

$$I_T = 10V + 5V^2 \text{ mA}$$

Hence, Table 11.4 can be obtained.

This data is plotted in Fig. 11.19. From the graph it is evident that for a maximum current of 100 mA the peak value of the applied voltage must be limited to 3.6 V (*Ans*)

Table 11.4

V (volts)	1	2	3	4	5
I (mA)	15	40	75	120	175

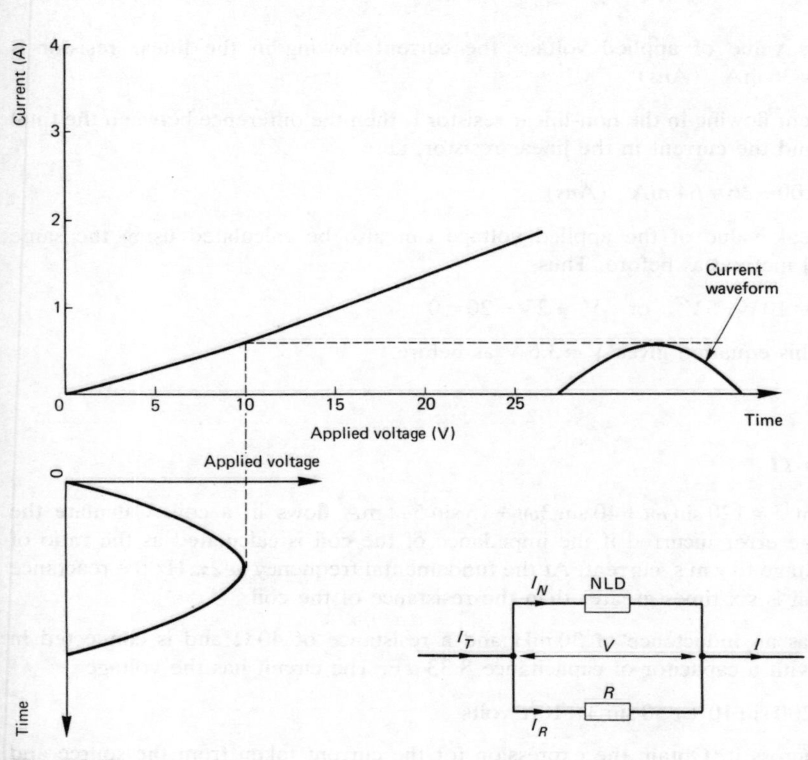

Fig. 11.17

Fig. 11.18 A non-linear device in parallel with a linear resistor

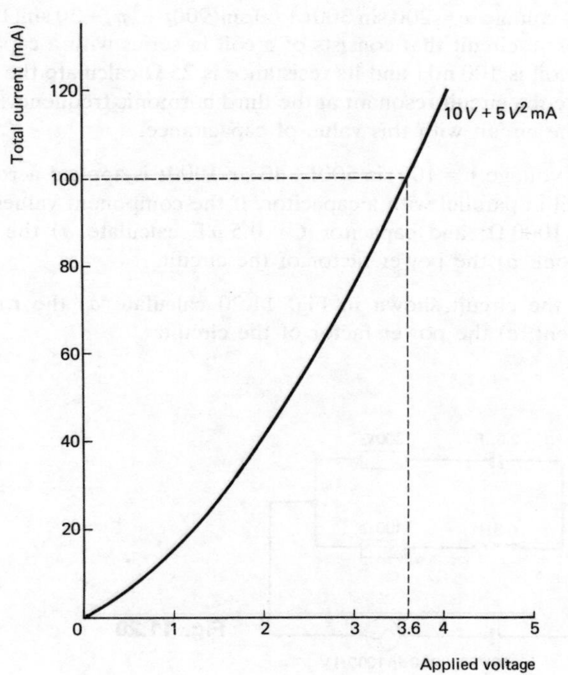

Fig. 11.19

With this value of applied voltage the current flowing in the linear resistor is $3.6/100 = 36$ mA (*Ans*)

The current flowing in the non-linear resistor is then the difference between the total current and the current in the linear resistor, i.e.

$$I = 100 - 36 = 64 \text{ mA} \quad (Ans)$$

The peak value of the applied voltage can also be calculated using the same analytical method as before. Thus,

$$100 = 10V + 5V^2 \quad \text{or} \quad V^2 + 2V - 20 = 0$$

Solving this equation gives $V = 3.6$ V as before.

Exercises 11

11.1 A current $i = 120 \sin \omega t + 40 \sin 3\omega t + 15 \sin 5\omega t$ mA flows in a coil. Calculate the percentage error incurred if the impedance of the coil is calculated as the ratio of r.m.s. voltage to r.m.s. current. At the fundamental frequency $\omega/2\pi$ Hz the reactance of the coil is six times greater than the resistance of the coil.

11.2 A coil has an inductance of 30 mH and a resistance of 40 Ω and is connected in parallel with a capacitor of capacitance 8.33 μF. The circuit has the voltage

$$v = 200 \sin 10^3 t + 50 \sin 3 \times 10^3 t \text{ volts}$$

applied across it. Obtain the expression for the current taken from the source and calculate the power factor of the circuit.

11.3 The voltage $v = 200 \sin 300t + 60 \sin(900t + \frac{1}{4}\pi) + 30 \sin(1500t - \frac{1}{4}\pi)$ volts is applied across a circuit that consists of a coil in series with a capacitor. If the inductance of the coil is 100 mH and its resistance is 25 Ω calculate the capacitance value that will make the circuit resonant at the third harmonic frequency. Also find the power factor of the circuit with this value of capacitance.

11.4 The voltage $v = 100 \sin 500t + 40 \sin 1000t$ is applied across a circuit that consists of a coil in parallel with a capacitor. If the component values are, for the coil, $L = 1$ H, $R = 1000$ Ω; and capacitor $C = 0.5$ μF calculate *a*) the expression for the circuit current, *b*) the power factor of the circuit.

11.5 For the circuit shown in Fig. 11.20 calculate *a*) the r.m.s. voltage, *b*) the r.m.s. current, *c*) the power factor of the circuit.

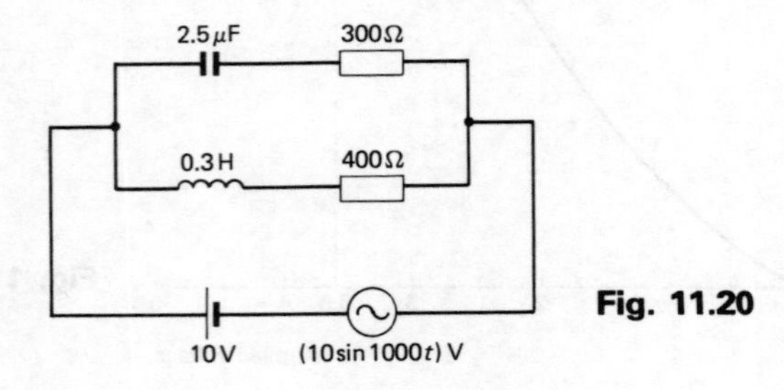

Fig. 11.20

11.6 Calculate the total power dissipated in the circuit shown in Fig. 11.21. Also find the power factor of the circuit.

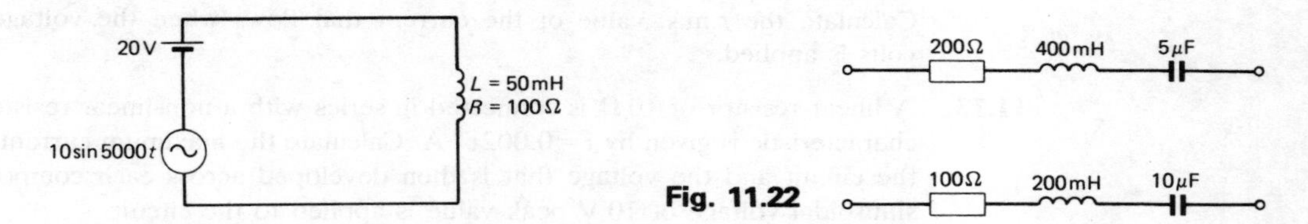

Fig. 11.21

Fig. 11.22

11.7 Calculate the admittance of both of the series circuits given in Fig. 11.22 at frequencies of (i) $500/2\pi$ Hz and (ii) $1000/2\pi$ Hz. The two circuits are connected in parallel. Calculate their total admittance at each frequency.

If the voltage $v = 10 \sin 500t + 4 \sin 1000t$ volts is applied across the circuit determine the r.m.s. value of the current supplied to the circuit.

11.8 The current/voltage characteristic of a non-linear resistor is given by $i = 0.1V + 0.02V^2$ amps. This non-linear resistor is connected in series with a linear resistor of 20 Ω and a d.c. voltage source of 9 V and negligible internal resistance. Calculate the current flowing in the circuit and the voltage across each component using a graphical method.

11.9 Repeat exercise 11.8 using an analytical method of solution.

11.10 Measurements on a non-linear device gave the data listed in Table 11.5 Assuming the resistor to have a square-law characteristic calculate *a*) the mean current

Table 11.5

Applied voltage (V)	1	3	5
Current (mA)	1	9	25

when a sinusoidal signal of 2 V peak value is applied, *b*) the amplitudes of the intermodulation components when the two voltages 1 sin ωt and 1 sin $2\omega t$ are applied.

11.11 The voltage applied to the circuit given in Fig. 11.23 is

$$v = 12(\sin 4000t + \tfrac{1}{3} \sin 12\,000t + \tfrac{1}{5} \sin 20\,000t) \text{ volts}$$

Calculate the r.m.s. value of the applied voltage. Determine the value of C that will cause the circuit to resonate at $20\,000/2\pi$ Hz. Calculate the peak value of the voltage at this frequency across the capacitor. Also find the impedance of the circuit at $12\,000/2\pi$ Hz.

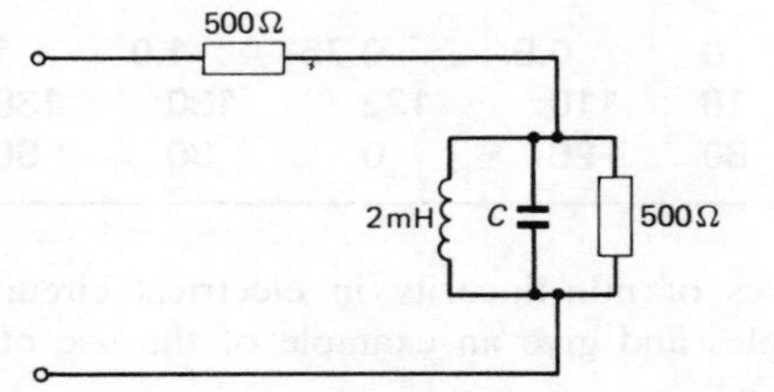

Fig. 11.23

11.12 The I/V characteristic of a non-linear circuit is

$$i = 4.5 + 0.5v + 0.03v^2 \text{ mA}$$

Calculate the r.m.s. value of the current that flows when the voltage $v = 6 \sin \omega t$ volts is applied.

11.13 A linear resistor of 10 Ω is connected in series with a non-linear resistor whose I/V characteristic is given by $i = 0.002v^3$ A. Calculate the maximum current that flows in the circuit and the voltage that is then developed across each component when a sinusoidal voltage of 10 V peak value is applied to the circuit.

11.14 A 78 Ω resistor is connected in parallel with a non-linear device whose current/voltage characteristic is given by

$$i = (4 \times 10^{-3} v) + (3 \times 10^{-4} v^3) \text{ A}$$

Calculate the current in each component and the voltage across them when the total current flowing is 200 mA.

Short Exercises

11.15 The total characteristic of a bridge rectifier circuit is given by Table 11.6. Determine the waveform of the current when the rectifier is connected in series with a 3300 Ω resistor and the applied voltage is 60 sin $100\pi t$ volts.

Table 11.6

Voltage (V)	0	6	8	10	12	14	16	18
Current (mA)	0	0	2.3	4.5	6.7	9.4	12.2	15.5

11.16 Two signals $V_1 \sin 1000t$ and $V_2 \sin 2000t$ volts are applied to a component that has a square-law I/V characteristic. Calculate the frequencies of the components of the output current.

11.17 A non-linear characteristic is given by

$$i = 3.8 + 1.2v + 0.6v^2 \text{ mA}$$

If $v = 2.5 \sin \omega t$ calculate the percentage second harmonic in the current.

11.18 Table 11.7 gives the values of flux produced in a magnetic material by various values of magnetizing current. Plot flux to a base of magnetizing current and use the graph to obtain the waveform of the current needed to produce a sinusoidal flux waveform.

Table 11.7

Magnetizing current (A)	0	0.5	0.75	1.0	1.5	2	2.5	3
Flux (Wb)	18	110	122	130	138	140	140	140
	−80	−26	0	30	80	110	128	140

11.19 List some examples of non-linearity in electrical circuits. Give both useful and undesirable examples and give an example of the use of each you consider to be useful.

11.20 A sinusoidal voltage of peak value 30 V is applied to a non-linear circuit whose I/V characteristic is given by the data of Table 11.8. Plot the half-cycle of the current that flows, on squared paper, and hence estimate the average value of the current.

Table 11.8

Applied voltage (V)	0	12	16	20	24	28	32	36
Current (mA)	0	0	2.6	7.0	11.4	16.8	22.4	30

11.21 Calculate the r.m.s. value of the voltage

$$v = 40 + 10 \sin 3000t + 5 \cos 15\,000t \text{ volts}$$

11.22 Calculate the mean value of the voltage

$$v = 1 \sin 5000t + 0.4 \sin(10\,000t + 36°) \text{ volts}$$

11.23 Draw a fundamental wave with 25% second harmonic content with the harmonic in anti-phase at time $t = 0$. Sketch the resultant waveform.

11.24 A capacitor having a reactance of 250 ohms at a frequency of $5000/2\pi$ Hz is connected in series with an inductance whose reactance at the same frequency is 150 ohms. The voltage

$$v = 100 \sin 5000t + 20 \sin 15\,000t \text{ volts}$$

is applied across the circuit. Obtain an expression for the current flowing in the circuit.

12 Transients

Introduction

When a non-sinusoidal voltage waveform is applied to a circuit that consists of one or more inductances and/or capacitances, the resultant current waveform will differ from that of the voltage. The current waveform will generally consist of two parts: a *steady state* component and a *transient* component. The determination of an expression for the current waveform can be carried out with the aid of a mathematical process known as the LAPLACE TRANSFORMATION.

The possible waveforms for the applied voltage and the resulting current are many and can be divided into two main groups. Waveforms are either *periodic* or they are *non-periodic* or *aperiodic*.

A **periodic waveform** is one that repeats itself at regular intervals of time, and some examples are given in Fig. 12.1. These are *a*) the sinusoidal, *b*) the square, *c*) the rectangular, *d*) the sawtooth and *e*) the triangular waveforms. The periodic time T of a periodic waveform is equal to the reciprocal of the fundamental frequency component of the waveform.

An **aperiodic waveform** does *not* repeat itself at regular intervals of time. Some examples of aperiodic waveforms are given in Fig. 12.1*f*, etc. These waves are known as *f*) the ramp, *g*) the delayed ramp, *h*) the exponential decay, *i*) the step, and *j*) the delayed step.

The ramp function increases linearly with increase in time; if the rate of increase is either 1 V/s or 1 A/s, it is said to be a *unit* ramp. The delayed ramp is similar to the ramp but it is delayed by a time of τ seconds. The step function (unit step for 1 V or 1 A) is an abrupt change of voltage from zero to V volts and it is often produced by the closing of a contact or switch connecting a battery to a circuit. The delayed step is merely a step delayed by τ seconds.

The Response of RC and RL Networks to Step and Ramp Waveforms

1 Fig. 12.2 shows a voltage step applied to a circuit that consists of a resistor R in series with a capacitor C. Assuming that the capacitor is initially discharged, all of the step voltage will initially appear across the resistor so that the initial current flowing in the circuit is $I = V/R$. This current flows through the capacitor to charge it up; as the capacitor voltage V_C rises, the voltage across the resistor V_R must fall because the sum of the two voltages is always equal to the applied voltage V. This means, of course, that the circuit current $I = V_R/R = (V - V_C)/R$ falls also. As the capacitor

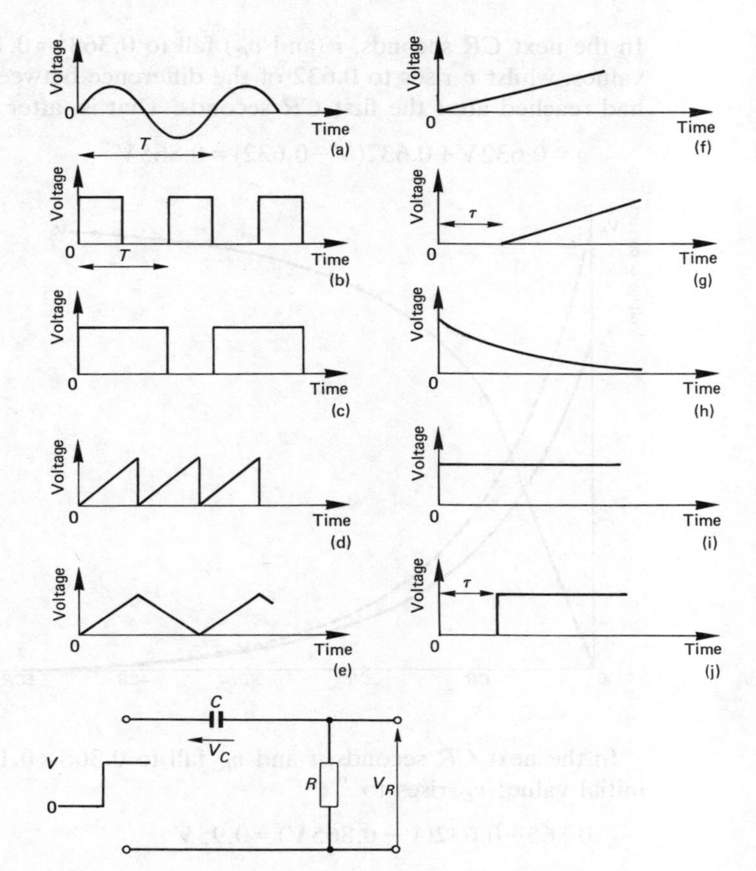

Fig. 12.1 *a*) to *e*) Periodic waveforms, *f*) to *j*) aperiodic waveforms

Fig. 12.2

voltage V_C increases, the current will fall and in so doing will reduce the rate at which the capacitor is charged.

The variations with time of the current and of the capacitor voltage are exponential in their nature and they are given by equations (12.1) and (12.2) respectively.

$$i = \frac{V}{R} e^{-t/CR} \tag{12.1}$$

$$v_C = V(1 - e^{-t/CR}) \tag{12.2}$$

where the product CR is known as the *time constant* of the circuit and has the dimensions of time (seconds). Equations (12.1) and (12.2) are derived on page 200.

The **time constant** of a circuit is the time in which the circuit current would fall to zero *if* the initial rate of decrease were to be maintained. When the time t is equal to the time constant,

$$i = \frac{V}{R} e^{-1} = 0.368I \quad \text{and} \quad v = (1 - 0.368)V = 0.632\ V$$

In the next CR seconds, i (and v_R) fall to $0.368^2 = 0.135$ times their initial values, whilst v rises to 0.632 of the difference between V and the value it had reached after the first CR seconds. That is, after $2CR$ seconds,

$$v = 0.632V + 0.632(V - 0.632) = 0.865V$$

Fig. 12.3

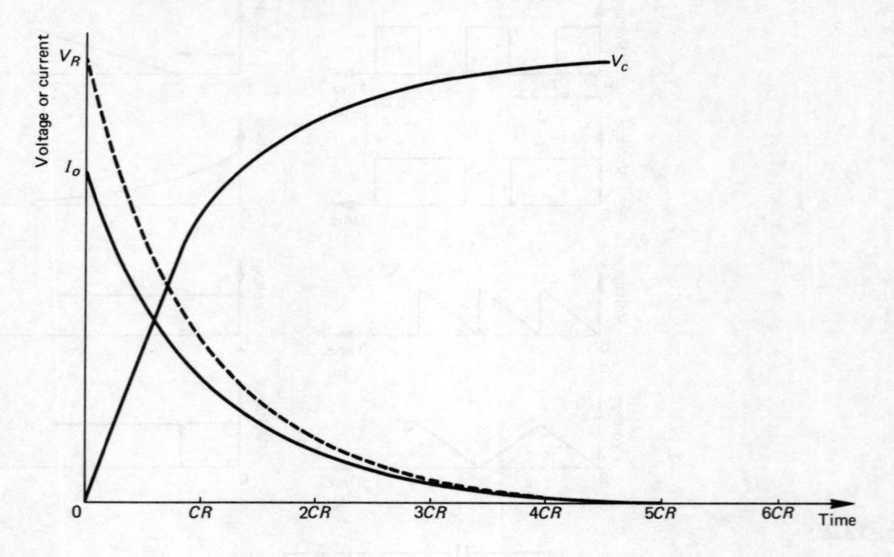

In the next CR seconds, i and v_R fall to $0.368 \times 0.135 = 0.05$ times their initial value; v_C rises to

$$0.865 + 0.632(V - 0.865V) = 0.95V$$

Similarly, after $4CR$ seconds, i and v_R are equal to 0.018 times their initial value and $v_C = 0.982V$ and so on. After $5CR$ seconds the current has very nearly fallen to zero and the capacitor voltage has almost risen to its maximum value of V volts (see Fig. 12.3). In practice, it is usual to assume that the transient is completed after $5CR$ seconds.

2 When a ramp voltage of V volts/second is applied to a series RC circuit (Fig. 12.4), the current in the circuit, and hence the voltage across R, increase with increase in time according to

$$i = CV(1 - e^{-t/CR}) \tag{12.3a}$$

$$v_R = CVR(1 - e^{-t/CR}) \tag{12.3b}$$

These two equations are derived on page 201.

Fig. 12.4

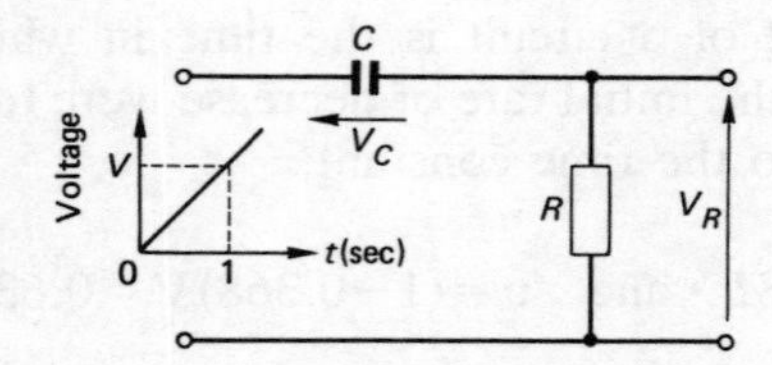

Fig. 12.5

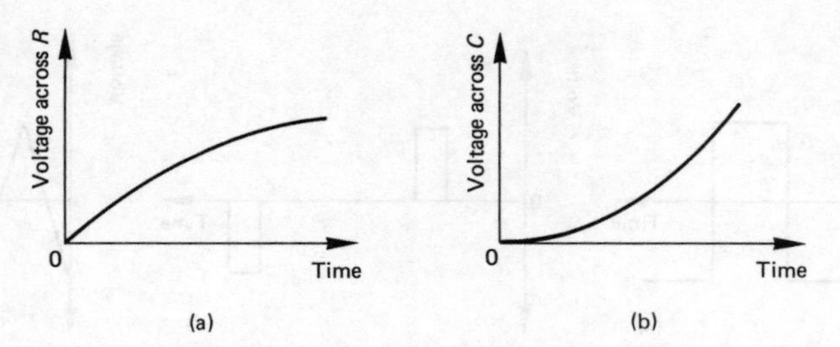

The voltage v_C across the capacitor is equal to the applied voltage minus the voltage drop across the resistor. Thus

$$v_C = Vt - CVR(1 - e^{-t/CR}) \tag{12.4}$$

The waveforms described by these equations are given in Figs. 12.5*a* and *b*.

3 When the step, or the ramp, voltage ends and is replaced by a short-circuit ($V = 0$), the capacitor will discharge through the resistor at a rate determined by the time constant. At any instant the circuit current is $i = (V/R)e^{-t/CR}$ and the voltage v_C is given by $v = Ve^{-t/CR}$.

The response of an *RC* circuit to an applied ramp or step voltage depends upon its time constant. Consider the case where $t/CR \ll 1$. Since

$$e^x = 1 + x + (x^2/2!) + (x^3/3!) + \cdots$$

equations (12.1) and (12.2) can be re-written in the form

$$i = \frac{V}{R}\left[1 - \frac{t}{CR} + \cdots\right] \simeq V/R$$

$$v_C = V\left[1 - \left(1 - \frac{t}{CR} + \cdots\right)\right] \simeq Vt/CR = \frac{1}{CR}\int_0^t V\,dt$$

Thus the voltage across the capacitor is the *integral* of the applied voltage. A similar result can be obtained from equations (12.3*a*) and (12.3*b*), (see Exercise 12.15).

The same result can be also obtained in another manner.

$$V = V_R + V_C \quad \text{but since} \quad CR/t \ll 1$$

$$V_C \ll V_R \quad \text{and so} \quad V \simeq V_R = iR$$

Therefore,

$$v_C = Q/C = \frac{1}{C}\int_0^t i\,dt \simeq \frac{1}{C}\int_0^t \frac{V}{R}\,dt$$

$$\text{or} \quad v_C = \frac{1}{CR}\int_0^t v\,dt \tag{12.5}$$

If the time constant of the circuit is large so that $CR/t \gg 1$, then

$$V = v_R + v_c \simeq v_c \quad \text{and} \quad i = dq/dt = C\,dv_C/dt \simeq C\,dV/dt$$

and hence

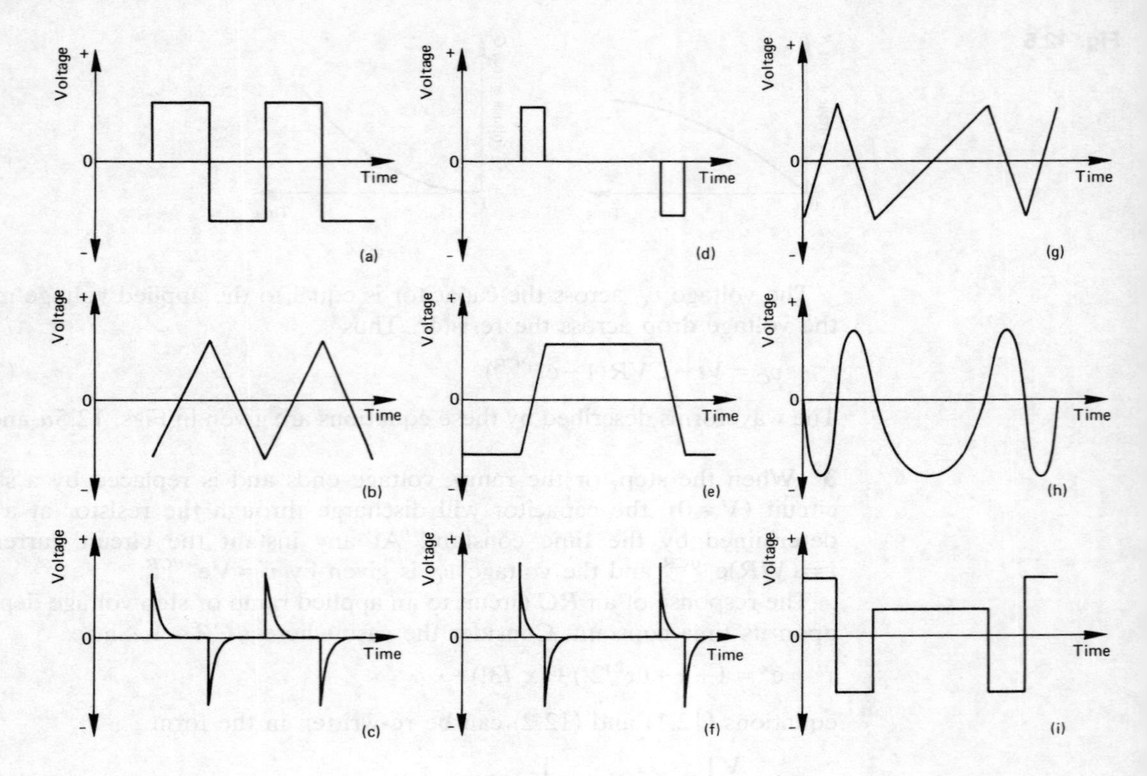

Fig. 12.6

$$v_R = iR = CR\, dV/dt \tag{12.6}$$

Equation (12.6) shows that the output voltage of the circuit is the differential with respect to time of the input voltage.

Some examples of the waveforms that can be produced by the processes of differentiation and integration are given in Fig. 12.6. Three waveforms *a*, *d* and *g* are shown integrated in *b*, *e* and *h* and differentiated in *c*, *f* and *i* respectively.

Response of RC Networks to Rectangular Pulses

1 Consider that the perfectly rectangular pulse of width τ seconds shown in Fig. 12.7*a* is applied to the input terminals of the *RC* network of Fig. 12.7*b*. Suppose that τ is very much less than the time constant, *CR* seconds, of the network and that the capacitor is initially discharged. The capacitor is unable to charge instantaneously and at the moment the pulse is first applied acts as a short-circuit. When the pulse is first applied the maximum current will flow in the circuit and the voltage across *R*, which is the output voltage, will jump to the same value as the input pulse. As the capacitor charges up, the voltage across *R* will fall but since $\tau \ll CR$ the reduction in the output voltage will be very little at the time the pulse ends. When the

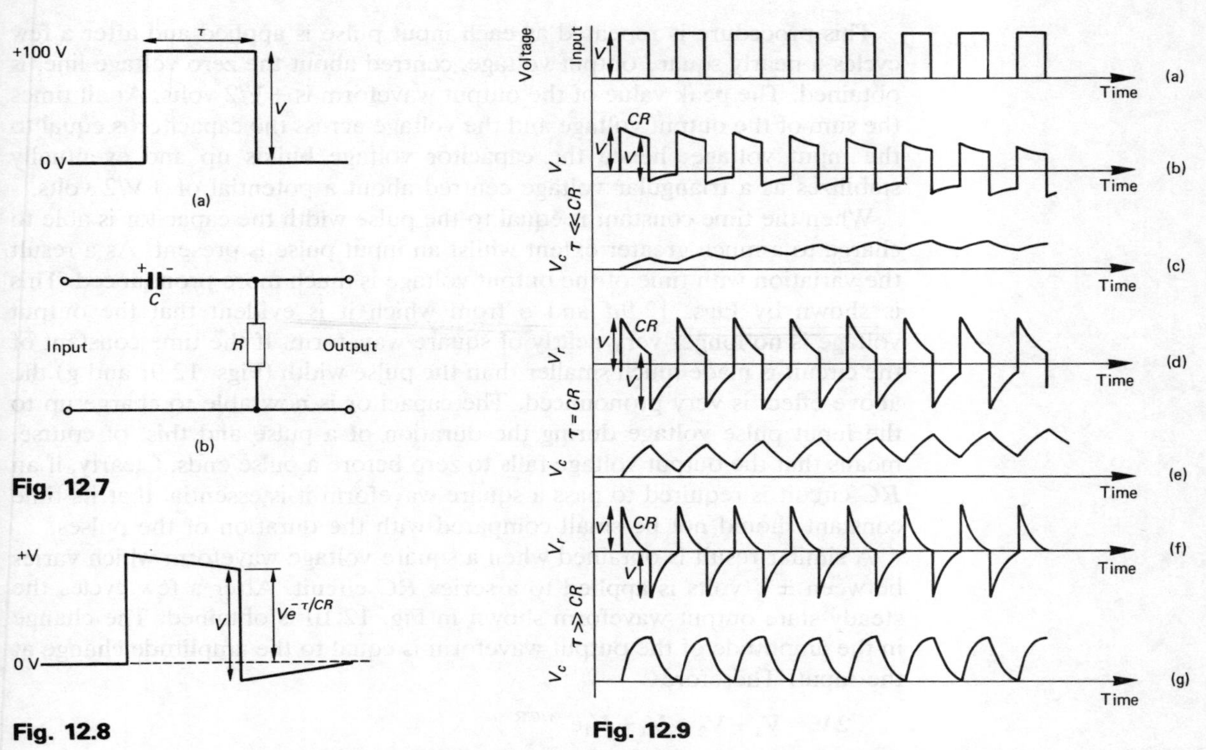

Fig. 12.7

Fig. 12.8

Fig. 12.9

input pulse ends, the capacitor will commence to discharge through resistor R with time constant CR seconds. The polarity of the capacitor voltage is shown in the diagram and so the discharge current will flow in the opposite direction to the charging current. Because of this the output voltage of the circuit suddenly falls by $-V$ volts, taking it negative, and then decays towards zero volts with time constant CR seconds. The output voltage waveform is shown in Fig. 12.8.

2 If a square waveform varying between 0 and $+V$ volts (Fig. 12.9*a*) is applied to the circuit of Fig. 12.7*b*, the output voltage waveform will take a few cycles to settle down to a *steady-state* value. Figs. 12.9*b* and *c* show how the voltages across the resistor and the capacitor vary when the pulse width τ is very much less than the time constant CR of the circuit. When the first pulse is applied, the voltage across R abruptly rises from zero to $+V$ volts and then starts to decay towards zero volts as the capacitor charges up. When the first pulse ends, the output voltage has fallen to $Ve^{-\tau/CR}$ volts and rapidly changes negatively by V volts. The output voltage is taken negative and discharges from this value towards zero but before it reaches zero the second input pulse is applied. The output voltage now rises in the positive direction by V volts but does not reach $+V$ volts because it has started from a small negative voltage. As the capacitor charges, the output voltage falls exponentially and at the end of the second pulse again falls suddenly by $-V$ volts.

This procedure is repeated as each input pulse is applied and after a few cycles a nearly square output voltage, centred about the zero voltage line, is obtained. The peak value of the output waveform is $+V/2$ volts. At all times the sum of the output voltage and the voltage across the capacitor is equal to the input voltage; hence the capacitor voltage builds up and eventually stabilizes as a triangular voltage centred about a potential of $+V/2$ volts.

When the time constant is equal to the pulse width the capacitor is able to charge to a much greater extent whilst an input pulse is present. As a result the variation with time of the output voltage is much more pronounced. This is shown by Figs. 12.9*d* and *e* from which it is evident that the output voltage is no longer very nearly of square waveform. If the time constant of the circuit is made much smaller than the pulse width (Figs. 12.9*f* and *g*) the above effect is very pronounced. The capacitor is now able to charge up to the input pulse voltage during the duration of a pulse and this, of course, means that the output voltage falls to zero before a pulse ends. Clearly, if an *RC* circuit is required to pass a square waveform it is essential that its time constant should not be small compared with the duration of the pulses.

A similar result is obtained when a square voltage waveform which varies between $\pm V$ volts is applied to a series *RC* circuit. After a few cycles the steady-state output waveform shown in Fig. 12.10 is obtained. The change in the amplitude of the output waveform is equal to the amplitude change at the input. Therefore

$$2V = V_1 + V_2 = V_1 + V_1 e^{-\tau/CR}$$

$$V_1 = \frac{2V}{1 + e^{-\tau/CR}} \tag{12.7}$$

Fig. 12.10

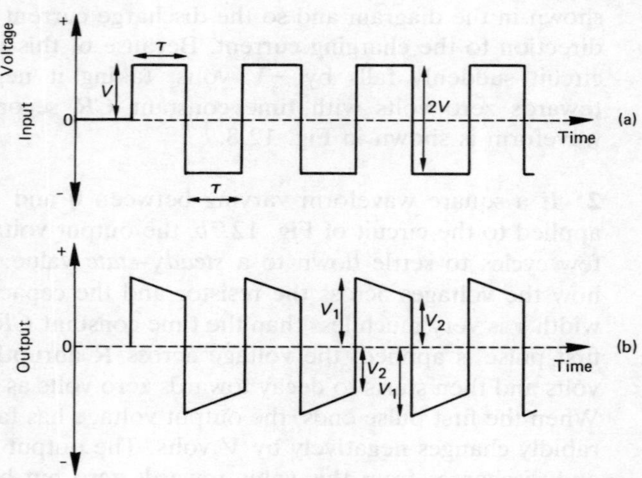

Example 12.1

A ±20 V square waveform is applied to the input terminals of a circuit that consists of a 100 kΩ resistor connected in series with a 0.01 μF capacitor. If the pulse duration is 0.05 ms sketch the output waveform.

Solution From equation (12.7)

$$V_1 = \frac{40}{1+e^{-0.05}} = 20.5\text{ V}$$

$$V_2 = 20.5e^{-0.05} = 19.5\text{ V}$$

The required waveform is shown in Fig. 12.11.

Fig. 12.11

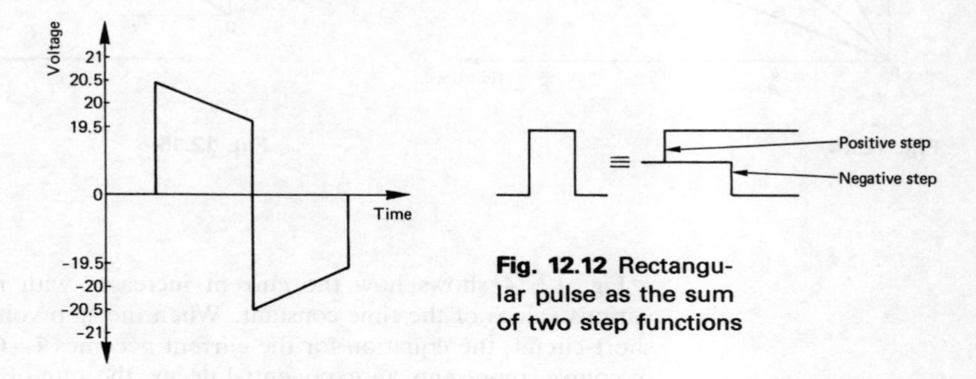

Fig. 12.12 Rectangular pulse as the sum of two step functions

A rectangular pulse can be considered to be the combination of a positive step followed, after a time equal to the pulse duration, by a negative step (see Fig. 12.12).

Response of RL Networks

When a step voltage is applied to a series RL circuit (Fig. 12.13), the sum of

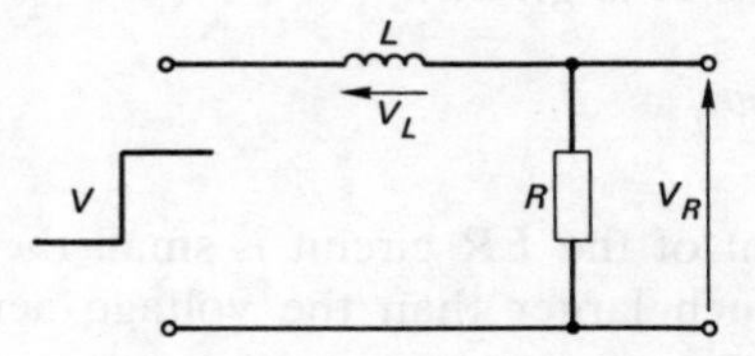

Fig. 12.13

the voltages across L and R is equal to the applied voltage V, i.e. $V = V_L + V_R$. The current i in the circuit, and hence the voltage V_R across the resistance, increase exponentially with time according to

$$i = \frac{V}{R}(1 - e^{-Rt/L}) \tag{12.8a}$$

$$v_R = V(1 - e^{-Rt/L}) \tag{12.8b}$$

where R/L is the *time constant* of the circuit and has the same meaning as before. The voltage v_L across the inductance is $v_L = V - V_R$ or

$$v_L = Ve^{-Rt/L} \tag{12.9}$$

Equations (12.8) and (12.9) are derived on page 202.

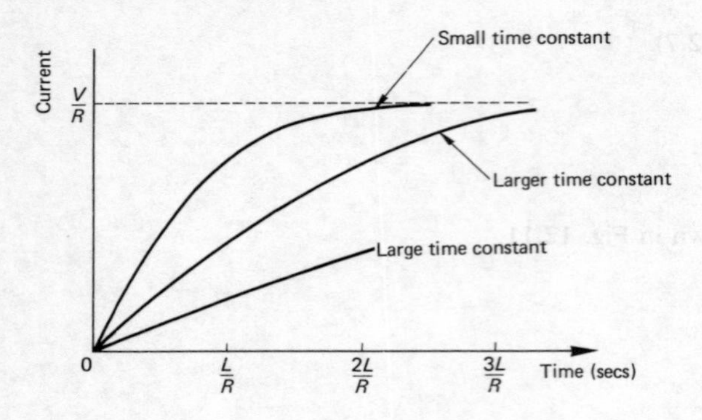

Fig. 12.14

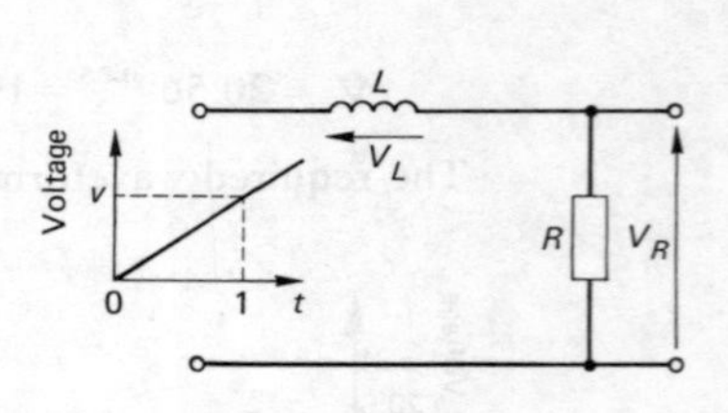

Fig. 12.15

Fig. 12.14 shows how the current increases with increase in time for various values of the time constant. When the step voltage is replaced by a short-circuit, the equation for the current becomes $i=(V/R)\mathrm{e}^{-Rt/L}$ and this, of course, represents an exponential decay, the rate of which is determined by the time constant of the circuit. The voltages v_L and v_R both decay exponentially at the same rate.

When a ramp voltage of V volts per second is applied to a series RL circuit (Fig. 12.15), the current in the circuit is given by

$$i=\frac{V}{R}t-\frac{VL}{R^2}(1-\mathrm{e}^{-Rt/L}) \tag{12.10}$$

and the voltage across L is given by

$$v_L=\frac{VL}{R}(1-\mathrm{e}^{-Rt/L}) \tag{12.11}$$

If the time constant of the LR circuit is small the voltage V_R across the resistance will be much larger than the voltage across the inductance L. Then

$$v=v_L+v_R\simeq v_R=iR$$

$$v_L=L\,di/dt=\frac{L}{R}\,dv/dt$$

and so the voltage across the inductance is the *differential* of the input voltage.

Conversely, if the time constant is large, $v\simeq v_L=L\,di/dt$. Then

$$i=\frac{1}{L}\int_0^t v\,dt \quad \text{and} \quad v_R=iR=\frac{R}{L}\int_0^t V\,dt$$

This means that the voltage developed across the resistance is the *integral* of the input voltage.

Laplace Transforms and Partial Fractions

When a voltage or a current is suddenly applied to a circuit containing capacitors and/or inductors, the currents and/or voltages produced at various points in the circuit will, in general, have two components. One of these is known as the *steady-state* response since it is the current or voltage that continues to exist for as long as the applied voltage or current is maintained. The other component is known as the *transient* and this will only exist for a short time after the application of the input voltage or current. Consider, for example, the series *RL* circuit shown in Fig. 12.16. When the switch is

Fig. 12.16

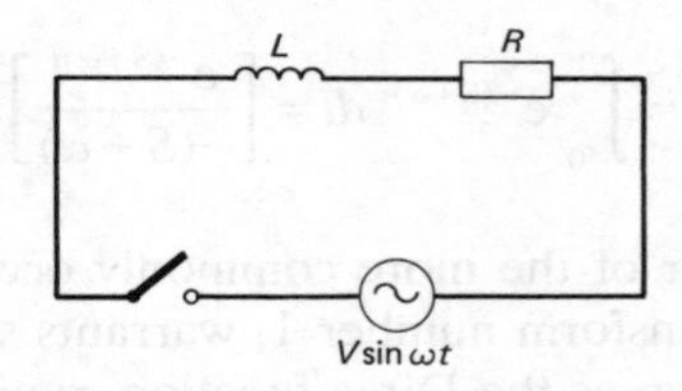

closed, the source voltage $v = V \sin \omega t$ is suddenly applied to the circuit and the current i that flows is

$$i = I_m \sin(\omega t + \theta) + I_t e^{-Rt/L} \tag{12.12}$$

The first term in equation (12.12) is, of course, the current that would be determined using ordinary a.c. theory and this is the *steady-state current.* The second, transient, term is an initial current, of peak value I_t, that flows at the instant the switch is closed and then decays at a rate determined by the time constant L/R of the circuit. Clearly, after some time t the transient term will be negligibly small and then the circuit current will consist solely of the steady-state component.

A convenient method for the calculation of the complete response of a circuit involves the use of both the LAPLACE TRANSFORMATION and of PARTIAL FRACTIONS. The theory and application of Laplace transforms is normally considered in a mathematics course and hence only a brief introduction will be provided here.

Laplace Transforms

The **Laplace transform** $f(S)$ of a function of time $f(t)$ is defined as

$$f(S) = \int_{-0}^{\infty} f(t) e^{-St}\, dt \tag{12.13}$$

S is a complex variable of the form $S = \alpha + j\omega$ where both α and ω have the dimensions of seconds^{-1}. The lower limit of integration is made -0 in order to include all functions that occur at time $t = 0$.

In the solution of an electrical problem it is usually not necessary to evaluate the integral given in equation (12.13) since tables giving the transforms of all the more commonly occurring waveforms are readily available. However, a few of the simpler cases will be evaluated as examples.

a) *Unit Step Function* $f(t)=1$

$$f(S)=\int_{-0}^{\infty} e^{-St}\,dt=\left[\frac{-e^{-St}}{S}\right]_{-0}^{\infty}=\frac{1}{S}$$

b) *Unit Ramp Function* $f(t)=t$

$$f(S)=\int_{-0}^{\infty} te^{-St}\,dt=\left[\frac{-e^{-St}t}{S}\right]_{-0}^{\infty}-\int_{-0}^{\infty}\frac{-e^{-St}}{S}\,dt=\left[\frac{-e^{-St}}{S^2}\right]_{-0}^{\infty}=\frac{1}{S^2}$$

c) *Exponential Delay* $f(t)=e^{-\alpha t}$

$$f(S)=\int_{-0}^{\infty} e^{-\alpha t}e^{-St}\,dt=\int_{-0}^{\infty} e^{-(S+\alpha)t}\,dt=\left[\frac{e^{-(S+\alpha)t}}{-(S+\alpha)}\right]_{-0}^{\infty}=\frac{1}{S+\alpha}$$

Table 12.1 lists a number of the more commonly occurring transforms.

The unit impulse δ, transform number 1, warrants some further mention. This function, often known as the Dirac function, represents a pulse of unit area and of width δt, where $\delta t \to 0$. Since the pulse width tends to zero, the height of the pulse must tend to infinity (see Fig. 12.17).

Two theorems are available that can often reduce considerably the work involved in solving an electrical circuit.

a) *Initial Value Theorem*

$$\lim_{t\to 0} f(t)=\lim_{S\to\infty}[Sf(S)]$$

b) *Final Value Theorem*

$$\lim_{t\to\infty} f(t)=\lim_{S\to 0}[Sf(S)]$$

Suppose, for example, that

$$f(t)=i=I_0e^{-t/RC}=1e^{-t/2}$$

The Laplace transform is, from Table 12.1 (number 6),

$$f(S)=\frac{1}{S+0.5} \qquad \text{hence } Sf(S)=\frac{S}{S+0.5}$$

The initial value of $f(t)$ is $\dfrac{\infty}{\infty+0.5}=1$

The final value of $f(t)=\dfrac{0}{0+0.5}=0$

Partial Fractions

Before a given $f(S)$ can be transformed into the time world it is very often necessary to manipulate the equation into a form that can be recognized in Table 12.1. For this step to be carried out the use of **partial fractions** is frequently essential.

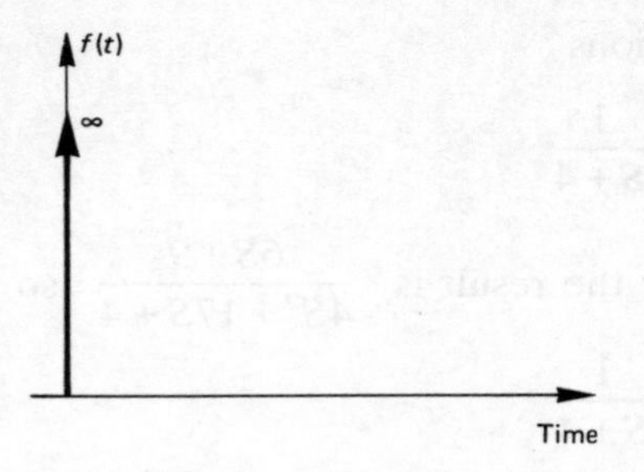

Fig. 12.17 The impulse function

Table 12.1 Laplace transforms

$f(t)$	$f(S)$
1 Unit impulse δ	1
2 Unit step function	$\dfrac{1}{S}$
3 Delayed unit step function	$\dfrac{e^{-ST}}{S}$
4 Rectangular pulse (of duration T)	$\dfrac{1-e^{-ST}}{S}$
5 Unit ramp function	$\dfrac{1}{S^2}$
6 $e^{-\alpha t}$	$\dfrac{1}{S+\alpha}$
7 $1-e^{-\alpha t}$	$\dfrac{\alpha}{S(S+\alpha)}$
8 $te^{-\alpha t}$	$\dfrac{1}{(S+\alpha)^2}$
9 $e^{-\alpha t}-e^{-\beta t}$	$\dfrac{\beta-\alpha}{(S+\alpha)(S+\beta)}$
10 $\sin \omega t$	$\dfrac{\omega}{S^2+\omega^2}$
11 $\cos \omega t$	$\dfrac{S}{S^2+\omega^2}$
12 $df(t)/dt=f'(t)$	$Sf(S)-f(0)$
13 $d^2f(t)/dt^2=f''(t)$	$S^2f(S)-Sf(0)-f'(0)$
14 $\int_0^t f(t)\,dt$	$\dfrac{f(S)}{S}+\dfrac{f(0)}{S}$

If the two equations

$$\frac{2}{4S+1} \quad \text{and} \quad \frac{1}{S+4}$$

are added together the result is $\dfrac{6S+9}{4S^2+17S+4}$ so

$$\frac{2}{4S+1} \quad \text{and} \quad \frac{1}{S+4}$$

are the partial fractions of $\dfrac{6S+9}{4S^2+17S+4}$

The problem is how the partial fractions of a given equation can be determined. One condition must be satisfied by an equation before it is possible to find its partial fractions. This is

The highest power of S in the numerator must be *less* than the highest power of S in the denominator. If this is not the case the numerator must be divided by the denominator to produce a remainder that does satisfy this requirement.

Thus $\dfrac{S+4}{S^2+3S+2}$

can have its partial fractions found but

$$\frac{S^2+4}{S^2+3S+2}$$

cannot. However, dividing the numerator of this latter equation by its denominator gives

$$\begin{array}{r|l} & \quad 1 \\ S^2+3S+2 & S^2+4 \\ & S^2+3S+2 \\ \hline & -3S+2 \end{array}$$

$$=1+\frac{-3S+2}{S^2+3S+2}$$

and the partial fractions of $\dfrac{-3S+2}{S^2+3S+2}$ can be found.

Partial fractions have denominators that fall into one of a number of different types:

A *Linear factors*

$$\frac{f(S)}{(S+a)(S+b)(S+c)}=\frac{A}{S+a}+\frac{B}{S+b}+\frac{C}{S+c}$$

where any one of a, b or c may be zero.

B *Repeated linear factors*

$$\frac{f(S)}{(S+a)^n} = \frac{A}{S+a} + \frac{B}{(S+a)^2} + \cdots + \frac{N}{(S+a)^n}$$

C *Quadratic*

$$\frac{f(S)}{aS^2+bS+c} = \frac{AS+B}{aS^2+bS+c}$$

Often it will prove necessary to modify the denominator by "completing the square", i.e.

$$aS^2+bS+c = S^2 + \frac{b}{a}S + \frac{c}{a}$$

$$= \left(S + \frac{b}{2a}\right)^2 + \frac{c}{a} - \frac{b}{4a^2}$$

The determination of the values for A, B, C, etc. can be carried out using either of two methods or perhaps a combination of each. These two methods will be illustrated by means of the following example.

Exercise 12.2

Determine the partial fractions of $\dfrac{S+4}{S^2+3S+2}$

Solution

$$\frac{S+4}{S^2+3S+2} = \frac{S+4}{(S+2)(S+1)} \equiv \frac{A}{S+2} + \frac{B}{S+1}$$

$$= \frac{A(S+1)+B(S+2)}{(S+2)(S+1)}$$

Therefore $S+4 = A(S+1)+B(S+2)$

Method 1 Make one of the unknowns zero by suitable choice of the value of S.

i) Let $S=-1$, then

$$-1+4=3=B(-1+2)=B$$

ii) Let $S=-2$, then

$$2=A(-2+1)=-A \quad \text{and so}$$

$$\frac{S+4}{(S+2)(S+1)} = \frac{-2}{S+2} + \frac{3}{S+1}$$

Method 2 Equate similar coefficients on either side of the identity. Thus

$$S+4 = AS+A+BS+2B$$

Equating like coefficients of S:

i) Constants $4 = A+2B$

ii) S $\quad 1 = A+B$

Hence, on subtracting, $3=B$ (as before) and $4=A+6$ or $A=-2$ (as before).

The Solution of Circuits Using Laplace Transforms

Consider the series *RC* circuit of Fig. 12.2 again when a step is applied. Applying Kirchhoff's voltage law to the circuit,

$$V = iR + (q/C) = iR + \frac{1}{C}\int_0^t i\,dt$$

Taking transforms

$$\frac{V}{S} = i(S)R + \frac{i(S)}{SC} = i(S)(R + 1/SC)$$

(no. 2) (no. 14)

The complex impedance $Z(S)$ of the circuit is

$$Z(S) = \frac{V(S)}{i(S)} = R + \frac{1}{SC}$$

It should be noted that this result could have been obtained by writing down the impedance of the circuit to a *sinusoidal* voltage and then replacing $j\omega$ with S throughout. This is true for any other circuit as well so that the "impedance" of an inductor can be written as SL.

Returning to the problem,

$$i(S) = \frac{V}{S(R + 1/SC)}$$

It is now necessary to put this equation into a form that can be recognized as one of the terms given in Table 12.1. Therefore,

$$i(S) = \frac{VSC}{S(SRC+1)} = \frac{VSC}{SRC(S+1/CR)} = \frac{V}{R}\cdot\frac{1}{S+1/CR}$$

This can be seen to correspond with Laplace transform number 6, where $\alpha = 1/CR$. Therefore,

$$i(t) = \frac{V}{R}e^{-t/CR} \qquad (12.1)\ (again)$$

The voltage across the capacitor is

$$v_C(S) = \frac{i(S)}{SC} = \frac{V}{SCR(S+1/CR)} = \frac{A}{S} + \frac{B}{S+1/CR}$$

or $$\frac{V}{CR} = AS + (A/CR) + BS$$

Equating like coefficients:

i) Constants $\frac{V}{CR} = \frac{A}{CR}$ or $A = V$

ii) S $0 = A + B$ or $B = -A = -V$ so that

$$v_C(S) = \frac{V}{S} - \frac{V}{S+1/CR}$$

$$v_C(t) = V(1 - e^{-t/CR}) \qquad (12.2)\ (again)$$

(no. 2 and no. 6)

The method adopted in the solution of the series RC circuit should be followed for all other circuits. The method may be summarized by the following listed steps:

A Apply Kirchhoff's laws to the circuit and write down the equation describing the operation of the circuit.
B Re-write the equations in terms of Laplace transforms (very often the analysis can commence with this step).
C Solve the equation(s) in terms of the unknown quantity(ies).
D Re-arrange the equations into a form that can be recognized in the available table of Laplace transforms.
E Use the appropriate parts of the table to write down the solution as a function of time.

Usually it is step **D** that presents the greatest difficulty encountered in a given problem. It should be noted from Table 12.1 that none of the given transforms have any coefficient of S other than unity and that $1/S$ does not occur in either the numerator or the denominator of any term.

When a ramp is applied to Fig. 12.4, equations (12.3), (12.4), (12.8), (12.10), (12.11) can now be derived.

$$\frac{V}{S^2} = i(S)(R + 1/SC)$$

$$i(S) = \frac{V}{S^2(R+1/SC)} = \frac{VC}{S(1+SCR)} = \frac{VC}{SCR(S+1/CR)}$$

$$\frac{V}{RS(S+1/CR)} = \frac{A}{S} + \frac{B}{S+1/CR}$$

$$\frac{V}{R} = AS + \frac{A}{CR} + BS$$

Equate the constants $V/R = A/CR$ or $A = VC$
Equate S $0 = A + B$ or $B = -A = -VC$

Hence $i(S) = \dfrac{VC}{S} - \dfrac{VC}{S+1/CR}$

and $i(t) = CV(1 - e^{-t/CR})$ (12.3a) (again)

$$v_C(S) = \frac{i(S)}{SC} = \frac{V}{S^2} - \frac{V}{S(S+1/CR)}$$

Now

$$\frac{V}{S(S+1/CR)} = \frac{A}{S} + \frac{B}{S+1/CR} = A(S+1/CR) + BS$$

Let $S = 0$ then $V = A/CR$ or $A = VCR$
Let $S = -1/CR$ then $V = -B/CR$ or $B = -VCR$
Therefore,

$$v_C(S) = \frac{V}{S^2} - \frac{VCR}{S} + \frac{VCR}{S+1/CR}$$

$$v_C(t) = Vt - VCR(1 - e^{-t/CR}) \qquad (12.4)\ \text{(again)}$$

When a step voltage is applied to the inductive circuit of Fig. 12.13, then $V = iR + L\,di/dt$. Taking transforms,

$$\frac{V}{S} = i(S)(R+SL)$$

Therefore,

$$i(S) = \frac{V}{S(R+SL)} = \frac{V/L}{S(S+R/L)} = \frac{A}{S} + \frac{B}{S+R/L} = AS + (AR/L) + BS$$

Equating constants $V/L = AR/L$ or $A = V/R$
Equating S $0 = A + B$ or $B = -A = -V/R$

Hence $i(S) = \dfrac{V}{RS} - \dfrac{V/R}{S+R/L}$

and $i(t) = \dfrac{V}{R}(1 - e^{-Rt/L})$ (12.8a) (again)

When a ramp voltage of V volt/sec is applied to the same circuit

$$V/S^2 = i(S)(R+SL)$$

and

$$i(S) = \frac{V/L}{S^2(S+R/L)}$$

$$= \frac{AS+B}{S^2} + \frac{C}{S+R/L}$$

$$= AS^2 + \frac{ASR}{L} + BS + CS^2 + \frac{BR}{L}$$

Equating constants $V/L = BR/L$ or $B = V/R$
Equating S $0 = B + (AR/L) = (V/R) + (AR/L)$ or $A = -VL/R^2$
Equating S^2 $0 = A + C$ or $C = VL/R^2$
Therefore,

$$i(S) = \frac{\dfrac{V}{R} - \dfrac{VSL}{R^2}}{S^2} + \frac{\dfrac{VL}{R^2}}{S+R/L}$$

$$= \frac{V/R}{S^2} - \frac{VL/R^2}{S} + \frac{VL/R^2}{S+R/L}$$

and $i(t) = \dfrac{V}{R}t - \dfrac{VL}{R^2}(1 - e^{-tR/L})$ (12.10) (again)

The voltage $v_L(t)$ across the inductance is obtained from

$$v_L(S) = i(S)SL = \frac{VL/R}{S} - \frac{VL^2}{R^2} - \frac{SVL^2/R^2}{S+R/L}$$

$$= \frac{VL/R}{S} - \frac{VL^2}{R^2}\left[1 - \left(1 - \frac{R/L}{S+R/L}\right)\right]$$

$$= \frac{VL}{SR} + \frac{VL}{R[S+R/L]}$$

and $v_L(t) = \frac{VL}{R}(1 - e^{-Rt/L})$ (12.11) (again)

When the step is removed and replaced by a short-circuit

$$0 = iR + L\,di/dt = i(S)(R + SL) - Li(0) \qquad \text{[see (12.17)]}$$
$$= i(S)(R + SL) - LV/R$$

and $i(S) = \frac{LV}{R(R+SL)} = \frac{LV}{LR[S + R/L]} = \frac{V/R}{S + R/L}$

and $i(t) = \frac{V}{R} e^{-Rt/L}$ (12.14)

Initial Conditions

Sometimes a capacitor has an initial charge, or an inductor is already carrying a current, at time $t = 0$.

A The voltage across the capacitor is then

$$v_C(t) = v_{C0} + \frac{1}{C}\int_0^t i(t)\,dt$$

or $v_C(S) = \frac{v_{C0}}{S} + \frac{i(S)}{SC}$ (12.15)

or the current in the capacitor is $i(t) = C\,dv_C(t)/dt$ and

$$i(S) = C[Sv_C(S) - v_{C0}] \qquad (12.16)$$

B The voltage across the inductor is then $v_L(t) = L\,di(t)/dt$

and $v_L(S) = L[Si(S) - i(0)]$ (12.17)

or the current in the inductor is then

$$i_L(t) = i_0 + \frac{1}{L}\int_0^t v_L(t)\,dt$$

or $i(S) = \frac{i_0}{S} + \frac{v_L(S)}{SL}$ (12.18)

Consider, as an example, the capacitive and inductive circuits given, respectively, in Figs. 12.2 and 12.13 when the applied voltage is removed and replaced by a short-circuit. For the CR circuit

$$0 = i(t)R + v_C = i(t)R + (q/C) = i(t)R + \frac{1}{C}\int_0^t i(t)\,dt + v_{C0}$$
$$= i(S)[R + 1/SC] - (v_{C0}/S) = i(S)[R + 1/SC] - (V/S)$$

assuming the capacitor is fully charged.

Therefore,

$$i(S) = \frac{V}{S(R+1/SC)} = \frac{V}{R(S+1/CR)}$$

$$\text{and} \quad i(t) = \frac{V}{R}\mathrm{e}^{-t/CR} \tag{12.18}$$

For the LR circuit,

$$0 = i(t)R + L\,di/dt$$

$$0 = i(S)(R+SL) - LI(0) = i(S)(R+SL) - (LV/R)$$

$$i(S) = \frac{LV}{R(R+SL)} = \frac{V}{R(S+R/L)}$$

$$\text{and} \quad i(t) = \frac{V}{R}\mathrm{e}^{-Rt/L} \tag{12.19}$$

Impedance Functions

It was shown on page 200 that the impedance of a capacitor can be written as $1/SC$ and the impedance of an inductor as SL. This means that the normal sinusoidal expressions can be used with S replacing $j\omega$. This step can be extended to the determination of the impedance, or the admittance of any circuit *provided the initial conditions are zero*. Some examples are given in Fig. 12.18.

Fig. 12.18

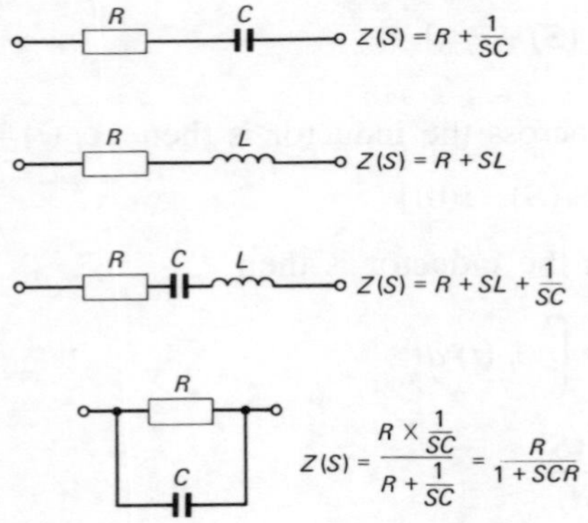

$Z(S) = R + \frac{1}{SC}$

$Z(S) = R + SL$

$Z(S) = R + SL + \frac{1}{SC}$

$$Z(S) = \frac{R \times \frac{1}{SC}}{R + \frac{1}{SC}} = \frac{R}{1+SCR}$$

Transfer Functions

The **transfer function** of a network is the ratio of the output quantity to the input quantity expressed using Laplace transforms. In the case of the voltage transfer function, the output terminals of the network are open-circuited, while for the current transfer function the output terminals are short-circuited.

Three simple examples follow:

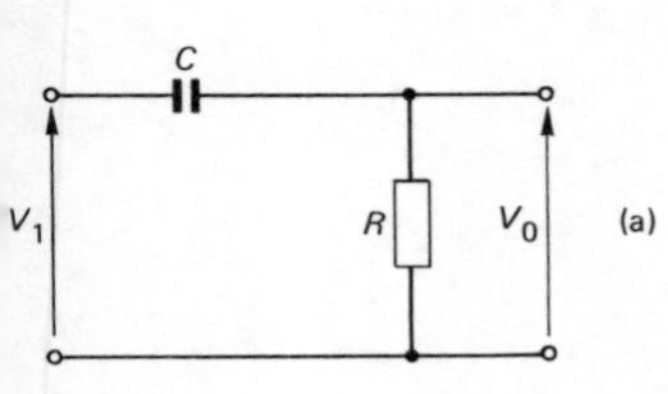

1 Fig. 12.19*a*

$$V_0(S)=\frac{V_1(S)R}{R+1/SC}=\frac{V_1(S)SCR}{1+SCR}$$

or $$\text{T.F.}=\frac{V_0(S)}{V_1(S)}=\frac{SCR}{1+SCR}$$

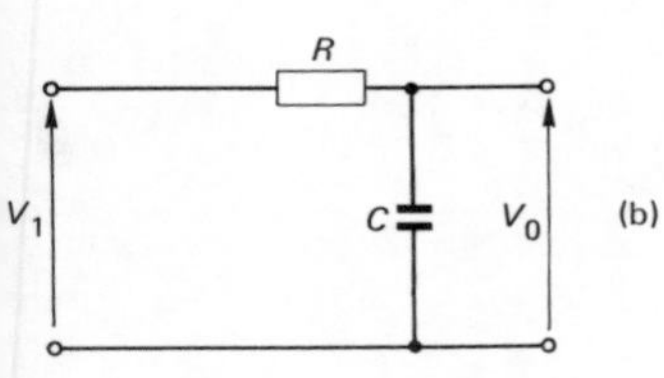

2 Fig. 12.19*b*

$$V_0(S)=\frac{V_1(S)\cdot 1/SC}{R+1/SC}=\frac{V_1(S)}{1+SCR}$$

or $$\text{T.F.}=\frac{V_0(S)}{V_1(S)}=\frac{1}{1+SCR}$$

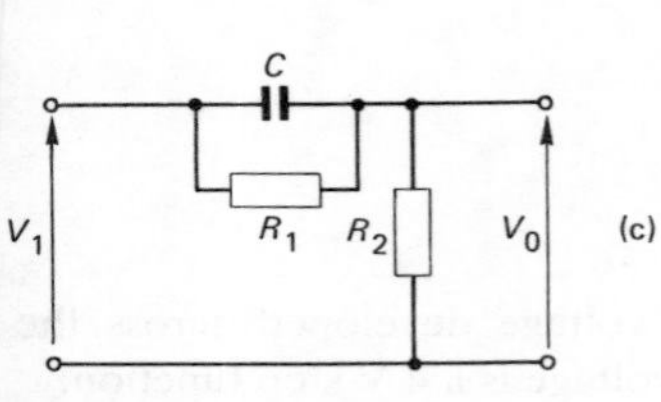

3 Fig. 12.19*c*

$$V_0(S)=\frac{V_1(S)R_2}{\dfrac{R_1}{1+SCR_1}+R_2}=\frac{V_1(S)R_2(1+SCR_1)}{R_1+R_2(1+SCR_1)}$$

or $$\text{T.F.}=\frac{V_0(S)}{V_1(S)}=\frac{R_2(1+SCR_1)}{R_1+R_2+SCR_1R_2}$$

Fig. 12.19

Several examples are now provided to illustrate the technique of solving transient problems using the Laplace transformation. Circuits containing both inductance and capacitance are *not* included nor are circuits subjected to a sinusoidal input voltage because the work involved in their solution is thought to be beyond the scope of this book. However the principles demonstrated are equally valid to such circuits.

1 MΩ

0.1 μF

Fig. 12.20

Example 12.3

Determine an expression for the variation with time of the output voltage of the circuit given in Fig. 12.20 when the input voltage is *a*) a 6 V step function, *b*) a 6 V/s ramp function.

Solution

$$V_0(S)=\frac{V_1(S)\cdot 1/SC}{R+1/SC}=\frac{V_1(S)}{1+SCR}$$

a) $V_1(S)=6/S$, $CR=0.1$

$$V_0(S)=\frac{6}{S(1+0.1S)}=\frac{6}{0.1S(S+1/0.1)}=\frac{60}{S(S+10)}$$

$$\frac{60}{S(S+10)}=\frac{A}{S}+\frac{B}{S+10}=A(S+10)+BS$$

Let $S=0$ then $60=10A$ or $A=6$

Let $S=-10$ then $60=-10B$ or $B=-6$.

Therefore $V_0(S) = \dfrac{6}{S} - \dfrac{6}{S+10}$

and $V_0(t) = 6(1 - e^{-10t})$ (*Ans*)

b) $V_1(S) = 6/S^2$, $CR = 0.1$

$$V_0(S) = \frac{6}{S^2(1+0.1S)} = \frac{60}{S^2(S+10)} = \frac{AS+B}{S^2} + \frac{C}{S+10}$$

$$60 = AS^2 + 10AS + BS + 10B + CS^2$$

Equate constants $60 = 10B$ or $B = 6$

Equate S $\quad 0 = 10A + B$ or $A = -0.6$

Equate S^2 $\quad 0 = A + C$ or $C = 0.6$

$$V_0(S) = \frac{6 - 0.6S}{S^2} + \frac{0.6}{S+10}$$

$$= \frac{6}{S^2} - \frac{0.6}{S} + \frac{0.6}{S+10}$$

Therefore $V_0(t) = 6t - 0.6(1 - e^{-10t})$ (*Ans*)

Example 12.4

Derive an expression for the time variation of the voltage developed across the resistor R_2 in the circuit of Fig. 12.21 when the input voltage is a 4 V step function.

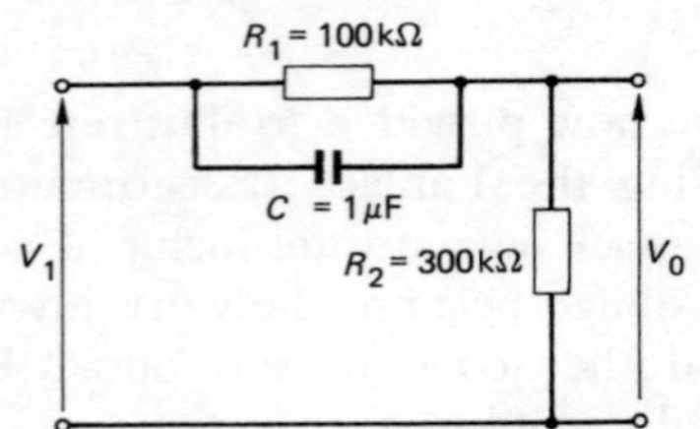

Fig. 12.21

Solution

$$V_0(S) = \frac{V_1(S)R_2}{\dfrac{R_1/SC}{R_1 + 1/SC} + R_2} = \frac{V_1(S)R_2(1 + SCR_1)}{R_1 + R_2 + SCR_1R_2}$$

$V_1(S) = 4/S$ and thence,

$$V_0(S) = \frac{1200 \times 10^3(1 + 0.1S)}{S[(400 \times 10^3) + (30 \times 10^3 S)]} = \frac{120(1 + 0.1S)}{S(40 + 3S)}$$

$$= \frac{120 + 12S}{3S(S + 40/3)} = \frac{40 + 4S}{S(S + 40/3)} = \frac{A}{S} + \frac{B}{S + 40/3}$$

Therefore $40 + 4S = AS + (40A/3) + BS$

Equating constants $40 = 40A/3$ or $A = 3$

Equating S $\quad 4 = A + B$ or $B = 1$

Hence $V_0(S) = \dfrac{3}{S} + \dfrac{1}{S + 40/3}$

and $V_0(t) = 3 + e^{-40t/3}$ (*Ans*)

Example 12.5

Determine an expression for the time variation of the current supplied to the circuit given in Fig. 12.22 when a 100 V step is applied to the terminals AB.

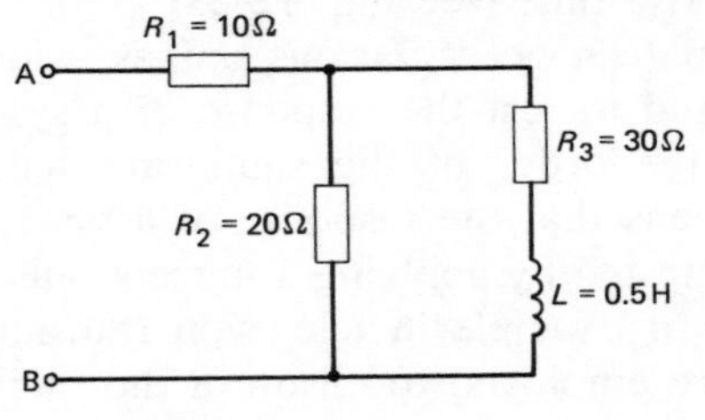

Fig. 12.22

Solution

$$Z_{in}(S) = R_1 + \frac{R_2(R_3 + SL)}{R_2 + R_3 + SL}$$

$$= \frac{R_1R_2 + R_1R_3 + R_2R_3 + SLR_1 + SLR_2}{R_2 + R_3 + SL}$$

or $$Z_{in}(S) = \frac{1100 + 15S}{50 + 0.5S}$$

$$I_{in}(S) = \frac{V_{in}(S)}{Z_{in}(S)} = \frac{100(50 + 0.5S)}{S(1100 + 15S)} = \frac{100(50 + 0.5S)}{15S(S + 73.3)}$$

$$= \frac{333.3 + 3.3S}{S(S + 73.3)} = \frac{A}{S} + \frac{B}{S + 73.3}$$

or $333.3 + 3.3S = AS + 73.3A + BS$

Equating constants $333.3 = 73.3A$ or $A = 4.55$

Equating S $3.3 = A + B$ or $B = -1.25$

Therefore $$I_{in}(S) = \frac{4.55}{S} - \frac{1.25}{S + 73.3}$$

and $I_{in}(t) = 4.55 - 1.25e^{-73.3t}$ (*Ans*)

Test Waveforms

In the measurement of many of the parameters of an electrical or an electronic circuit, a test signal voltage or current is applied to one pair of terminals of the circuit and the response at some other point in the circuit is measured. The test signal waveform is most often one of the following:

A Sinusoidal
B Square
C Pulse (many versions)
D Ramp

A A *sinusoidal* voltage source such as a laboratory oscillator or a signal generator is commonly used, in conjunction with a c.r.o., to measure a number of amplifier parameters. The more important of these include *a*) the measurement of gain, *b*) the measurement of distortion, *c*) the measurement of input and/or output impedance, and *d*) the measurement of efficiency.

The methods used for these tests are described elsewhere [EIII].

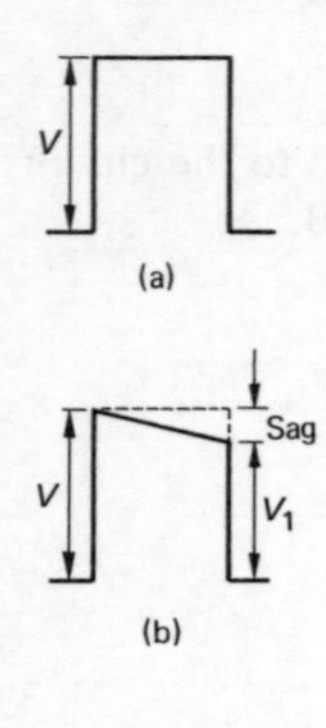

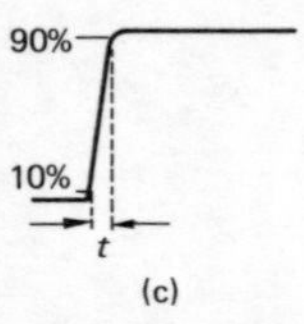

Fig. 12.23 Sag of a pulse waveform

B *Square* waveforms can also be used to test the performance of an amplifier. If the square waveform is applied to the input of an amplifier, any imperfections in the circuit will cause the output waveform to exhibit *sag* and/or a non-zero rise-time (see Fig. 12.23).

C *Pulse* waveforms can be of various shapes, widths, and periodic times and can also be used to test the response of a system. The narrower the pulse, the higher the order of the significant harmonics present in the waveform. This means that the response of a system over a wide band of frequencies can be tested by applying a narrow pulse to its input terminals. In some cases, as for example in television transmission circuits, complex pulse waveforms are employed to ensure a thorough test of the important parameters of the circuit.

D The *ramp* waveform is less often used to test the performance of a system. Clearly its main application arises whenever the response of a system to a linearly changing voltage is to be measured.

Exercises 12

12.1 A V volt/sec ramp voltage is applied to the circuit of Fig. 12.24. Derive an expression for the subsequent variation with time of the current flowing in the circuit when the initial voltage stored in the capacitor is zero.

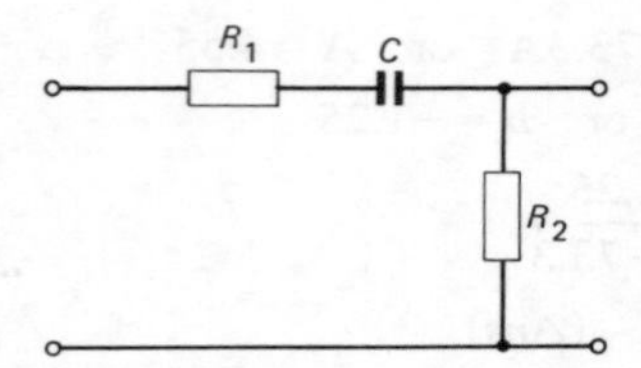

Fig. 12.24

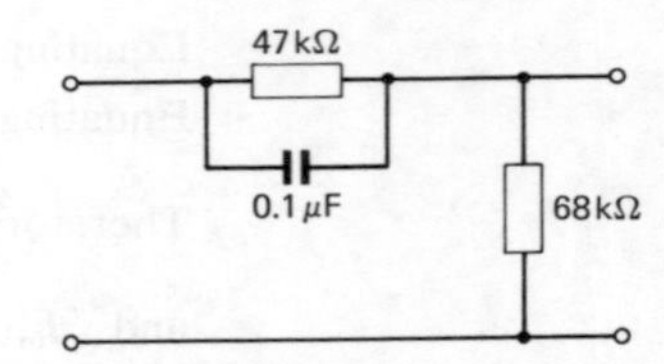

Fig. 12.25

12.2 Repeat **12.1** for the case when the initial capacitor voltage is $V/4$ volts.

12.3 Derive an expression for the time variation of the voltage across the output terminals of the circuit in Fig. 12.25 if the input voltage is a 6 V step function.

12.4 A 1 V step voltage is applied to the circuit shown in Fig. 12.26. Derive an expression for the variation with time of the current supplied to the circuit.

12.5 A rectangular pulse train is applied to the circuit of Fig. 12.26. The pulse repetition frequency is 500 Hz, the pulse width is 1 ms, and the pulse amplitude is 6 V. Sketch the waveform of the currents flowing in both the resistors and hence obtain the waveform of the total current.

12.6 Obtain an expression for the output voltage of the circuit given in Fig. 12.27, if R_1 = 160 kΩ, R_2 = 350 kΩ and C = 0.3 µF when the input voltage is a 6 V step.

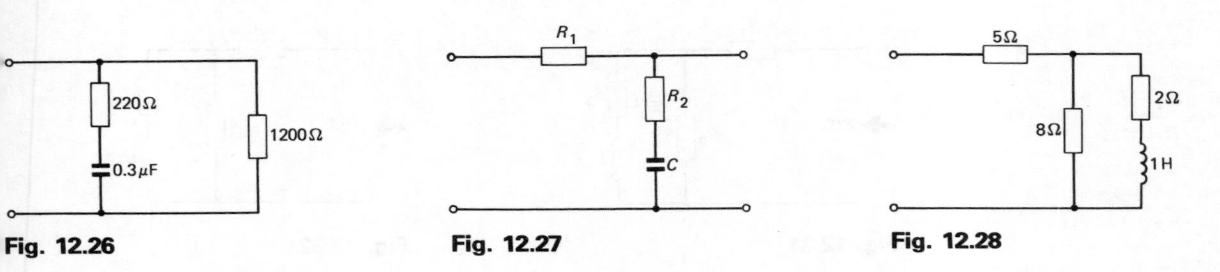

Fig. 12.26 Fig. 12.27 Fig. 12.28

12.7 Derive an expression for the current flowing in the 5 Ω resistor of Fig. 12.28 when the input voltage is a unity step function.

12.8 For the circuit given in Fig. 12.28 obtain an expression for the current flowing in the inductor when the applied voltage is a 3 V/s ramp function.

Short Exercises

12.9 A circuit consists of a capacitor and a resistor connected in series with one another and then connected in series with the parallel combination of an identical capacitor and resistor. Determine an expression for the complex impedance of the circuit.

12.10 Show that the transfer function of the circuit of Fig. 12.29 is

$$\frac{V_0(S)}{V_1(S)}=\frac{SR_2}{\left[S+\dfrac{R_1R_2}{L(R_1+R_2)}\right][R_1+R_2]}$$

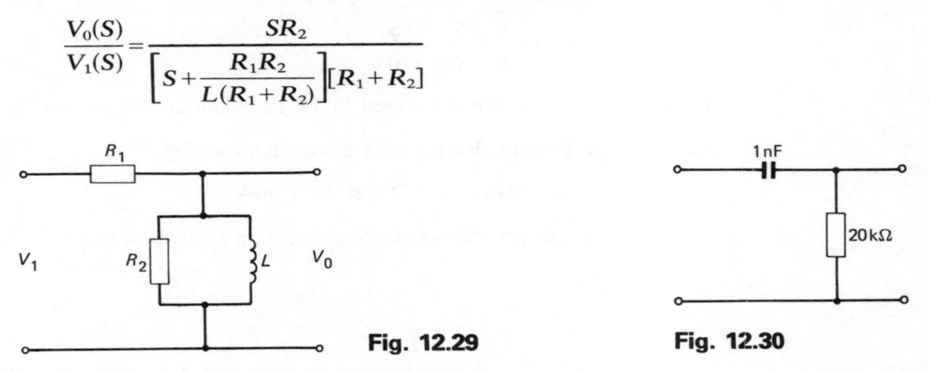

Fig. 12.29 Fig. 12.30

12.11 A train of rectangular pulses is applied to the circuit of Fig. 12.30. The pulse repetition frequency is 1 kHz, the pulse duration is 10 μs, and the pulse amplitude is 6 V. Sketch the output waveform.

12.12 A square-wave generator has an e.m.f. of 25 V p-p and an internal impedance of 1000 Ω. The generator is connected to a circuit that consists merely of a 10 μF capacitor. If the generator is set to have a p.r.f. of 200 Hz sketch the waveform that appears across the capacitor's terminals.

12.13 The internal impedance of a pulse generator consists of a 150 Ω resistor in parallel with a 1 H inductor. The generator is set to produce a rectangular pulse of 12 V amplitude and 10 ms duration. Sketch the waveform of the inductor voltage.

12.14 Write down the complex impedance $Z(S)$ of *a*) R, L and C in series, *b*) R, L and C in parallel, *c*) R and C in parallel connected in series with L and R in series.

12.15 Show that a series RC circuit will integrate an applied voltage ramp provided $t/CR \gg 1$.

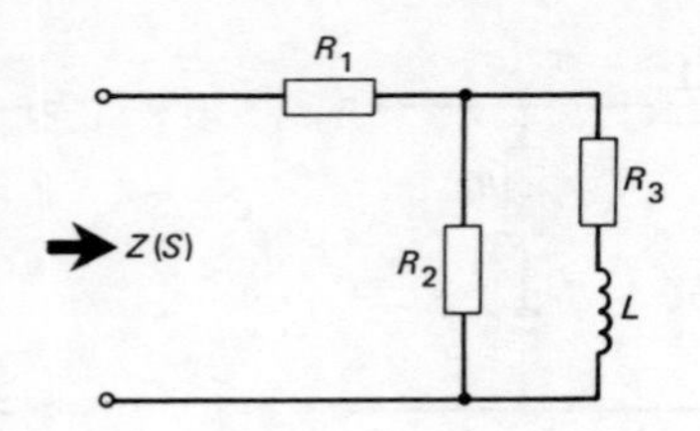

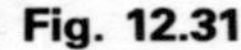

Fig. 12.31

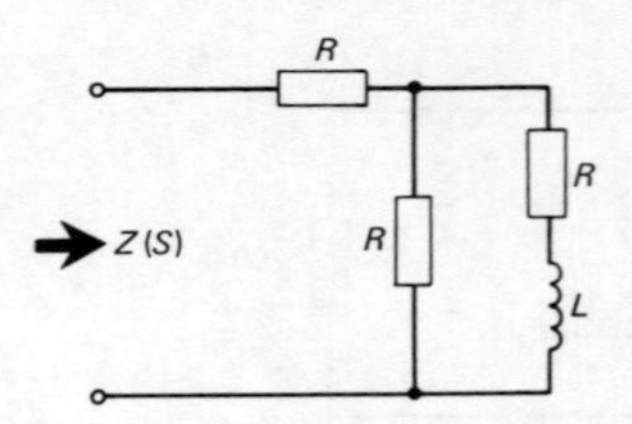

Fig. 12.32

12.16 Show that for the circuit given in Fig. 12.31

$$Z(S)=\frac{R_1R_2+R_1R_3+R_2R_3+SL(R_1+R_2)}{R_2+R_3+SL}$$

12.17 Determine the impedance function $Z(S)$ for the circuit shown in Fig. 12.32.

12.18 Find the expression for the current $i(t)$ flowing in a circuit if

$$i(S)=\frac{25}{S}+\frac{20S+12}{S+4}$$

12.19 Find the expression for the current $i(t)$ in a circuit if

$$i(S)=\frac{6}{S}+\frac{5S+16}{S(S+6)}$$

12.20 For the circuit described in **12.18** calculate the current after 1.427s.

12.21 The current flowing in a circuit is given by

$$i(t)=50(1-e^{-40t})+200e^{-20t} \text{ mA}$$

Write down the transformed version of this equation.

13 Filters

Introduction

In both line and radio systems the need often arises for a group of frequencies contained within a wider frequency band to be transmitted while all other frequencies are suppressed. A **filter** is a circuit designed to pass a certain band of frequencies whilst introducing considerable attenuation at all other frequencies.

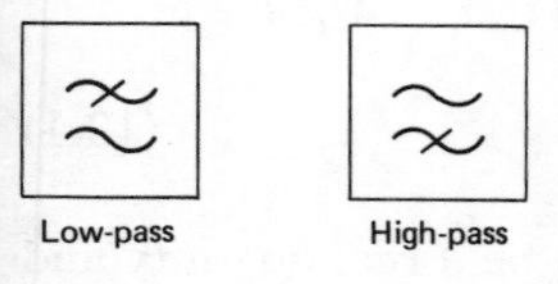

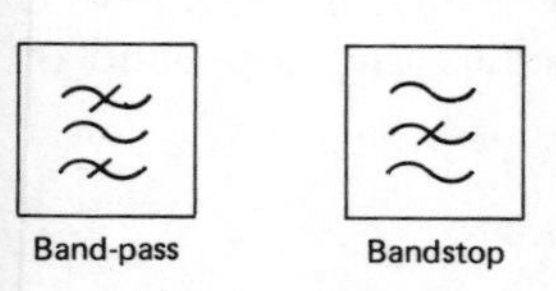

Fig. 13.1 Symbols for *a*) low-pass, *b*) high-pass, *c*) band-pass, and *d*) band-stop filters

Four basic types of filter are available: the low-pass, the high-pass, the band-pass, and the band-stop, and Fig. 13.1 shows the circuit symbols for each of these filters. Filters can be designed using inductors and capacitors, piezo-electric crystals, and active networks. The traditional approach to the design of a filter is an adaption of the network theory described in Chapter 8. This method assumes the filter to be terminated in its characteristic impedance at all frequencies and that lossless components are used. Since neither of these assumptions are satisfied in practice the end product does not completely meet the specifications of the filter.

An alternative method of filter design, using modern network synthesis techniques, has become increasingly popular and leads to filter designs that more nearly meet their specification.

Most of the emphasis in this chapter will be on the classical method of filter design but a brief introduction to the modern techniques will also be given.

Prototype or Constant-k Filters

A **prototype** or **constant-k filter** consists of a number of inductors and capacitors connected as either a T or a π network. The term "constant-k" indicates that the product of the series impedance and the shunt impedance is a constant at all frequencies. Assuming that the components employed in the filter have zero resistance, the characteristic impedance of the network must always be either wholly real or wholly imaginary. When the characteristic impedance is real the filter will be able to accept power from a source and, since it is supposed to contain zero resistance, *none* of this power will be dissipated within the filter. This means that all the input power is transmitted through the filter and delivered to the load. The filter therefore has zero attenuation in its pass-band. On the other hand, when the characteristic impedance of the filter is imaginary, the filter will not accept power from the source and so there can be no output power. Thus, the pass- and stop-bands of a filter are determined by whether or not its characteristic impedance is real or imaginary.

The T Low-pass Filter

A **low-pass filter** should be able to transmit, with the minimum attenuation, all frequencies from 0 Hz up to the **cut-off frequency** f_{co}. At frequencies greater than f_{co} the attenuation of the filter will increase with increase in frequency up to a high value. The circuit of a T constant-k low-pass filter is given in Fig. 13.2. The total series impedance is $Z_1 = j\omega L$ and the total shunt impedance is $Z_2 = 1/j\omega C$. The product of the series and shunt impedances is to be constant at all frequencies, i.e. $Z_1Z_2 = R_0^2$ where R_0 is the *design impedance*. Therefore,

$$R_0^2 = j\omega L \times 1/j\omega C = L/C \quad \text{and} \quad R_0 = \sqrt{(L/C)}\ \Omega$$

The characteristic impedance Z_0 of the filter is, from equation (8.4).

$$Z_{0T} = \sqrt{\left(\frac{Z_1^2}{4} + Z_1Z_2\right)} = \sqrt{\left[Z_1Z_2\left(1 + \frac{Z_1}{4Z_2}\right)\right]} = \sqrt{\left[\frac{L}{C}\left(1 + \frac{j\omega L}{4/j\omega C}\right)\right]}$$

$$\text{or} \quad Z_{0T} = R_0\sqrt{\left(1 - \frac{\omega^2 LC}{4}\right)} \qquad (13.1)$$

At all frequencies where $1 > LC\omega^2/4$, then Z_{0T} will be a real quantity and the frequencies lie in the pass band of the filter. Conversely, if $\omega^2 LC/4 > 1$, then Z_{0T} will be an imaginary quantity and this specifies the stop-band of the filter.

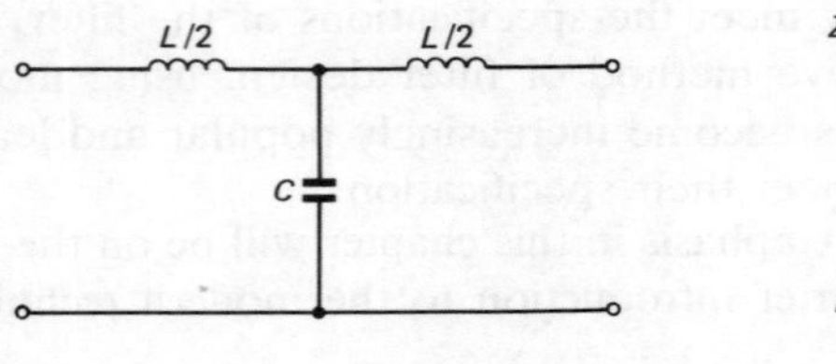

Fig. 13.2 T constant-k low-pass filter

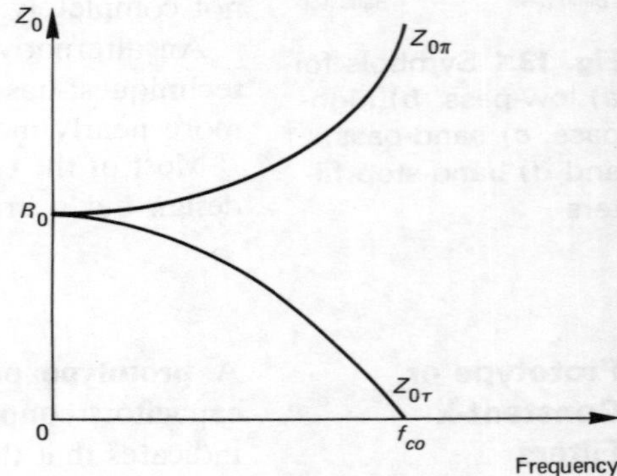

Fig. 13.3 Characteristic impedance/frequency characteristic of a constant-k low-pass filter

The cut-off point of the filter occurs at the frequency at which the changeover from the pass-band to the stop-band takes place. This is the frequency f_{co} at which $1 = \omega_{co}^2 LC/4$ or

$$f_{co} = \frac{1}{\pi\sqrt{(LC)}}\ \text{Hz} \qquad (13.2)$$

When the frequency is 0 Hz the input impedance of the filter is equal to the design impedance R_0, but at the cut-off frequency Z_{0T} is zero. In between these two frequencies Z_{0T} varies in the manner shown by Fig. 13.3.

The π Low-pass Filter

The circuit of a constant-k π low-pass filter is shown by Fig. 13.4. Once again the total series impedance is $Z_1 = j\omega L$ and the total shunt impedance is $Z_2 = 1/j\omega C$ so that the design impedance is unchanged at $R_0 = \sqrt{(L/C)}\ \Omega$. From equation (8.6)

$$Z_{0\pi} = \frac{Z_1 Z_2}{Z_{0T}} = \frac{R_0^2}{Z_{0T}} = \left(\frac{R_0}{\sqrt{(1-\omega^2 LC/4)}}\right) \qquad (13.3)$$

At zero frequency $Z_{0\pi} = R_0$, as for the T filter, but as the frequency is increased $Z_{0\pi}$ rises and becomes equal to $R_0/0$ or ∞ at the cut-off frequency (see Fig. 13.3).

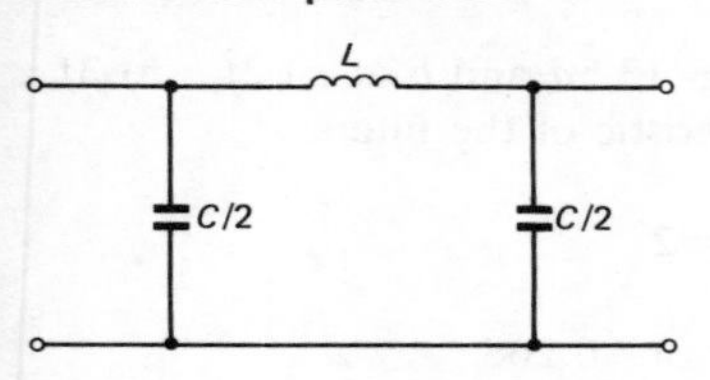

Fig. 13.4 The constant-k low-pass filter

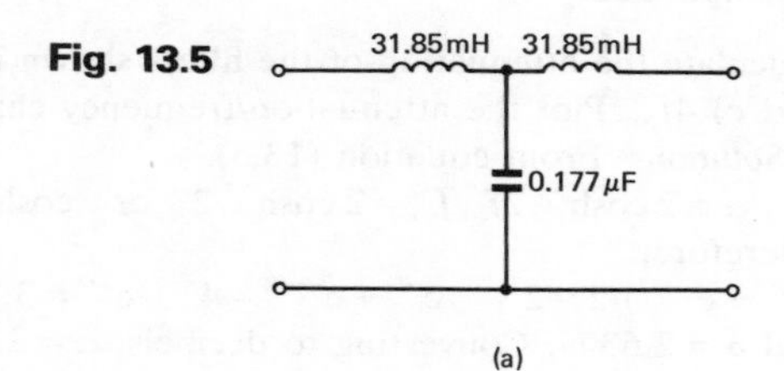

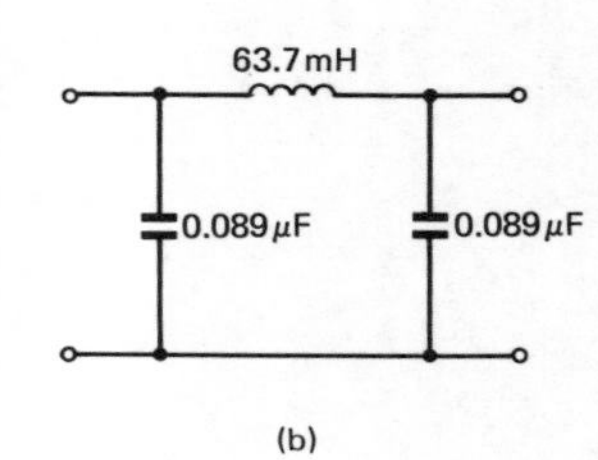

Fig. 13.5

Example 13.1

Calculate the component values for a constant-k low-pass filter using *a*) a T section and *b*) a π section if the design impedance is to be 600 Ω and the cut-off frequency is 3000 Hz.

Solution

$$600 = R_0 = \sqrt{(L/C)} \quad \text{and} \quad 3000 = 1/\pi\sqrt{(LC)}$$

Therefore,

$$600 \times 3000 = 1/\pi C \quad \text{or} \quad C = 1/600 \times 3000 \times \pi = 0.177\ \mu\text{F} \quad (Ans)$$

Also $600/3000 = \pi L$ or $L = 600/3000\pi = 63.7$ mH (*Ans*)
The required filters are shown in Fig. 13.5.

The attenuation of a low-pass filter at any frequency in its stop-band can be determined by using the expression

$$\cosh \gamma = 1 + (Z_1/2Z_2)$$

For such a filter this expression becomes

$$\cosh \gamma = 1 + \frac{j\omega L}{2/j\omega C} = 1 - (\omega^2 LC/2)$$

Therefore,

$$\cosh(\alpha + j\beta) = 1 - (\omega^2 LC/2)$$
$$\cosh \alpha \cos \beta + j \sinh \alpha \sin \beta = 1 - (\omega^2 LC/2)$$

In the pass-band, $\alpha = 0$. Hence $\cosh \alpha = 1$ and $\sinh \alpha = 0$. Hence,

$$\cos \beta = 1 - (\omega^2 LC/2)$$

or the phase shift β in the pass-band is

$$\beta = \cos^{-1}[1 - 2f^2/f_{co}^2] \tag{13.4}$$

In the stop-band, $\beta = 180°$, $\sin \beta = 0$ and $\cos \beta = -1$. Hence

$$-\cosh \alpha = 1 - (\omega LC/2) \qquad \cosh \alpha = \frac{2\omega^2}{\omega_{co}^2} - 1 \qquad \cosh^2 \alpha/2 = \omega^2/\omega_{co}^2$$

Therefore $\alpha = 2 \cosh^{-1}(f/f_{co})$ N (13.5)

Example 13.2

Calculate the attenuation of the filters shown in Fig. 13.5*a* and *b* at *a*) $2f_{co}$, *b*) $3f_{co}$, and *c*) $4f_{co}$. Plot the attenuation/frequency characteristic of the filters.

Solution From equation (13.5),

(i) $\alpha = 2 \cosh^{-1} 2f_{co}/f_{co} = 2 \cosh^{-1} 2$ or $\cosh \alpha/2 = 2$

Therefore,

$(e^{\alpha/2} + e^{-\alpha/2})/2 = 2 \qquad e^{\alpha/2} + e^{-\alpha/2} = 4 \qquad e^{\alpha/2} = 3.732$ or 0.268

and $\alpha = 2.63$ N. Converting to decibels, $\alpha = 22.8$ dB.

ii) At frequency $f = 3f_{co}$, $\cosh \alpha/2 = 3$ and $\alpha = 3.53$ N or 30.6 dB.

iii) At frequency $f = 4f_{co}$, $\cosh \alpha/2 = 4$ and $\alpha = 4.13$ N or 35.8 dB.

Plotting these values gives the theoretical attenuation/frequency characteristic of the filters (see Fig. 13.6).

Fig. 13.6

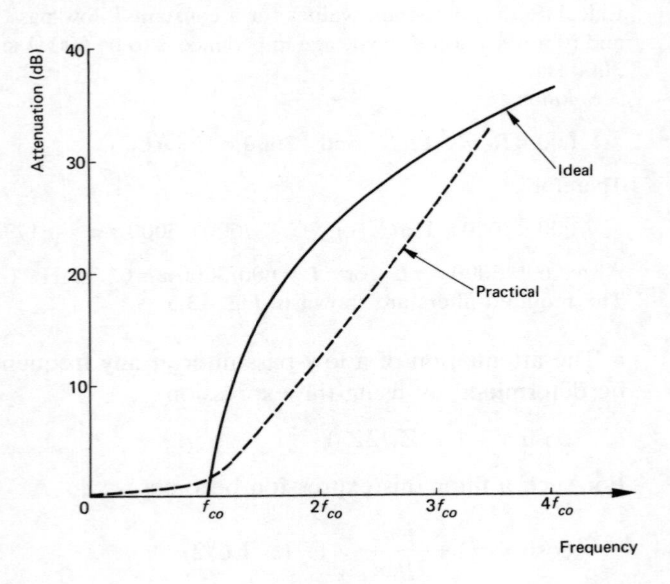

Practical attenuation/frequency characteristics differ from the theoretical in two ways: first, the loss in the pass-band is not zero because the inductances employed possess inherent self-resistance; secondly, the input and output terminals of the filter are only matched to the source and to the load impedances (which are equal to the design impedance) at zero hertz. At all other frequencies, a mismatch exists at both terminals, and the effects of this manifest themselves in the loss of the filter starting to increase at a lower frequency than the cut-off frequency. The difference between the theoretical and the practical characteristics is shown by the dotted line in Fig. 13.6.

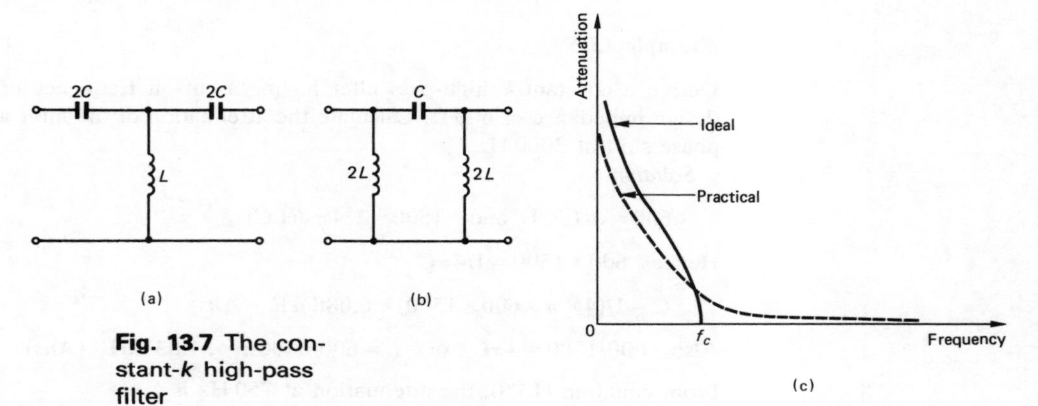

Fig. 13.7 The constant-*k* high-pass filter

The High-pass Filter

Fig. 13.7 shows the arrangements of a T and a π **high-pass filter**. In each case the total series impedance is $Z_1 = 1/j\omega C$ and the total shunt impedance is $Z_2 = j\omega L$. The design impedance is $R_0 = \sqrt{(Z_1 Z_2)} = \sqrt{(L/C)}$, as for the low-pass filter. The characteristic impedance Z_{0T} of the T filter is

$$Z_{0T} = \sqrt{\left(\frac{Z_1^2}{4} + Z_1 Z_2\right)} = \sqrt{\left[Z_1 Z_2\left(1 + \frac{Z_1}{4Z_2}\right)\right]} = \sqrt{\left[\frac{L}{C}\left(1 - \frac{1}{4\omega^2 LC}\right)\right]}$$

or $$Z_{0T} = R_0\sqrt{\left(1 - \frac{1}{4\omega^2 LC}\right)} \qquad (13.5)$$

In this case Z_{0T} will be imaginary for all frequencies low enough to ensure that $1/4\omega^2 LC > 1$ and real for all high frequencies at which $1 > 1/4\omega^2 LC$. The cut-off frequency f_{co} is the frequency at which Z_{0T} changes over from being imaginary to being real. At this point $1 = 1/4\omega_{co}^2 LC$ or

$$f_{co} = \frac{1}{4\pi\sqrt{(LC)}} \text{ Hz} \qquad (13.6)$$

At the cut-off frequency, $Z_{0T} = 0\ \Omega$ and then increases with increase in frequency, becoming equal to the design impedance R_0 when the frequency is very high. Since $Z_{0T} = Z_1 Z_2 / Z_{0T}$ the impedance $Z_{0\pi}$ of a constant-*k* π high-pass filter varies from infinity $(Z_1 Z_2/0)$ at the cut-off frequency to R_0 at very high frequencies.

The phase-shift in the pass-band and the attenuation in the stop-band can be determined from the equation $\cosh \gamma = 1 + (Z_1/2Z_2)$. Following the same steps as before gives

$$\beta = \cos^{-1}[1 - 2f_{co}^2/f^2] \tag{13.7}$$

$$\alpha = 2\cosh^{-1}(f_{co}/f) \tag{13.8}$$

Fig. 13.18 shows how the attenuation of a constant-k high-pass filter varies with frequency for both the ideal filter and a practical filter.

Example 13.3

Design a constant-k high-pass filter having a cut-off frequency of 1500 Hz and a design impedance of 600 Ω. Calculate the attenuation of the filter at 750 Hz and its phase shift at 3000 Hz.

Solution

$600 = \sqrt{(L/C)}$ and $1500 = 1/4\pi\sqrt{(LC)}$

Hence $600 \times 1500 = 1/4\pi C$

$C = 1/(4 \times \pi \times 600 \times 1500) = 0.088\ \mu\text{F}$ (*Ans*)

Also $600/1500 = 4\pi L$ or $L = 600/6000\pi = 31.83$ mH (*Ans*)

From equation (13.8), the attenuation at 750 Hz is

$\alpha = 2\cosh^{-1}(2) = 22.8$ dB (*Ans*)

From equation (13.7) the phase shift at 3000 Hz is

$\beta = \cos^{-1}[1 - (\frac{1}{2})^2] = 41.4°$ (*Ans*)

Band-stop and Band-pass Filters

The **band-pass filter** is one that passes, with (theoretically) zero attenuation, a particular band of frequencies and offers considerable attenuation to all frequencies outside of this bandwidth. The band-pass effect can be achieved by the cascade connection of a low-pass and a high-pass filter, but more usually by the use of a T or a π network whose arms consist of series- and/or parallel-tuned circuits. Fig. 13.8*a* shows the circuit of a constant-k band-pass filter and Fig. 13.8*b* shows its practical and theoretical attenuation/frequency characteristics.

Fig. 13.8 The constant-k band-pass filter

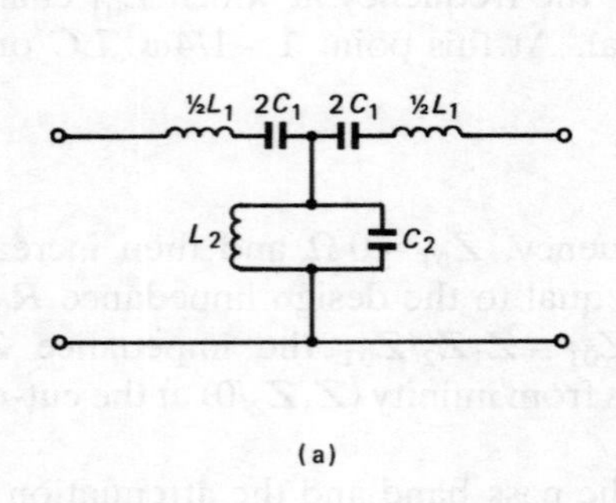

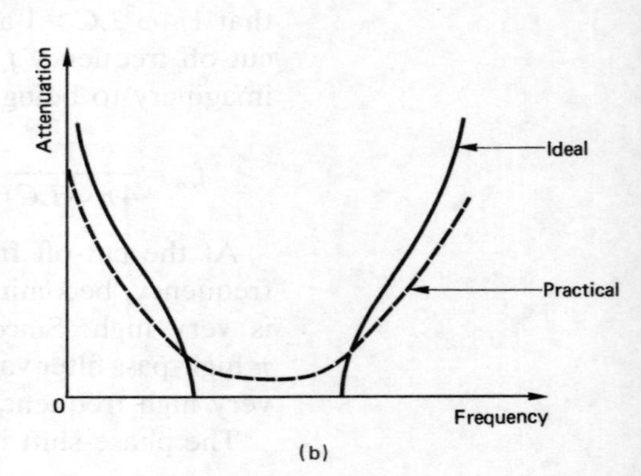

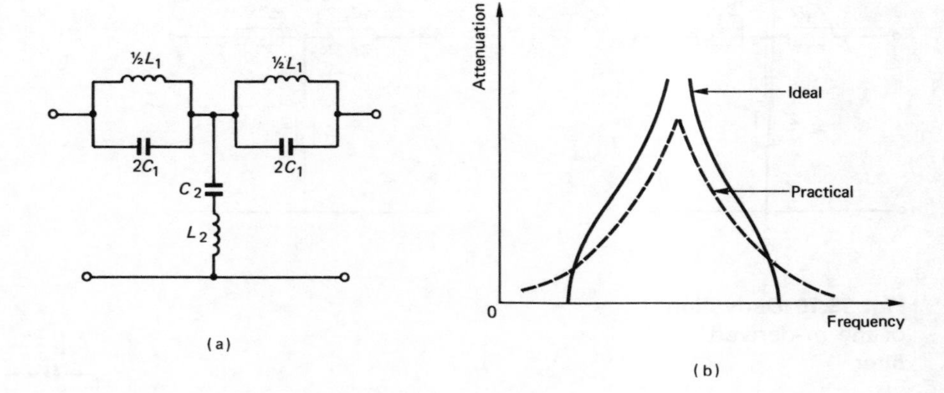

Fig. 13.9 The band-stop filter

The design of the filter follows a similar course to that described for the low- and high-pass filters but since considerably more algebra is involved it will not be attempted here.

Fig. 13.9*a* shows a T **band-stop filter** whose function is to pass all frequencies except those within a particular frequency band. The ideal and practical attenutation/frequency characteristics of a band stop filter are shown in Fig. 13.9*b*.

m-derived Filters

The constant-k filters possess the disadvantages that

a) their attenuation does not increase very rapidly at frequencies above (or below for a high-pass filter) the cut-off frequency.

b) the input impedance varies with frequency and at most frequencies it is not equal to the design impedance as is wanted.

These disadvantages can be overcome by the use of m-derived filter sections connected in cascade with the constant-k basic section.

An ***m*-derived filter** is obtained from a basic constant-k filter section by connecting an extra component in series with the shunt arm in order to make it resonate at some particular frequency. Thus a low-pass filter will have an inductance connected in series with the shunt capacitor. The input impedance and the cut-off frequency of the filter section will be altered by the inclusion of this component, and to restore these quantities to their original values the series and shunt components must have their values modified.

Consider Fig. 13.10*a* and *b*. If the series impedance is multiplied by a factor m, where $0<m<1$, and if the input impedances of the two networks are to be equal to one another,

$$\frac{Z_1^2}{4}+Z_1Z_2=\frac{m^2Z_1^2}{4}+mZ_1Z_2'$$

$$Z_2'=\frac{Z_2}{m}+Z_1\left[\frac{1-m^2}{4m}\right] \qquad (13.9)$$

The m-derived section is shown in Fig. 13.10*c*.

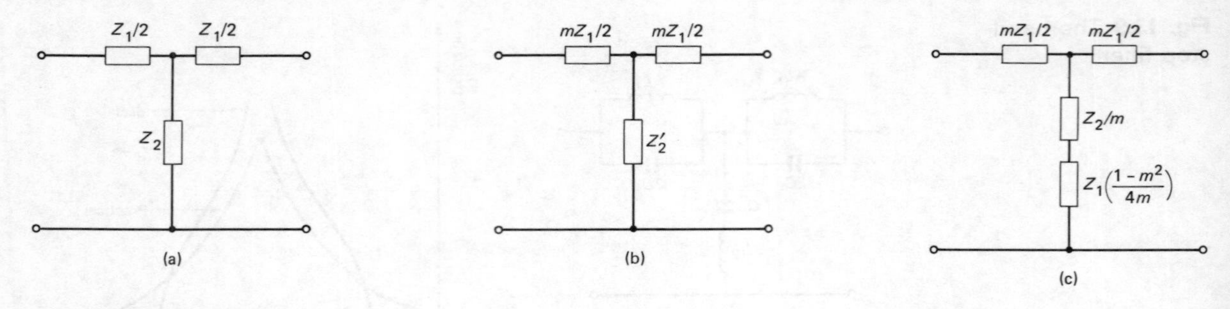

Fig. 13.10 Derivation of the m-derived filter

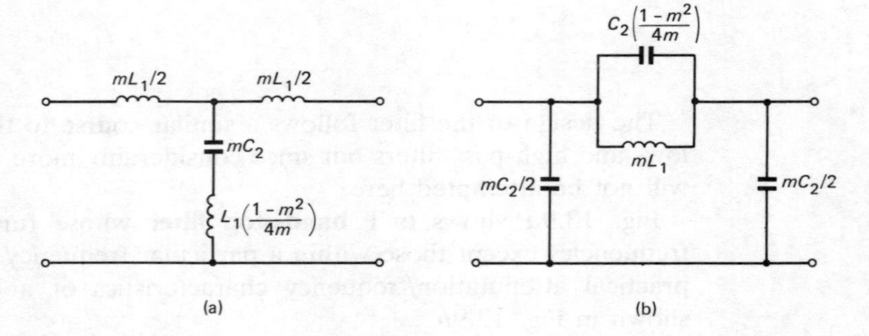

Fig. 13.11 m-derived filter

The m-derived Low-pass Filter

Applying equation (13.9) to the constant-k section shown in Fig. 13.2 gives the m-derived low-pass section of Fig. 13.11*a*. For the T section, at the resonant frequency f_∞ of the shunt path, the attenuation of the filter will be very high, theoretically infinite, since the impedance of the shunt path will be very low. At this frequency,

$$1/\omega_\infty mC_2 = \omega_\infty L_1(1-m^2)/4m$$

$$\omega_\infty^2 = \frac{4}{(1-m^2)L_1C_2}$$

Hence

$$f_\infty = \frac{1}{\pi\sqrt{(L_1C_2)}\sqrt{(1-m^2)}} = \frac{f_{co}}{\sqrt{(1-m^2)}} \qquad (13.10)$$

For the π section, "infinite" loss occurs when

$$m\omega_\infty L_1 = \frac{4m}{\omega_\infty C_2(1-m^2)}$$

$$f_\infty = \frac{f_{co}}{\sqrt{(1-m^2)}} \qquad \text{(again)} \qquad (13.10)$$

Re-arranging equation (13.10) in terms of m gives

$$m = \sqrt{[1-(f_{co}/f_\infty)^2]} \qquad (13.11)$$

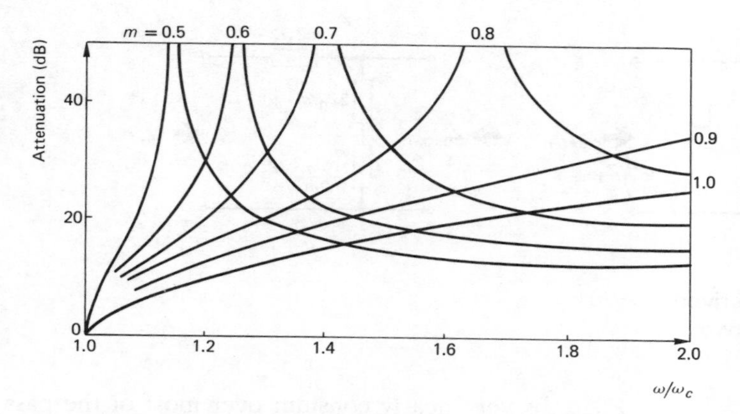

Fig. 13.12 The relationship between m and the frequency of maximum attenuation

Unfortunately, at frequencies above the cut-off frequency the attenuation of the filter falls and eventually reaches quite a low value. Graphs showing the attenuation of an m-derived low-pass section plotted against frequency ratio f/f_{co} for various values of m are readily available and can be used in the design of an m-derived section. Fig. 13.12 shows the relationship between m and the attenuation/frequency characteristic of an m-derived low-pass section.

Example 13.4

Design an m-derived T section low-pass filter to have a cut-off frequency of 4000 Hz, a design impedance of 600 Ω, and maximum attenuation at 6000 Hz.

Solution The component values of the constant-k section must first be obtained.

$600 = \sqrt{(L/C)}$ $\quad 4000 = 1/\pi\sqrt{(LC)}$ $\quad$ Hence,

$600 \times 4000 = 1/\pi C$ or $C = 1/600 \times 4000\pi = 0.133\ \mu\text{F}$ (*Ans*)

Also $600/4000 = \pi L$ or $L = 600/4000\pi = 47.75$ mH (*Ans*)

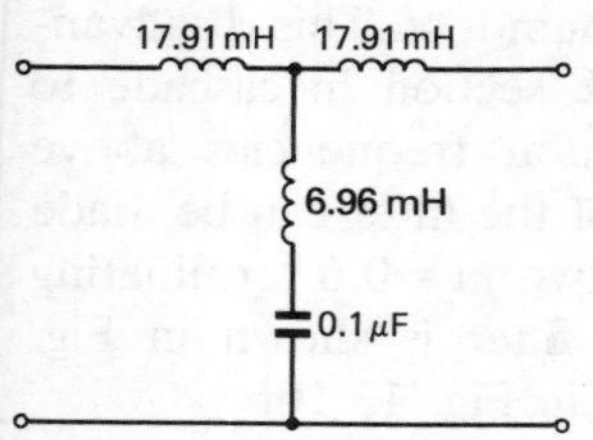

Fig. 13.13

Now $f/f_{co} = 6000/4000 = 1.5$ and from Fig. 13.12 the required value of m is 0.75. Therefore,

$$mL_1/2 = (0.75 \times 47.75)/2 = 17.91 \text{ mH}$$

$$\frac{(1-m^2)L_1}{4m} = \frac{(1-0.75^2)47.75}{4 \times 0.75} = 6.96 \text{ mH}$$

$$mC = \frac{0.75 \times 0.133}{2 \times 0.75} = 0.1\ \mu\text{F}$$

The required filter section is shown in Fig. 13.13.

The Input Impedance of an m-derived Half-section

If the m-derived T section of Fig. 13.10 is split into two halves, as in Fig. 13.14, each half will have input and output impedances of Z_{0T} and $Z_{0\pi(m)}$ respectively. Z_{0T} is the input impedance of a constant-k T section and $Z_{o\pi(m)}$ is a modified version of $Z_{0\pi}$, where $Z_{0\pi}$ is the input impedance of a π constant-k section. When the value of m is chosen to be 0.6, the impedance

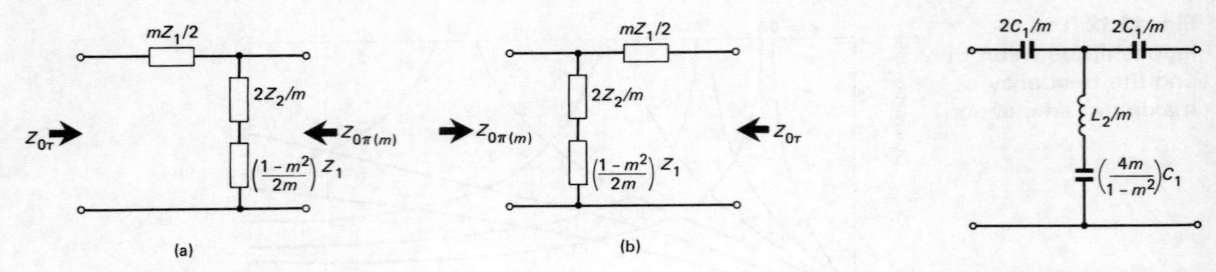

Fig. 13.14 *m*-derived filter split into two equal parts

Fig. 13.15 *m*-derived high-pass filter

$Z_{0\pi(m)}$ is very nearly constant over most of the pass bandwidth of the half section. This statement is also true when a π half-section is given the value $m = 0.6$. The impedance seen at the other pair of terminals, i.e. Z_{0T} (or $Z_{0\pi}$), varies with frequency in the same way as the impedance of a whole section, constant-k or m-derived. This means that half-sections with $m = 0.6$ can be used to terminate both the input and the output terminals of a filter to ensure that the filter presents constant values of impedance at both of its terminals.

The m-derived High-pass Filter

An m-derived T high-pass filter section is shown by Fig 13.15. The graph given in Fig. 13.12 can also be used in the design of an m-derived high-pass filter except that the labelling of the horizontal axis should be altered to read ω_{co}/ω.

Composite Filters

The attenuation of an m-derived low-pass filter section decreases with increase in frequency above the frequency of maximum loss. This disadvantage can be overcome by connecting a constant-k section in cascade to ensure that the overall attenuation remains high at frequencies above cut-off. Further, the input and output impedances of the filter can be made very nearly constant with frequency by the use of two $m = 0.6$ terminating half-sections. The block diagram of a **composite filter** is shown in Fig. 13.16*a* and its attenuation/frequency characteristic in Fig. 13.16*b*.

Example 13.5

Design a high-pass filter to work between 600 Ω impedances with a cut-off frequency of 10 kHz. The filter should have (i) infinite attenuation at 6 kHz, (ii) a constant-k section to ensure a high loss at frequencies below 6 kHz, and (iii) $m = 0.6$ matching half-sections.

Solution

a) Constant-k

$$600 = \sqrt{(L/C)} \qquad 10\,000 = 1/4\pi\sqrt{(LC)}$$

$$C_1 = 1/(600 \times 10\,000 \times 4\pi) = 0.013\ \mu\text{F}$$

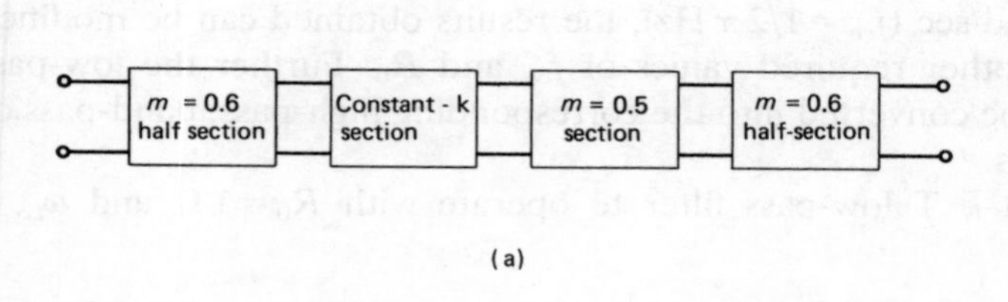

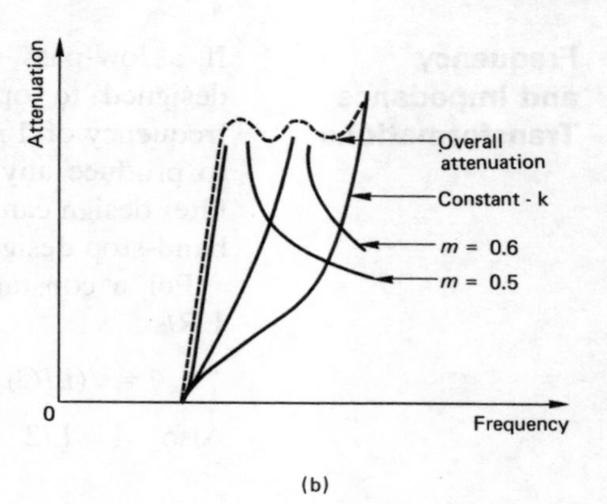

Fig. 13.16 Composite filter

and $L = 600/(10\,000 \times 4\pi) = 4.78\text{ mH}$

b) $m = 0.6$ half-section

$$C_1/m = 0.0133/0.6 = 0.022\ \mu\text{F}$$

$$\frac{C_1 \times 4m}{1-m^2} = \frac{0.013 \times 4 \times 0.6}{1-0.36} = 0.049\ \mu\text{F}$$

$$L_2/m = 4.78/0.6 = 7.96\text{ mH}$$

c) Infinite loss at 6000 Hz

From Fig. 13.12, $m = 0.8$ and so

$$C_1/m = 0.0133/0.8 = 0.016\ \mu\text{F}$$

$$\frac{C_1 4m}{1-m^2} = \frac{0.0133 \times 4 \times 0.8}{1-0.64} = 0.116\ \mu\text{F}$$

$$L_2/m = 4.78/0.8 = 5.98\text{ mH}$$

The required filter is shown in Fig. 13.17*a* and this circuit can be reduced to the simpler form shown in Fig. 13.17*b*.

Fig. 13.17

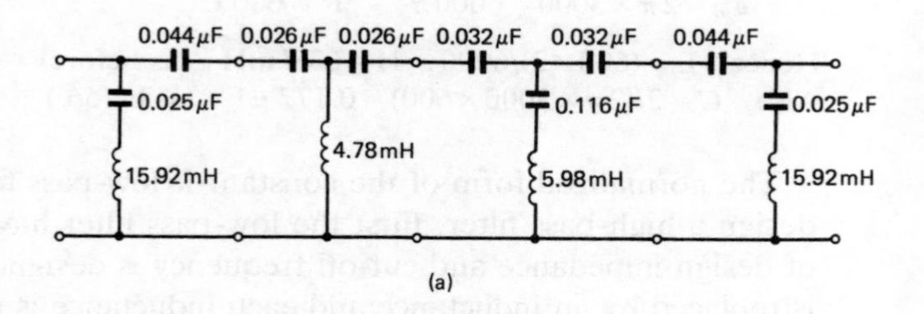

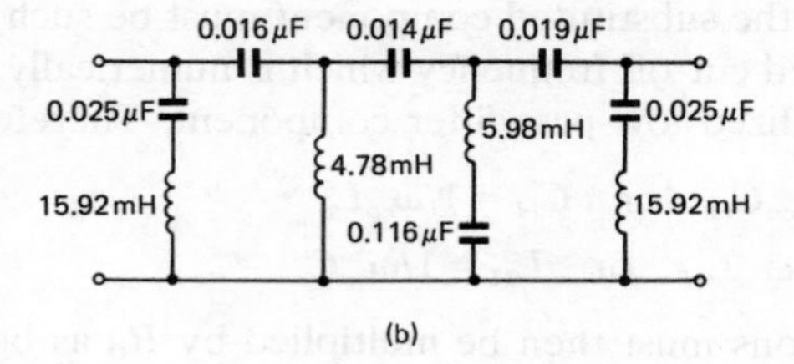

Frequency and Impedance Transformations

If a low-pass constant-k filter is designed in its *normalized* form, i.e. designed to operate with a design impedance of 1 ohm and a cut-off frequency of 1 rad/sec ($f_{co} = 1/2\pi$ Hz), the results obtained can be modified to produce any other required values of f_{co} and R_0. Further the low-pass filter design can be converted into the corresponding high-pass, band-pass or band-stop designs.

For a constant-k T low-pass filter to operate with $R_0 = 1\,\Omega$ and $\omega_{co} = 1$ R/s

$$1 = \sqrt{(L/C)} \quad \text{and} \quad 1 = 2/\sqrt{(LC)} \qquad \text{Hence} \quad 1 = 2/C \quad \text{or} \quad C = 2\,\text{F}$$

Also $1 = L/2$ or $L = 2\,\text{H}$

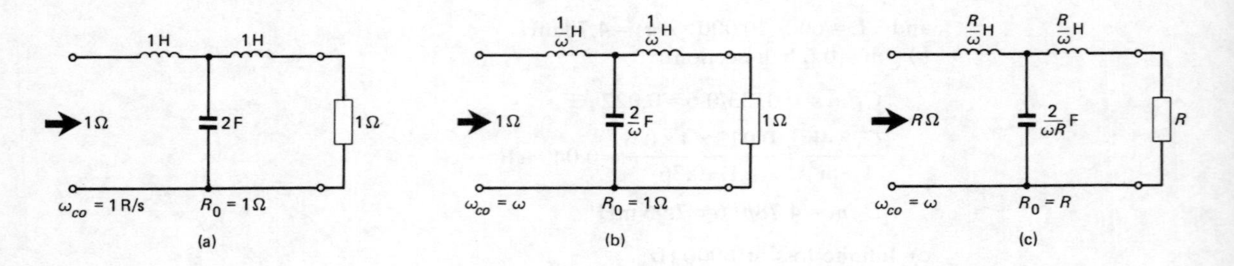

Fig. 13.18 Frequency and impedance transformation

The normalized T low-pass filter is shown in Fig. 13.18. To convert the filter to have a cut-off *angular frequency* of ω R/s, DIVIDE both L and C by ω (see Fig. 13.18*b*). To convert the filter in Fig. 13.18*b* to have a design impedance of R_0, MULTIPLY the *impedance* of each component by R_0. This is shown by Fig. 13.18*c*.

Example 13.6

Re-design the low-pass filter of Example 13.1, i.e. $R_0 = 600\,\Omega$ and $f_{co} = 3000$ Hz.

Solution

$$\omega_{co} = 2\pi \times 3000 = 6000\pi \qquad R = 600\,\Omega$$

Hence $L = (600 \times 2)/6000\pi\ \text{H} = 63.7\ \text{mH}$ (as before)
Also $C = 2/(2\pi \times 3000 \times 600) = 0.177\ \mu\text{F}$ (as before)

The normalized form of the constant-k low-pass filter can also be used to design a high-pass filter. First the low-pass filter having the required values of design impedance and cut-off frequency is designed. Then each capacitor is replaced by an inductance and each inductance is replaced by a capacitor. The value of the substituted component must be such that it has a reactance at the required cut-off frequency which is numerically equal to the reactance of the normalized low-pass filter component. Therefore,

$$L_n = 1/\omega_{co}C_H \quad \text{or} \quad C_H = 1/\omega_{co}L_n \qquad (13.11)$$

and $$1/C_n = \omega_{co}L_H \quad \text{or} \quad L_H = 1/\omega_{co}C_n \qquad (13.12)$$

These equations must then be multiplied by R_0 as before,

Example 13.7

Re-design the high-pass filter given in Example 13.3, i.e. $R_0 = 600$ and $f_{co} = 1500$ Hz, *a*) using the frequency transformation method, *b*) by converting from low-pass to high-pass.

Solution

a) For a constant-*k* T high-pass filter section to operate at $\omega_{co} = 1$ R/s with $R_0 = 1\ \Omega$.

$$1 = \sqrt{(L/C)} \quad \text{and} \quad 1 = 1/2\sqrt{(LC)}$$

Hence

$$1 = 1/2C \quad \text{or} \quad C = 1/2\ \text{F}$$

and $1 = 2L$ or $L = 1/2$ H

(see Fig. 13.19*a*). Therefore,

$$L = 600/(2\times 2\pi\times 1500) = 31.83\ \text{mH} \quad \text{(as before)} \quad (Ans)$$

Also $C = 1/(2\times 600\times 2\pi\times 1500) = 0.088\ \mu\text{F}$ (as before) (*Ans*)
The filter is shown by Fig. 13.19*c*.

Fig. 13.19

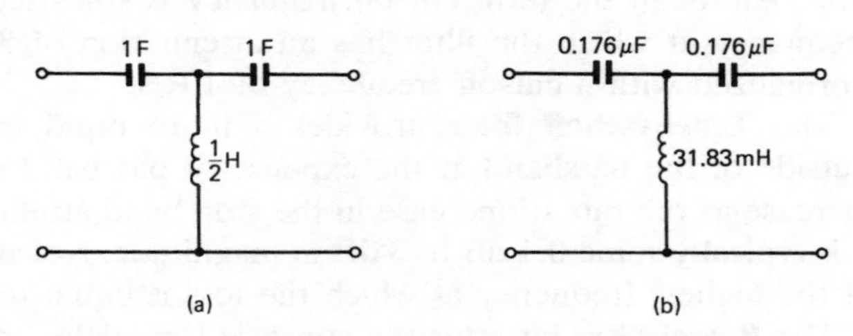

b) The normalized low-pass filter (Fig. 13.18*a*) has a capacitive reactance of $X_C = 1/2\ \Omega$. Hence, the required inductance for the high-pass filter is

$$L = \frac{1\times 600}{2\times 2\pi\times 1500} = 31.83\ \text{mH}$$

The normalized filter has an inductive reactance of $2\ \Omega$. Hence, the required high-pass filter capacitance is

$$1/(2\times 2\pi\times 1500\times 600) = 0.088\ \mu\text{F} \quad (Ans)$$

A band-pass filter can also be derived from the low-pass filter by scaling the normalized frequency up to a frequency equal to the desired bandwidth. Then a capacitor is placed in series with each inductor and an inductor is placed in parallel with each capacitor. The values of the new components are chosen so that resonance occurs at the wanted centre frequency.

Modern Filter Design

Because the conventional method of filter design assumes (incorrectly) that a filter operates between matched impedances, the measured characteristics of the filter are rarely as good as anticipated.

There are three main types of modern filter known respectively as the *Butterworth*, the *Bessel*, and the *Tchebyscheff* filters.

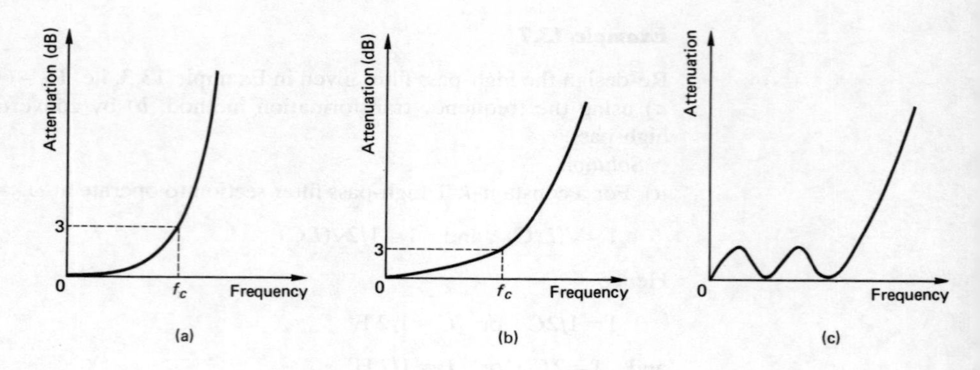

Fig. 13.20 Attenuation/frequency characteristics for *a*) Butterworth, *b*) Bessel, and *c*) Tchebyscheff low-pass filters

The **Butterworth filter** has a *maximally-flat* response in the passband; this means that the response is as flat as is possible, with no ripple occurring at any point. This can be alternatively expressed by saying that the filter is *monotonic*, i.e. an increase in frequency always gives an increase in attenuation. Although the term *cut-off frequency* is still used, it now refers to the frequency at which the filter has an attenuation of 3 dB. The filter can be normalized with a cut-off frequency of 1 R/s.

The **Tchebyscheff filter** provides a more rapid increase in attenuation outside of the passband at the expense of passband ripple. This ripple will increase as the rate of increase in the stop band attenuation is increased and it is typically some 0.1 dB to 3 dB in magnitude. Normalization is carried out at the highest frequency at which the loss is equal to the passband ripple.

The **Bessel filter** introduces a constant time delay to all frequencies in the passband. Outside of the passband the attenuation increases at a slower rate than for either the Butterworth or the Tchebyscheff filters.

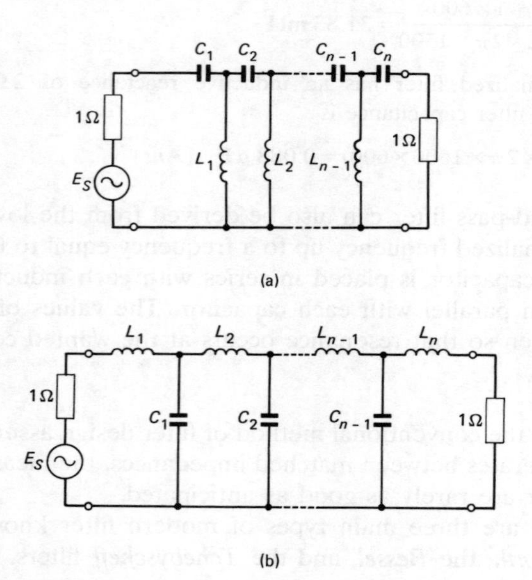

Fig. 13.21 *a*) Minimum-inductance filter, *b*) minimum-capacitance filter

With all the modern filter designs the source and load impedances need not be of the same value. Any filter may be either of the minimum inductance type (Fig. 13.21*a*) or of the minimum capacitance type (Fig. 13.21*b*). The number of components used in a filter is known as the *order* of the filter, and the higher the order the more selective its attenuation/frequency characteristic. The calculation of values for any of these filters is beyond the scope of this book.

Active Filters

The use of inductors in conventional *LC* filters is opposed to the requirement for electronic equipment to be as physically small and light as possible. Inductors, particularly at the lower frequencies where fairly large values of inductance may be required, are relatively bulky and heavy and, because of their inherent self-resistance, introduce unwanted losses into the passband. It is therefore desirable to be able to realise a filter design using only resistors and capacitors. It is fairly obvious that a non-active *RC* network is able to act as a low-pass filter—and it is frequently so employed—but the selectivity provided is poor. Fig. 13.22*a* shows a first-order *RC* low-pass filter. The designed-for performance of the circuit will be modified immediately it is connected to a load and so very often an op-amp voltage follower is used as a buffer as shown by Fig. 13.22*b*. The 3 dB frequency of the filter is easily shown to be $1/RC$ Hz.

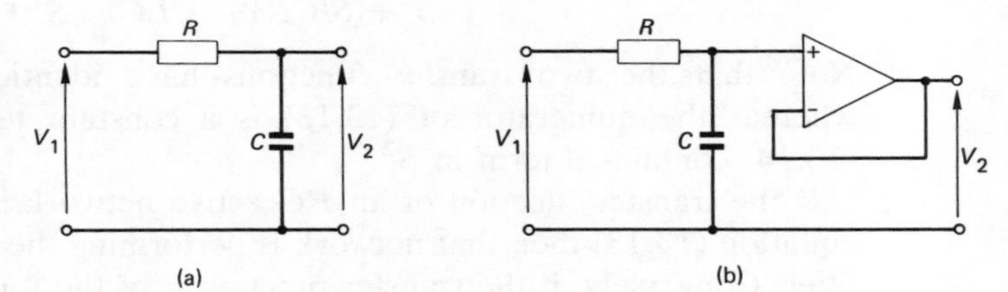

Fig. 13.22 *a*) First-order *RC* filter, *b*) second-order *RC* filter

Improved selectivity can be obtained, by the use of higher-order *RC* active networks in which the active element is usually provided by an op-amp. A large number of different active filter designs have been proposed in the literature and these can be broadly divided into two main classes:

a) Circuits where an *LC* filter design is used but all the inductances needed are simulated by *gyrators*.

b) An op-amp circuit is used to directly produce the required filter attenuation/frequency characteristic. This is normally expressed in terms of the required *transfer function*.

One of the more commonly used active filter circuits is that due to **Sallen and Key** and the discussion in the rest of this chapter will be restricted to this circuit.

Fig. 13.23

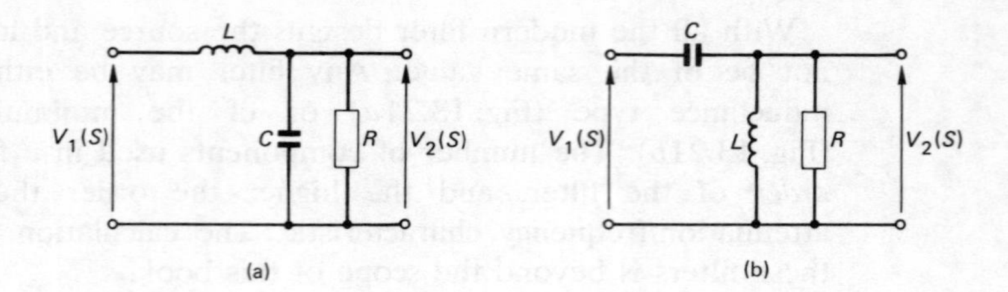

Transfer Functions of Second-order Filters

The transfer function of the low-pass filter shown in Fig. 13.23*a* is

$$V_2(S)/V_1(S) = \frac{R/(1+SCR)}{SL+[R/(1+SCR)]} = \frac{R}{R+SL+S^2LCR}$$

$$V_2(S)/V_1(S) = T(S) = \frac{1}{LC\left(S^2+\dfrac{S}{CR}+\dfrac{1}{LC}\right)} = \frac{N}{S^2+bS+c} \tag{13.13}$$

For the second-order high-pass filter of Fig. 13.23*b*,

$$V_2(S)/V_1(S) = \frac{SLR/(R+SL)}{\dfrac{1}{SC}+\dfrac{SLR}{R+SL}} = \frac{SLR\times SC}{R+SL+S^2LCR}$$

$$V_2(S)/V_1(S) = T(S) = \frac{S^2}{S^2+(S/CR)+(1/LC)} = \frac{NS^2}{S^2+bS+c} \tag{13.14}$$

Note that the two transfer functions have identical denominators, but whereas the numerator of (13.13) is a constant term the numerator of (13.14) contains a term in S^2.

If the transfer function of an *RC* active network is of the same form as equation (13.13) then that network is performing the function of a low-pass filter. Conversely, if the transfer function is of the form of equation (13.14) the active network will act as a high-pass filter.

Similar analysis shows that for a band-pass filter

$$T(S) = \frac{NS}{S^2+bS+c} \tag{13.15}$$

The final term $1/LC$ in these equations is the square of the *angular resonant frequency* of the circuit, i.e. $\omega_0^2 = 1/LC$. Also, applying equation (1.25) to Fig. 13.23*a*,

$$Q = \frac{2\pi CV}{V^2/Rf_0} = \omega_0 CR = \frac{CR}{\sqrt{(LC)}} = R\sqrt{\frac{C}{L}}$$

Hence

$$\frac{1}{CR} = \frac{L}{LCR} = \frac{\omega_0^2}{LR} = \frac{L\omega_0}{R\sqrt{(LC)}} = \frac{\omega_0}{R\sqrt{(C/L)}} = \omega_0/Q$$

This means that the denominator of each of the transfer functions can be written in the form

$$D(S) = S^2 + \frac{S\omega_0}{Q} + \omega_0^2 \tag{13.16}$$

Generalized Sallen and Key Circuit

The general form of the Sallen and Key filter network is shown in Fig. 13.24. The resistors R_{f1} and R_{f2} are provided to set the voltage gain of the op-amp to the desired value

$$A_{v(F)} = (R_{f1} + R_{f2})/R_{f1}$$

Often the required gain is unity and it is rarely more than 3.

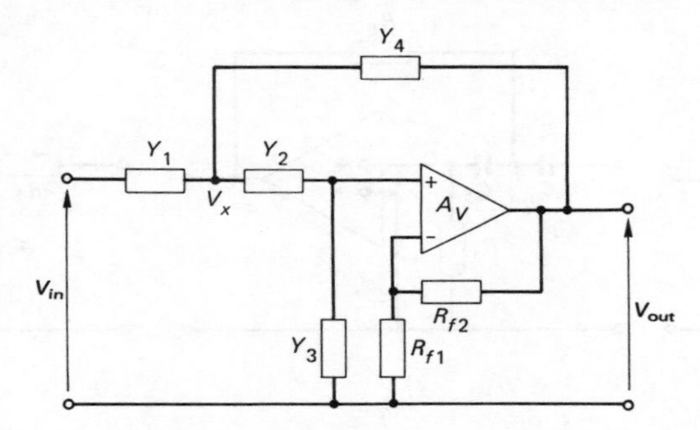

Fig. 13.24 Active filter of the Sallen and Key type

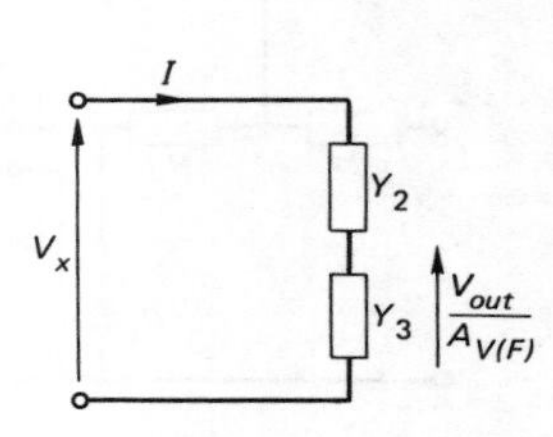

Fig. 13.25

Summing the currents at the node A:

$$(V_{in} - V_x)Y_1 + (V_{out} - V_x)Y_4 - \frac{V_{out}Y_3}{A_{v(F)}} = 0$$

$$V_{in}Y_1 = (Y_1 + Y_4)V_x - V_{out}Y_4 + \frac{V_{out}Y_3}{A_{v(F)}}$$

Referring to Fig. 13.25, $I = V_{out}Y_3/A_{v(F)}$ and so

$$V_x = \frac{V_{out}Y_3}{A_{v(F)}}\left(\frac{Y_2 + Y_3}{Y_2Y_3}\right) = \frac{V_{out}}{A_{v(F)}}\left(\frac{Y_2 + Y_3}{Y_2}\right)$$

Hence

$$V_{in}Y_1 = \left[(Y_1 + Y_4)\left(\frac{Y_2 + Y_3}{Y_2}\right) + Y_3\right]\frac{V_{out}}{A_{v(F)}} - V_{out}Y_4$$

or $$T(S) = \frac{A_{v(F)}Y_1Y_2}{(Y_1 + Y_4)(Y_2 + Y_3) + Y_2Y_3 - A_{v(F)}Y_2Y_4}$$

$$= \frac{A_{v(F)}Y_1Y_2}{Y_2[Y_1 + Y_4(1 - A_{v(F)})] + Y_3(Y_1 + Y_2 + Y_4)} \tag{13.17}$$

For the network to act as a low-pass, or a high-pass, or a band-pass filter the admittances Y_1, Y_2, etc. must be provided by the appropriate components (R or C) so that the transfer function has the desired form.

Suppose a low-pass filter is wanted. Then, since the numerator of equation (13.13) does not contain a term in S, both Y_1 and Y_2 must be provided by resistors. Equation (13.17) then reduces to

$$T(S)=\frac{A_{v(F)}G_1G_2}{G_2[G_1+Y_4(1-A_{v(F)})]+Y_3(G_1+G_2+Y_4)} \tag{13.18}$$

and from this it can be seen that for terms in both S and S^2 to appear in the denominator both Y_3 and Y_4 must be provided by capacitors. The required circuit is shown by Fig. 13.26. The choice of voltage gain $A_{v(F)}$ is complex and beyond the scope of this book but if, as is frequently the case, it is chosen to be unity, the op-amp should be connected to act as a voltage follower.

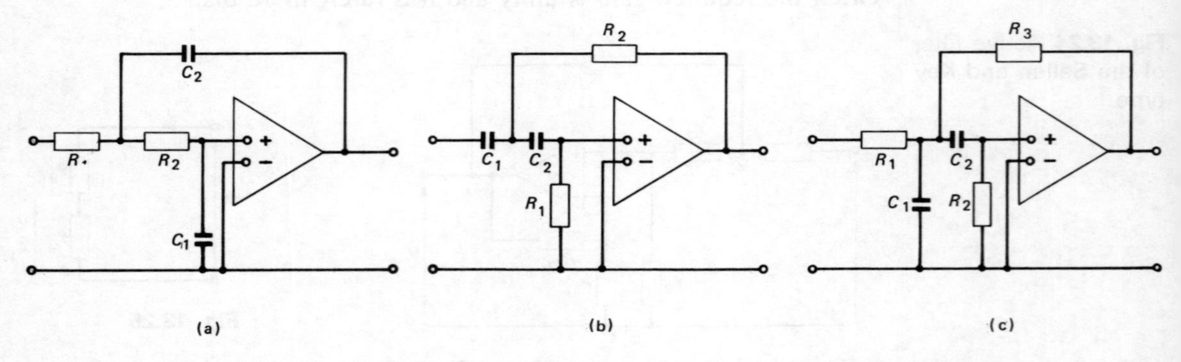

Fig. 13.26 Active filters: *a*) low-pass, *b*) high-pass, and *c*) band-pass

The transfer function of Fig. 13.26*a* can be derived directly or by substituting into equation (13.18), thus

$$T(S)=\frac{A_{v(F)}G_1G_2}{G_1G_2+SC_2G_2(1-A_{v(F)})+SC_1(G_1+G_2+SC_2)}$$

$$=\frac{A_{v(F)}}{1+SC_2R_1(1-A_{v(F)})+SC_2[R_1+R_2+(SC_2/G_1G_2)]}$$

or $$T(S)=\frac{A_{v(F)}}{S^2C_1C_2R_1R_2+S[C_2R_1(1-A_{v(F)})+C_1(R_1+R_2)]+1} \tag{13.19}$$

The calculation of the components values required to produce a given filter characteristic is also beyond the scope of this book. Values have been computed, however, and are available in tabular form from a number of sources.

The equivalent high-pass filter can be obtained by merely (i) replacing each resistor in the low-pass circuit by a capacitor of reciprocal value and (ii) replacing each capacitor by a resistor of the reciprocal value. The circuit of a high-pass filter is given in Fig. 13.26*b*, while Fig. 13.26*c* shows the band-pass filter.

Greater selectivity can be obtained by cascading *a*) a first-order filter and a second-order filter to obtain a third-order filter, *b*) two second-order filters may be cascaded to give a fourth-order filter and so on.

Exercises 13

13.1 Derive an expression for the characteristic impedance of the filter section shown in Fig. 13.27. Sketch a graph to show how the characteristic impedance varies with frequency. Determine the type of filter and obtain an expression for its cut-off frequencies.

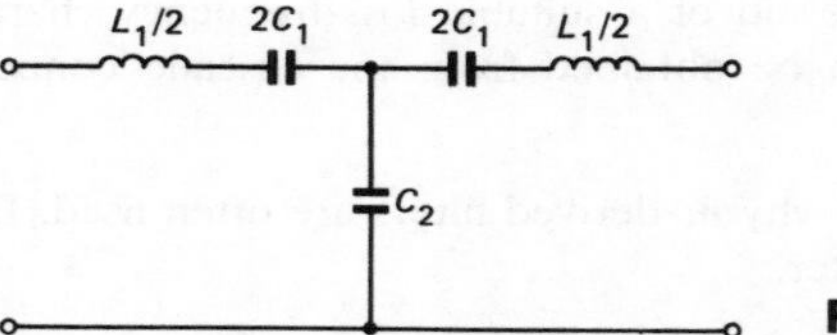

Fig. 13.27

13.2 Calculate the components for an m-derived low-pass half-section with $m = 0.6$ if $R_0 = 1000$ ohms and the cut-off frequency is 4000 Hz and the frequency of maximum attenuation is to be 3200 Hz.

13.3 A low-pass filter section has a total series inductance of 2 H and a total shunt capacitance of 2 μF. Calculate the loss of the filter at $2500/2\pi$ Hz when it is connected between a source and load impedance equal to the design impedance.

13.4 Show that the general π m-derived section is given by Fig. 13.28.

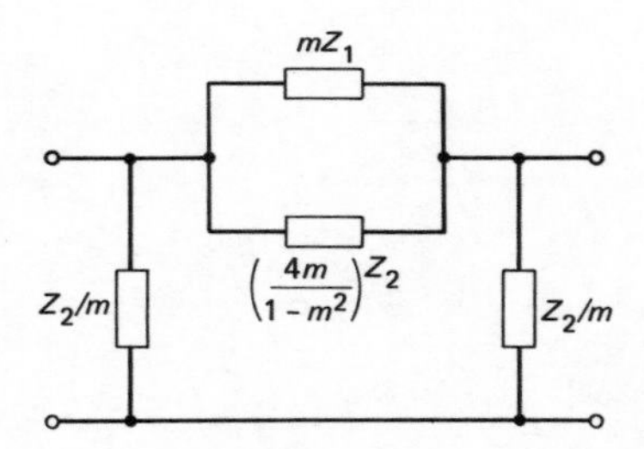

Fig. 13.28

13.5 Use nodal analysis to derive the transfer function of the active filter shown in Fig. 13.26*b*.

13.6 Design a low-pass filter to operate between a source and a load impedance of 600 ohms. The cut-off frequency is to be $25\,000/2\pi$ Hz and the frequency of maximum attenuation is to be $32\,500/2\pi$ Hz.

13.7 A T low-pass filter has a total series inductance of 0.2 H and a total shunt capacitance of 0.6 μF. Calculate its characteristic impedance at *a*) the cut-off frequency f_{co}, *b*) at 0.5 f_{co}, and *c*) at 0.2 f_{co}.

13.8 Derive an expression for the cut-off frequency of a T low-pass filter. Calculate component values for the filter to have a cut-off frequency of 10 kHz and a design impedance of 600 ohms. Calculate the attenuation of the filter at twice the cut-off frequency.

13.9 A T low-pass filter has a total series inductance of 60 mH and a total shunt capacitance of 0.2 μF. Calculate the phase shift through the filter at 1 kHz.

Short Exercises

13.10 Draw the practical attenuation/frequency characteristic of a high-pass constant-k filter. What are the disadvantages of the filter and how may they be overcome?

13.11 What are the advantages of using an active filter in place of a traditional LC design? Draw the circuit of an active high-pass filter.

13.12 Explain, with the aid of a suitable loss-frequency characteristic, how a band-pass characteristic can be obtained from the cascade connection of a low-pass and a high-pass filter.

13.13 State the reasons why m-derived filters are often used. Draw the m-derived version of a high-pass filter.

13.14 A low-pass filter has a cut-off frequency of 5000 Hz. What will be its frequency of maximum attenuation if $m = 0.75$?

Answers to Exercises

1.1 1.58 V, 14.14 mA, 26.6°, 19.98 mW, 0.89
1.2 29.058 kHz, 80, 2.74 mA, 6.37 V
1.3 2037 Hz, 63.4 Ω
1.4 8.66 kHz, 15 kHz, 49.75 kHz
1.5 200 nF, 1 mH, 11.254 kHz, 7.07, 200 nF, 1 mH, 11.254 kHz, 3.54, 500 μH, 400 nF, 11.254 kHz, 3.54, 1592 Hz, 3184 Hz, 3184 Hz, 11.254 kHz, 2546 Hz
1.6 $11.03\angle 53^\circ\ \Omega$, 10.88 A
1.7 $0.277\angle -33.7^\circ$ A, 23.0−2 W, 0.977 $(2.31 - j1.54)\times 10^{-3}$ S
1.8 $1.23\angle 12.4^\circ$ mA, 14.4 mW, $(9538 - j2098)\ \Omega$
1.9 $(55.1 - j8.8)\ \Omega$, $(17.71 + j2.82)\times 10^{-3}$ S, $1.79\angle 9.1^\circ$ A, 177 W
1.10 $1201\angle 87.6^\circ\ \Omega$, $832.6\angle -87.6^\circ\ \mu$S, 4.99 mW
1.11 $(1951 + j2439)\ \Omega$, 0.5 W, $4841\angle -14.5^\circ\ \Omega$
1.12 $163.8\times 10^{-6}\angle -55^\circ$ S, 26.8 nF
1.13 $9.82\angle -78.67^\circ$ mA
1.14 $(17 - j6)\times 10^{-3}$ S, 2.45 W, $0.216\angle 19.4^\circ$ A
1.15 46.15
1.16 20 kΩ, 79.577 kHz, 1
1.17 200 W
1.18 1125 Hz, 56.55 Ω, 23.5
1.19 353.4 Ω, 2278 Ω
1.20 $0.8\angle -46.7^\circ$ A
1.21 $18.39\angle -8.2^\circ\ \Omega$
1.22 2373 Hz, 55.91 kΩ, 25, 54.22 kΩ
1.23 142.9 kHz, 50 mA, 50
1.25 405 pF, 24.1 Ω, 32.6
1.26 29.058 kHz, 32.9 μA, 3.29 mA
1.27 253 pF, 12.22 kΩ, 38.9
2.1 $(90.9 + j200)\ \Omega$, 0.23 W
2.2 7.913 : 1, 0.112 W
2.3 $30.3\angle -5.3^\circ$ mA, $2.37\angle 44.9^\circ$ V
2.4 100 Ω and 2 μF, 125 Ω and 0.4 μF
2.6 8.2 Ω, 76.5 mW
2.7 324 Ω, 18.3 mW
2.8 35.4 mW
2.9 $0.51\angle -123.2^\circ$ A
2.10 $9.17\angle 3.5^\circ$ A, $5.9\angle -24.3^\circ$ A, $5.26\angle 52^\circ$ A

2.11 $0.434\angle{-16^\circ}\ \Omega$, 3.26 A, $6.92\angle{60.8^\circ}$ A, $6.04\angle{89^\circ}$ A
2.14 0.51 W
2.15 3.33 Ω each
2.16 30 Ω each
2.17 3.36 : 1
2.18 125 mH
2.20 $57.78\angle{-8.5^\circ}$ mA
2.23 $(37.1+j21.8)\ \Omega$, $(42.5+j12.5)\ \Omega$, $(55-j30)\ \Omega$
2.24 $(1000+j100)\ \Omega$, 59.3 mW
3.1 71.32 Ω, $j799\ \Omega$
3.2 0.2 μF, 0.8 μF, 4×10^{-4}, 4×10^{-4}
3.3 1250 Ω, 0.3125 Ω, 5 MΩ
3.4 2 MΩ, 2.5×10^{-4}, 3996
3.5 33.2 nF, 10 mA, 0.48 Ω, 48 μW
3.6 3.46×10^{-5}, 28 902, 3.46×10^{-5}, 41.82 MΩ
3.7 312.4 Ω, 0.4925 H, 3.124 W
3.8 200 kV/m, 500 kV/m
3.9 626.24 W
3.10 1183, 349 W
3.11 760 W
3.15 500 kΩ, 125 kΩ
4.1 767 V
4.2 97.56%, 97.32%, 355 A, 97.61%
4.3 3.13 V, 8.97 V
4.4 1 mH, 4 nF, 15 μH, 0.015, 0.014
4.5 100 kHz, 13.95 Ω
4.6 12.2 mA, 68.42
4.7 0.02 μF, 0.08 μF, 20 Ω
4.8 33.33 V
4.9 216 mA, 161 mA, 0.167
4.10 0.433 Ω, 0.423 Ω
4.11 0.37%, 0.19%
4.12 0.04 Ω, 0.066 $100\angle{120^\circ}$ *VV*
4.13 540.5 W, 2.1 Ω, 7.03 Ω
4.14 367 V, 425.3 V
4.15 $R_{p(eff)}=2\ \Omega$, $X_{p(eff)}=3.58\ \Omega$, $R_c=6116\ \Omega$, $L_m=514.2$ H, $n=5$
4.16 49.2 Ω, 150 Ω, 320.2 V, 333.8 V
4.17 1 kW, 17 kW
4.18 5.753 Ω
4.20 0.012, 5.5 μH, 84.85, 38.4 pF
4.21 $(19.2+j2480)\ \Omega$
4.22 480 mA, 7.24 V
4.23 0.38 Ω, 1.61 Ω
4.24 592.1 W
5.1 $1\angle{53.1^\circ}$, $34.13\angle{58.2^\circ}\ \Omega$, $0.06\angle{71.6^\circ}$ S, $1.7\angle{49.8^\circ}$
5.2 $\begin{bmatrix} 7 & 1950 \\ 0.03 & 8.5 \end{bmatrix}$

5.3 $50\underline{/36.8^\circ}\ \Omega$, $50\underline{/50^\circ}\ \Omega$

5.4 $1.77+j3.06\ \Omega$, 0.071

5.5 $\begin{bmatrix} 0.010\ \text{S} & 0.01\ \text{S} \\ 0.01\ \text{S} & 0.012\ \text{S} \end{bmatrix}, \begin{bmatrix} 100\ \Omega & 1 \\ 1 & 2\times10^{-3}\ \text{S} \end{bmatrix}$

5.6 $\begin{bmatrix} 29.5 & -2000\ \Omega \\ -0.03 & -2 \end{bmatrix}$

5.8 $\begin{bmatrix} 1.36 & 420\ \Omega \\ 3.03\times10^{-3}\ \text{S} & 1.67 \end{bmatrix}$, 0.585

5.9 $\begin{bmatrix} 1-j1 & 100-j300\ \Omega \\ -j2.5\times10^{-3}\ \text{S} & 0.25 \end{bmatrix}$, $200\underline{/53^\circ}$

5.10 $\begin{bmatrix} -0.94+j1.39 & -471.5+j368.5\ \Omega \\ j5.56\times10^{-3}\ \text{S} & -0.94+j1.39 \end{bmatrix}$

5.11 $161.4\underline{/35.6^\circ}$ V, $0.296\underline{/80.3^\circ}$ A

5.12 $\begin{bmatrix} 8.75 & 2300\ \Omega \\ 12.5\times10^{-3}\ \text{S} & 3.4 \end{bmatrix}$, 256 V

5.13 3 V

5.15 $\begin{bmatrix} 50 & 40 \\ 10 & 40 \end{bmatrix}$

5.17 19.998 S

5.18 $\begin{bmatrix} 10^{-3}\ \text{S} & -1\times10^{-9}\ \text{S} \\ 120\times10^{-3}\ \text{S} & 4\times10^{-4}\ \text{S} \end{bmatrix}$

5.19 $\begin{bmatrix} 3.5 & 120\ \Omega \\ 0.094\text{S} & 3.5 \end{bmatrix}$

5.20 $\begin{bmatrix} 1 & 47\underline{/62^\circ}\ \Omega \\ 1.2\times10^{-3}\ \text{S} & 1 \end{bmatrix}$

6.1 12.88 N

6.2 0.52 N

6.3 1.4 A

6.4 84.4%

6.5 67.8 μF, 295 V, 257 V

6.6 0.377 T, 0.992 T, 195.85 mH, 515.4 mH, 2.56 J

6.7 4.87 N

6.9 1.04 μF, 1733 V, 1.8×10^{-3} C, 1733 V, 900 V, 900 V

6.10 0.345 J

6.11 0.1 J

6.12 5 H

6.13 8 J

6.14 0.9 W

6.15 300 V, 0.075 J

6.16 0, 2.75 J, 11 J

6.17 2.64 mJ, 2.88 mJ

7.1 181 mm, 262 mm

7.4 0.89 μJ 1.77pF

7.6 9.44pF/m, 1.28μH/m

7.7 115.5 pF/m

7.9 1.86×10^{12} Ω/m
7.11 40.55 kV
7.12 17.28×10^{6} A/m^2
8.1 77.8 Ω, 86.8 Ω, 6.47 dB
8.2 $Z_1 = 147.37$ Ω, $Z_2 = 952.63$ Ω, $Z_3 = 697.37$ Ω
8.3 1087 Ω, 1398 Ω, 10.1dB, 1049 Ω, 1449 Ω
8.4 3493 Ω, 0.59 N, 0.52 mA
8.5 8.25 dB
8.6 4272.5 Ω, 10 223 Ω, 12.5 dB, 3774 Ω, 11 574 Ω, 13 dB
8.7 438.2 Ω, 34.24 dB
8.8 2324 Ω, 1704 Ω, 4.86 dB
8.9 196 Ω, 1004 Ω, 392 Ω
8.10 $R_1 = 359$ Ω, $R_2 = 302$ Ω
8.11 $R_1 = 97.7$ Ω, $R_2 = 51.4$ Ω
8.13 $R_1 = 563$ Ω, $R_2 = 38$ Ω
8.14 0.766 N
8.16 350 Ω, No, 8.57 mA, 16.14 dB, 16.14 dB
8.17 1.73 N, 1250 Ω, 2500 Ω
9.1 7.134 km, 20 km, 4 mW
9.2 131 mW, $-60°$
9.4 0.2 W, 1.8×10^{8} m/s
9.4 5.02 mA
9.5 $605\angle{-14°}$ Ω, $0.1362\angle{83°}$, $9.25\angle{-73°}$ mA
9.6 3 m, 1.33 W, $133.3\angle{-120°}$ mA
9.7 3 V, 3 V, 0.18 W, $\pi/100$ R/m
9.8 $232\angle{-34.1°}$ Ω, 0.065 N/km, 0.096 R/km
9.9 0.419 mA, 0.197 λ
9.10 126.6 km/s
9.11 3.16 W, $-240°$, $-220°$
9.12 $0.109 + j0.183$ per km
9.13 $160\angle{-10°}$ Ω, $0.233 + j0.119$ per km
9.14 2.564 W, 2.338×10^{8} m/s
9.15 193 mW, 25.71 dB
9.16 37.5 Ω
9.17 8.686 dB, 73.3°
9.18 1.4 dB/km
9.19 12 km, 180°
9.20 167 mW
9.21 0.23 μH, 90.9 pF
10.1 1.056 m
10.2 3.37, $360.6\angle{-56.3°}$ Ω
10.3 $0.528\angle{10.5°}$, 3.24
10.4 $687.5\angle{7.6°}$ Ω
10.5 -0.5, 3, 36 V, 12 V
10.6 600 Ω
10.7 $284\angle{18.9°}$ Ω

10.8 5.66 V, 7 V
10.9 −0.5, 200 Ω
10.10 225 Ω, 45 Ω
10.11 13.32 V, 22.2 mA
10.12 $455.3\underline{/-24^\circ}$ V
10.17 693 Ω
10.20 75 Ω, 37.5Ω, 150 Ω
10.21 640 Ω, 250 Ω
11.1 +44.7%
11.2 $i = 3.28\sin(1000t - 13^\circ) + 0.812\sin(3000t + 75.3^\circ)$, 0.963
11.3 12.35 μF, 0.304
11.4 $i = 0.082\sin(500t - 10.6^\circ) + 0.02\sin(1000t + 90^\circ)$, 0.973
11.5 12.25 V, 32 mA, 0.996
11.6 4.069 W, 0.95
11.7 $7.5 + j7.5$ mS, $7.5 - j7.5$ mS, 80.8 mA
11.8 0.335 A, 2.3 V, 6.7 V
11.10 2 mA, 1 mA, 0.335 A, 2.3 V, 6.7 V
11.11 9.1 V, 1.25 μF, 1.2 V, $529\underline{/4^\circ}$ Ω
11.12 5.48 mA
11.13 410 mA, 5.9 V, 4.1 V
11.14 85 mA, 115 mA, 6.63 V
11.16 159.2 Hz, 318.3 Hz, 477.5 Hz, 636.6 Hz
11.17 62.5%
11.20 8.5 mA
11.21 40.77 V
11.22 0.267 V
11.24 $i = 1\sin(5000t + \pi/2) + 0.055\sin(15\,000t - \pi/2)$ A
12.1 $i(t) = CV(1 - e^{-t/C(R_1+R_2)})$
12.2 $i(t) = CV\left(1 - e^{-t/C(R_1+R_2)} - \dfrac{1}{4(R_1+R_2)}e^{-t/C(R_1+R_2)}\right)$
12.3 $v(t) = 3.55 + 2.45e^{-360t}$
12.4 $i(t) = (8.33 + 45.45e^{-1.52\times10^4 t}) \times 10^{-4}$ A
12.6 $v(t) = (6 - 1.88e^{-6.54t})$ V
12.7 $i(t) = (0.152 - 0.075e^{-5.08t})$ A
12.8 $i(t) = 0.456t - 0.044(1 - e^{-5.08t})$ A
12.16 $z(S) = \dfrac{3R^2 + S2LR}{2R + SL}$
12.18 $i(t) = 25 - 68e^{-4t} + 20\delta$
12.19 $i(t) = 8.67 + 2.33e^{-6t}$
12.20 8.676 mA
13.2 24 mH, 42.7 mH, 0.024 μF
13.3 1.82 dB
13.6 154 mH, 221 mH, 85 nF
13.7 0 Ω, 500 Ω, 565 Ω
13.8 0.053 μF, 19.1 mH, 22.8 dB
13.9 40.2°
13.14 7500 Hz

Electrical Principles IV and V: Learning Objectives (TEC)

(A) Complex Quantities

page 1 (1) *Applies complex quantities to single-phase a.c. series circuits.*

6 1.1 Calculates equivalent impedance and admittance of a series a.c. circuit in Cartesian and polar form.

11 1.2 Applies sinusoidal voltage and current driving functions in Cartesian form and polar form to the equivalent impedance and admittance of a series a.c. circuit, and derives the appropriate response.

14 1.3 Solves problems involving series a.c. circuits in Cartesian and polar form, including resonance and Q-factor.

(2) *Applies complex quantities to single-phase a.c. parallel circuits.*

10 2.1 Calculates equivalent impedance and admittance of a parallel a.c. circuit in Cartesian and polar form.

12 2.2 Applies sinusoidal voltage and current driving functions in Cartesian form and polar form to the equivalent impedance and admittance of a parallel single-phase a.c. circuit, and derives the appropriate response.

14 2.3 Solves problems involving parallel a.c. circuits in Cartesian and polar form including resonance and Q-factor.

(3) *Determines power in an a.c. circuit.*

9 3.1 Calculates power in a.c. circuits from parameters in Cartesian and polar forms.

(B) Circuit Theory

24 (4) *Applies various laws and theorems to the solution of problems involving linear d.c. and a.c. circuits.*

26 4.1 Derives the Thevenin equivalent circuit for simple a.c./d.c. sources.

27 4.2 Solves linear network problems using Thevenin's theorem.

28 4.3 Derives the Norton equivalent circuit for simple a.c./d.c. sources.

28 4.4 Solves linear network problems using Norton's theorem.

29 4.5 States the superposition principle.

29 4.6 Solves linear network problems using the superposition principle.

30 4.7 Derives the conversion formulae for star-to-delta connection and delta-to-star connection.

33 4.8 Simplifies and solves linear network problems using star-to-delta and delta-to-star transformations.

35 4.9 States the maximum power transfer theorem for resistive and complex impedance loads.

36	4.10	Calculates for a given source the load for maximum transfer and the power transferred.
37	4.11	Specifies the matching transformer for maximum power transfer.
37	4.12	Identifies practical problems to which the maximum power transfer theorem would be applied.

(C) Magnetic and Dielectric Materials

40	(5)	*Analyses the properties of magnetic and dielectric materials used in electrical engineering.*
48	5.1	Analyses the losses in ferro-magnetic materials.
70	5.2	Shows how hysteresis and eddy current losses may be separated.
48	5.3	Solves problems involving losses in ferro-magnetic materials.
49	5.4	Outlines the properties of modern non-permanent magnetic materials.
49	5.5	Outlines the properties of modern permanent magnetic materials.
40	5.6	Explains and justifies the importance of permittivity with respect to dielectrics and insulating materials.
41, 42	5.7	Compares the characteristics of dielectric materials including permittivity, dielectric hysteresis and breakdown strength.
45	5.8	Distinguishes between dielectric materials with respect to their mechanical and thermal characteristics.
42	5.9	States the losses incurred in dielectric materials.
42, 43	5.10	Derives the equivalent circuit and phasor diagram for a capacitor to show dielectric loss and loss angle.

(D) Complex Waveforms

165	(6)	*Synthesizes a complex wave from fundamental and harmonic components.*
166	6.1	States that the resultant instantaneous value of voltage for a complex wave is $v = V_{1M}(\sin \omega t + \phi_1) + V_{2M}(\sin 2\omega t + \phi_2) + \cdots + V_{nM}(\sin n\omega t + \phi_n)$
169	6.2	States an expression for the instantaneous value of current.
165	6.3	Carries out a harmonic synthesis for 6.1 graphically and/or practically.
165, 185	6.4	Shows the effect of phase shift of second and third harmonics.
170	6.5	Calculates power due to complex waveforms in simple cases.
174	6.6	Lists devices which produce complex waveforms.
175, 176	6.7	Explains how the devices in 6.6 produce complex waveforms.
167	6.8	Recognises complex waveforms which contain i) only even-order harmonics, ii) only odd-order harmonics.

(E) Transmission Lines

132	(7)	*Understands transmission line theory.*
136	7.1	Derives an expression for the characteristic impedance in terms of primary constants.
138	7.2	Explains attenuation and phase change per unit length.
147	7.3	Describes the effect of a mis-matched load.

149, 152 7.4 Represents incident and reflected voltage and current waves, mathematically or graphically.
150 7.5 Defines reflection coefficient.
157 7.6 Defines standing wave ratio.
159, 161 7.7 Relates standing wave ratio to the load impedance.
154 7.8 Calculates the input impedance of a finite line with various terminations.
140, 141 7.9 Defines phase delay and group delay.
7.10 Explains the use of phase delay equalization.
159 7.11 Explains the practical importance of the measurement of standing wave ratio for the correct operation of a transmission system.
159 7.12 Relates power transfer to standing wave ratio.
156 7.13 Explains the action of the quarter wave line as a matching transformer.
162 7.14 Explains the action of single stub matching.

(F) Attenuators

125 (8) *Solves problems on passive attenuators.*
125 8.1 States the need for attenuators.
8.2 Defines the terms:
118 i) characteristic impedance and
123 ii) insertion loss
when applied to an attenuator.
115 8.3 Draws diagrams of T, L and π symmetrical and non-symmetrical attenuators.
8.4 Calculates the insertion loss and characteristic impedance of attenuators:
123 i) individually and
127 ii) cascaded.

(G) Filters

211 (9) *Understands the purpose and types of filter.*
211 9.1 States the need for filters.
212 9.2 Defines the terms cut-off frequency, bandwidth, attenuation and phase characteristics in respect of a filter.
9.3 Explains what is meant by a:
216 i) band-pass
217 ii) band-stop
212 iii) low-pass and
213 iv) high-pass filter.
214–217 9.4 Draws the ideal characteristics of the filters in 9.3.
214–217 9.5 Draws the practical characteristics of the filters in 9.3.
214–217 9.6 Draws the practical circuit configuration of filters in 9.3.
225 9.7 Defines the term "active filter".

(H) Systems Response

186 (10) *Interprets observed system responses.*
186, 188, 193, 194 10.1 States the expected response of *RL* and *RC* combinations to square and ramp functions.

207 10.2 Lists forms of test signal as:
i) sinusoidal waves
ii) square waves
iii) pulses and
iv) ramp functions.

EIII, RSIII 10.3 Draws a block diagram of the test equipment needed to obtain a system response.

(J) Field Theory

107 (11) *Understands the concepts associated with analogous field plotting techniques.*

106 11.1 Describes the analogies of the electric, magnetic and thermal fields.

103 11.2 Explains that for an electric system equipotentials are parallel to the plates and at right-angles to the lines of flux.

103 11.3 Sketches a diagram to show the distribution of flux and equipotentials.

104 11.4 Sketches the electric field distribution in a parallel plate capacitor.

104 11.5 Sketches the electric field distribution for a concentric cylinder capacitor.

107 11.6 Explains the concept of "curvilinear squares".

110 (12) *Derives expressions for parameters of common cable/line configurations.*

111 12.1 Derives the expression for the capacitance/metre length for a pair of concentric cylinders

$$C = 2\varepsilon\pi / \ln(b/a) \text{ farads/metre}$$

111 12.2 Derives the expression for the electric stress for a pair of concentric cylinders

$$E = V/x \ln(b/a) \text{ volts/metre}$$

112 12.3 Derives the expression for the inductance/metre length for a pair of concentric cylinders

$$L = \frac{1}{4}\frac{\mu}{2\pi} + \frac{\mu \ln(R/r)}{2\pi} \text{ henrys/metre}$$

110 12.4 Derives the expression for the capacitance/metre length for an isolated twin line.

$$C = \frac{\pi\varepsilon}{\ln[(D-r)/r]} \text{ farads/metre}$$

112 12.5 Derives the expression for the inductance/metre length of an isolated twin line

$$L = \frac{\mu}{4\pi} + \frac{\mu \ln[(D-r)/r]}{\pi} \text{ henrys/loop metre}$$

89, 103 (13) *Understands that energy is stored in an electric field.*

105 13.1 Derives the formula for the energy stored in an electric field as $\frac{1}{2}DE$ joules/m^3.

89	13.2	Shows that energy in 13.1 can be expressed in terms of circuit parameters as $\frac{1}{2}CV^2$ joules.
89, 105	13.3	Solves more advanced problems on stored energy in an electric field.
	(14)	*Understands that energy is stored in an electromagnetic field.*
91	14.1	Derives the formula for the magnetic energy stored in a non-magnetic medium as $\frac{1}{2}BH$ joules/m^3.
91, 106	14.2	Derives from the formula in 14.1 the expression for total stored energy in terms of circuit parameters as $\frac{1}{2}LI^2$ joules.
91, 106	14.3	Solves more advanced problems on stored energy in a magnetic field.

(K) Transformers

52, 59	(15)	*Understands that a transformer may be treated as a mutually coupled circuit.*
52, 53	15.1	Explains how self and mutually induced e.m.f.s occur in coils which are magnetically coupled.
55	15.2	Defines the term coefficient of coupling (k).
52	15.3	Defines the term coefficient of mutual inductance (M).
55	15.4	Derives the expression $M = k\sqrt{(L_1L_2)}$.
56, 59	15.5	Explains how the choice of core material affects the value of k.
59	15.6	Explains that the core material chosen depends upon the frequency of supply.

(L) Energy Transfer

96	(16)	*Understands the principles of energy transfer with particular reference to electro-mechanical conversion.*
88	16.1	Reviews the various energy conversion systems applied in industry e.g. electrical generating plant.
88–100	16.2	Integrates the expressions for energy dealt with in previous levels e.g. Fd, $\frac{1}{2}mv^2$, $\frac{1}{2}LI^2$, $B^2A/2\mu_0$, $\frac{1}{2}QV$, $\frac{1}{2}CV^2$, $mC(\theta_2-\theta_1)$, etc.
88	16.3	States the principle of energy conservation.
94	16.4	Identifies the various energy changes in the operation of a simple electromagnetic relay or lifting magnet.
97	16.5	States that the change in *a)* electrical energy input is represented by δW_e, *b)* mechanical energy output δW_m, *c)* stored mechanical energy δW_s, *d)* stored magnetic field energy δW_p, *e)* losses δW_c.
97	16.6	Draws the energy balance diagram for the system in 16.4.
97	16.7	Sets up the energy balance equation $\delta W_E + \delta W_M = \delta W_F + \delta W_S + \delta W_C$
97	16.8	Writes down a power balance equation $P_e + P_m = \frac{\delta W_F}{\delta t} + \frac{\delta W_S}{\delta t} + \frac{\delta W_C}{\delta t}$
98	16.9	Identifies and explains the significance of the transient terms in 16.8.
98	16.10	Writes down the simplified power balance expression for steady state conditions.

98	16.11	Identifies the power losses which occur in the following electrical motors *a*) d.c. shunt motor, *b*) cage rotor induction motor, *c*) a.c. generator.
99–100	16.12	Draws a flow diagram showing power input to power output including the various losses for each of the examples in 16.11.
99–100	16.13	Identifies those losses which vary with load and those which remain relatively constant for each of the examples in 16.11.
99	16.14	Shows that maximum efficiency occurs when the load is such that the variable losses = fixed losses.

Index